Applications of Remote Sensing Data in Mapping of Forest Growing Stock and Biomass

Applications of Remote Sensing Data in Mapping of Forest Growing Stock and Biomass

Editor

José Aranha

MDPI • Basel • Beijing • Wuhan • Barcelona • Belgrade • Manchester • Tokyo • Cluj • Tianjin

Editor
José Aranha
University of Trás-os-Montes and Alto Douro
Portugal

Editorial Office
MDPI
St. Alban-Anlage 66
4052 Basel, Switzerland

This is a reprint of articles from the Special Issue published online in the open access journal *Forests* (ISSN 1999-4907) (available at: https://www.mdpi.com/journal/forests/special_issues/Remote_Sensing_Growing_Stock_Biomass).

For citation purposes, cite each article independently as indicated on the article page online and as indicated below:

LastName, A.A.; LastName, B.B.; LastName, C.C. Article Title. *Journal Name* **Year**, *Volume Number*, Page Range.

ISBN 978-3-0365-0568-8 (Hbk)
ISBN 978-3-0365-0569-5 (PDF)

Contents

About the Editor

José Aranha is a full professor of Geographical Information Systems and and Remote Sensing (GIS—RS) in agro-forestry applications, in the "Forestry Department" at the University of Trás-os-Montes e Alto Douro, Vila Real, Portugal. He earned his Ph.D. in GIS-RS (1998) from Kingston University, England (thesis title: An Integrated Geographical Information Systems for Vale do Alto Tâmega - GISVAT). His research interests are multidisciplinary, and include forest fire hazard indices, forest biomass calculation, forest inventory, satellite image processing and classification, land use land cover classification and mapping, spatial analysis, map Algebra modeling in agro-forestry, ecosystem services modeling and biological struggle, wildlife survey. He is the author (2021) of one book, five book chapters, seven books published by UTAD, 24 papers in international journals (ISI JCR), six papers in national journals (Latindex, scoupus), 48 papers in International Congress Proceedings Books, 72 papers in National Conference Proceedings Books, and 30 papers in the dissemination and transmission of science. He is the supervisor (2021) of PhD students (six with their thesis concluded and three ongoing), master's students (26 with their disseratation concluded and four ongoing), undergarduate students (40 with their final report concluded (UTAD), and 14 concluded (Erasmus)). His fellowship supervision includes: 1 concluded postdoc, 1 concluded Phd, 2 concluded Master's, 1 ongoing PhD and 1 ongoing Master's. He is a full member of the Centre for the Research and Technology of Agro-Environmental and Biological Sciences (CITAB) (http://www.citab.utad.pt/).

Preface to "Applications of Remote Sensing Data in Mapping of Forest Growing Stock and Biomass"

Forested areas and stock biomass are of interest in a climatic change scenario, with increasing carbon emissions. The forest canopy plays an important role in ecosystem services, which can provide soil protection against erosion, water management cycles, biomass production and carbon stock.

Surveying regular forested areas for biophysical measurement and inventory purposes is time-consuming and very expensive. In this way, remote sensing has proven to be a very important and useful tool in the cartography of the forested areas in recent years, for canopy changes analysis and biophysical variable modelling, such as canopy density, basal area growing and biomass stocking. Based on satellite image bands, it is possible to derive vegetation indices and allometric models to estimate forest cover characteristics and variations in the amount of biomass.

Nowadays, several spatial agencies provide regular and free satellite images, with high quality and resolution, over almost all continents and countries. Due to improvements in technology, which are followed by price reductions, local image capture is increasingly accessible to researchers, giving them the autonomy to develop and share a wide variety of projects. Several original manuscripts from different research teams and countries were submitted to this Special Issue, which led to 13 published papers. With regard to study area, the published papers range from purely forested areas to agroforestry and urban areas. Regarding remote sensing methods and techniques, both free and commercial satellite images, and LiDAR images collected using UAV and classified were processed.

José Aranha
Editor

Article

Finer Resolution Estimation and Mapping of Mangrove Biomass Using UAV LiDAR and WorldView-2 Data

Penghua Qiu [1], Dezhi Wang [2,*], Xinqing Zou [3], Xing Yang [1], Genzong Xie [1], Songjun Xu [4] and Zunqian Zhong [1]

1   College of Geography and Environmental Science, Hainan Normal University, Haikou 571158, China; cph6688@hainnu.edu.cn (P.Q.); yangxinggeo@163.com (X.Y.); 152262552@163.com (G.X.); a1193132383@163.com (Z.Z.)
2   Faculty of Information Engineering, China University of Geosciences (Wuhan), Wuhan 430074, China
3   School of Geography and Ocean Science, Nanjing University, Nanjing 210023, China; zouxq@nju.edu.cn
4   School of Geography, South China Normal University, Guangzhou 510631, China; xusj@scnu.edu.cn
*   Correspondence: dzwang@cug.edu.cn; Tel.: +86-027-6788-3728

Received: 31 July 2019; Accepted: 26 September 2019; Published: 4 October 2019

Abstract: To estimate mangrove biomass at finer resolution, such as at an individual tree or clump level, there is a crucial need for elaborate management of mangrove forest in a local area. However, there are few studies estimating mangrove biomass at finer resolution partly due to the limitation of remote sensing data. Using WorldView-2 imagery, unmanned aerial vehicle (UAV) light detection and ranging (LiDAR) data, and field survey datasets, we proposed a novel method for the estimation of mangrove aboveground biomass (AGB) at individual tree level, i.e., individual tree-based inference method. The performance of the individual tree-based inference method was compared with the grid-based random forest model method, which directly links the field samples with the UAV LiDAR metrics. We discussed the feasibility of the individual tree-based inference method and the influence of diameter at breast height (DBH) on individual segmentation accuracy. The results indicated that (1) The overall classification accuracy of six mangrove species at individual tree level was 86.08%. (2) The position and number matching accuracies of individual tree segmentation were 87.43% and 51.11%, respectively. The number matching accuracy of individual tree segmentation was relatively satisfying within 8 cm $\leq$ DBH $\leq$ 30 cm. (3) The individual tree-based inference method produced lower accuracy than the grid-based RF model method with R^2 of 0.49 vs. 0.67 and RMSE of 48.42 Mg ha^{-1} vs. 38.95 Mg ha^{-1}. However, the individual tree-based inference method can show more detail of spatial distribution of mangrove AGB. The resultant AGB maps of this method are more beneficial to the fine and differentiated management of mangrove forests.

Keywords: AGB estimation and mapping; mangroves; UAV LiDAR; WorldView-2

---

1. Introduction

Mangroves have attracted considerable attention due to their unique morphological characteristics and diverse eco-environmental service functions [1]. These services include coastal protection, biodiversity maintenance, and carbon sequestration [2–5]. The organic carbon in mangrove forests per unit area is four times higher than that of other terrestrial forest ecosystems [6]. Mangroves are therefore considered a strong candidate for the United Nations Framework Convention on Climate Change (UNFCCC), the payments for ecosystem services (PES) program [7], and the policymaking and implementation in blue carbon. However, all these initiatives require accurate biomass and carbon stock estimations. The aboveground biomass (AGB) of mangroves is one of the fundamental parameters describing a mangrove ecosystem's functioning and is essential for determining its storage of carbon. Accurate estimate of AGB is critical for mangrove monitoring and management. To build a beautiful China and protect the ecological marine environment, local governments of coastal areas in China have been asked to fully implement the bay chief system (A leader responsible system in governance of bays began in 2017, which was proposed by the State Oceanic Administration, China.) and conduct pilot work on carbon sinks in marine ecosystems. Strengthening the research on mangrove biomass will help to support this work.

Biomass estimation methods can be categorized into direct and indirect methods. The direct approach is a traditional field harvesting method. Among the indirect methods, the allometric estimation is the most common approach and has become a standard tool for biomass prediction [8]. The basic theory of allometric relationships suggests that in many organisms, the growth rate of one part of an organism is proportional to the growth rate of growth of another [9]. Thus, researchers have often used allometric regression equations based on available and measurable woody plant parameters such as stem diameter, tree height, or crown diameter to estimate biomass [10,11].

To improve the accuracy of mangrove biomass estimation and mapping, more remote sensing data and methods have been tested and applied to mangroves [4,12–20]. Very high resolution images, such as WorldView-2 (Digitalglobe, Westminster, CO, USA), GeoEye (GeoEye, Herndon, VA, USA), IKONOS (Spacing Imaging, Herndon, VA, USA), QuickBird (Digitalglobe, Westminster, CO, USA), digital photographs, and light detection and ranging (LiDAR) point cloud data, can be used for extracting individual trees [21–26], as well as the accurate estimation of AGB or carbon stocks at the individual tree level [27–34]. Yin and Wang (2019) used UAV LiDAR to study the individual tree detection and delineation (ITDD) of mangroves and proposed a rule-of-thumb that the spatial resolution should be finer than one-fourth of the crown diameter for ITDD [26]. To date, to the best of our knowledge, no study has discriminated mangrove species and estimated mangrove biomass in a finer resolution, such as individual tree or clump level. In addition, despite the wide use of remote sensing techniques to obtain and retrieve mangrove information, there is no general consensus approach [35]. Because of the ability to provide spectral information, optical images are considered to be very useful in distinguishing plant species [36]. While, LiDAR data can provide more information about forest structure [32,33]. Therefore, the integration use of very high spatial resolution WorldView-2 imagery and UAV LiDAR data may provide the potential for mangrove biomass estimation and mapping at finer resolution.

The overall aim of this study was to use optical remote sensing, UAV LiDAR, field survey data sets, and allometric equations to estimate and map the AGB of mangroves in Qinglan harbor, Hainan province, China. In this study, we aimed to (1) compare the merits and demerits of AGB estimation based on the newly proposed individual tree-based inference method and the grid-based RF model method, (2) examine the accuracies of AGB estimation for two methods, and (3) map the AGB of mangroves.

2. Materials and Methods

2.1. Study Site

The study was conducted in the core area of the Qinglan Harbor Provincial Nature Reserve, located on the northeastern part of Hainan Island, China (Figure 1). The Qinglan Harbor Nature Reserve is the reserve that has the most abundant mangrove species in China, including 24 true mangrove species and 10 semi-mangrove species, which account for 88.89% and 100% of the total number of true- and semi-mangrove species in China, respectively. In China, mangroves in the Qinglan Harbor Nature Reserve are distributed with the tallest plants, the oldest forests, the most complete community preservation, and many rare and endangered mangrove species. The total mangrove area of the Qinglan Harbor Nature Reserve is about 835.5 ha.

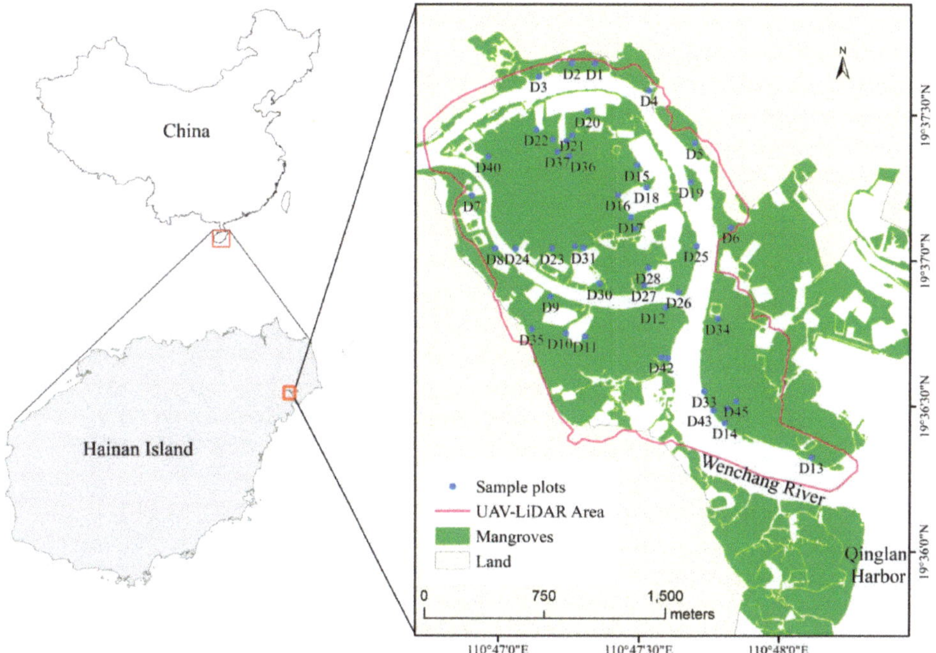

Figure 1. Location of the study area and the sampling plots.

The study area was 386.32 ha, with a mangrove area of 209.99 ha. The landforms are mainly sea alluvial plain, lagoon plain, sea platform, and mangrove beach. The soil type is mostly alluvial soil with a smooth soil texture (silty clay). The mean annual temperature is about 24.1 °C, the average annual rainfall is around 1650 mm and the average tidal range is about 0.89 m. The salinity of its water ranges between 3‰ and 25‰.

2.2. Field Data Collection

The field surveys were conducted from July to September 2018 and March 2019. A total of 45 plots were surveyed. The base plot area was 10×10 m, which was expanded up to 600 m^2 depending on the specific field situation. All individual trees with a diameter at breast height (DBH) of ≥1.5cm and height that was ≥1.5 m within the plots were counted and measured for species, number, height, stem girth at breast height (GBH, cm; DBH = GBH/π), and crown cover. A real-time kinetic global positioning system (RTK-GPS) was used to measure the geographical position of each plot and most of measured trees. A total of 3832 individual trees and 13 species (*Bruguiera sexangula* (Lour.) Poir., *Excoecaria agallocha* Linn., *Rhizophora apiculata* Blume, *Aegiceras corniculatum* (Linn.) Blanco, *Kandelia candel* (Linn.) Druce, *Xylocarpus granatum* Koenig, *Ceriops tagal* (Perr.) C. B. Rob., *Lumnitzera racemosa* Willd., *Heritiera littoralis* Dryand., *Hibiscus tiliaceus* Linn., *Sonneratia apetala* Buch-Ham., *Sonneratia alba* J. Smith, *Sonneratia ovata* Backer) were recorded in the 45 field plots. 68.89% of the plots had at least 3 types of mangrove species. The mangrove forest was dominated by *B. sexangula*. The tree density of *B. sexangula* was highest among all mangrove species, which was 1.5 times and 2.0 times higher than that of *E. agallocha* and *L. racemosa*, respectively. The summary of the community structure of the mangrove forest in Qinglan Harbor Nature Reserve was presented in Table 1.

Table 1. Summary of the community structure of the mangrove forest in 45 sample plots.

Species	Count	Density (Individuals·100 m^{-2})	RD (%)	F (%)	ReF (%)	BA (m^2)	Dominance (m^2 ha^{-1})	RDo (%)	IVI
B. sexangula	1063	8.93	27.74	67.31	20.02	16.514	13.877	46.75	31.50
E. agallocha	694	5.83	18.11	71.11	21.15	4.462	3.750	12.63	17.30
H. tiliaceus	250	2.10	6.53	35.56	10.58	1.371	1.152	3.88	7.00
H. littoralis	18	0.15	0.47	4.44	1.32	1.187	0.997	3.36	1.72
X. granatum	98	0.82	2.56	31.11	9.25	0.887	0.745	2.51	4.77
R. apiculata	402	3.38	10.49	35.56	10.58	3.938	3.309	11.15	10.74
A. corniculatum	469	3.94	12.24	8.89	2.64	0.655	0.550	1.85	5.58
K. candel	4	0.03	0.10	6.67	1.98	0.009	0.007	0.03	0.70
L. racemosa	543	4.56	14.17	28.89	8.59	1.516	1.274	4.29	9.02
C. tagal	112	0.94	2.92	2.22	0.66	0.213	0.179	0.60	1.39
S. ovata	50	0.42	1.31	13.33	3.97	1.453	1.221	4.12	3.13
S. apetala	48	0.40	1.25	13.33	3.97	1.859	1.562	5.26	3.49
S. alba	81	0.68	2.11	17.78	5.29	1.259	1.058	3.57	3.66
Total	3832	32.20	100.00	336.20	100.00	35.322	29.682	100.00	100.00

BA is basal area, BA = $\sum$(DBH$^2 \times \pi$)/(4 $\times$ 10,000); RD is relative density; F is frequency, which calculated as: F = number of sample plots occurring in some kind of plant/total number of sample plots $\times$ 100; ReF is relative frequency; RDo is relative dominance; IVI is importance value index, IVI = (relative density + relative frequency + relative dominance)/3 [37].

The AGB in each field plot was estimated by the allometric equation method. The selection principles were as follows: (1) Domestic models take precedence over foreign models; (2) the closer the study area of the model to Hainan, the more preferred it is; (3) the closest forest age takes precedence; and (4) species with similar DBH are preferred. When no specific allometric equation was available, a common allometric equation was used (Table 2). We first calculated the AGB of each tree using the species-specific allometric equations presented in Table 2 and then summed the AGBs in each field plot. To obtain the AGB density for each plot, we divided the summed AGB of a plot by the plot's area.

Table 2. Allometric equations for mangrove species used in this study.

No.	Species	Allometric Equations	References
1	*B. sexangula*	AGB = 0.168 $\times$ DBH$^{2.42}$	[38]
2	*E. agallocha*	LogAGB = 1.0996 $\times$ logDBH^2 − 0.8572	[39]
3	*R. apiculata*	AGB = 0.235 $\times$ DBH$^{2.420}$	[40]
4	*C. tagal*	AGB = 0.1885 $\times$ DBH$^{2.3379}$	[38]
5	*L. racemosa*	AGB = 0.1023 $\times$ DBH$^{2.50}$	[41]
6	*A. corniculatum*	LogAGB = 1.496 + 0.465 $\times$ log (DBH$^2 \times$ H)	[42]
7	*X. granatum*	AGB = 0.0823 $\times$ DBH$^{2.5883}$	[38]
8	*K. candel*	LogAGB = 2.814 + 1.053 $\times$ log (DBH$^2 \times$ H)	[42]
9	*Sonneratia* spp. (*S. ovata, S. apetala*, and *S. alba*)	AGB = 0.258 $\times$ DBH$^{2.287}$	[43]
10	Others (*H. littoralis, H. tiliaceus*)	AGB = 0.251 $\times$ ρ $\times$ DBH$^{2.46}$	[44]

2.3. Remote Sensing Data and Processing

2.3.1. WorldView-2 Imagery

The WorldView-2 optical images were acquired on 5 October 2018. They included a 0.5 m panchromatic band and eight 2.0 m multispectral bands.

First the images were radiometrically corrected. Then, the multispectral images were pan-sharpened by the panchromatic image using the NNDiffuse tool in ENVI 5.3 (Harris Geospatial, Melbourne, FL, USA). Finally, this pan-sharpened imagery was co-registered to the UAV LiDAR data. The precision was controlled within 0.5 pixels. The mangrove extent of the study area was extracted using this processed imagery. The detail process of how to extract mangrove extent using high spatial resolution imagery can refer to our previous work [45]. The main flowchart of remote sensing data processing and AGB estimation was presented in Figure 2.

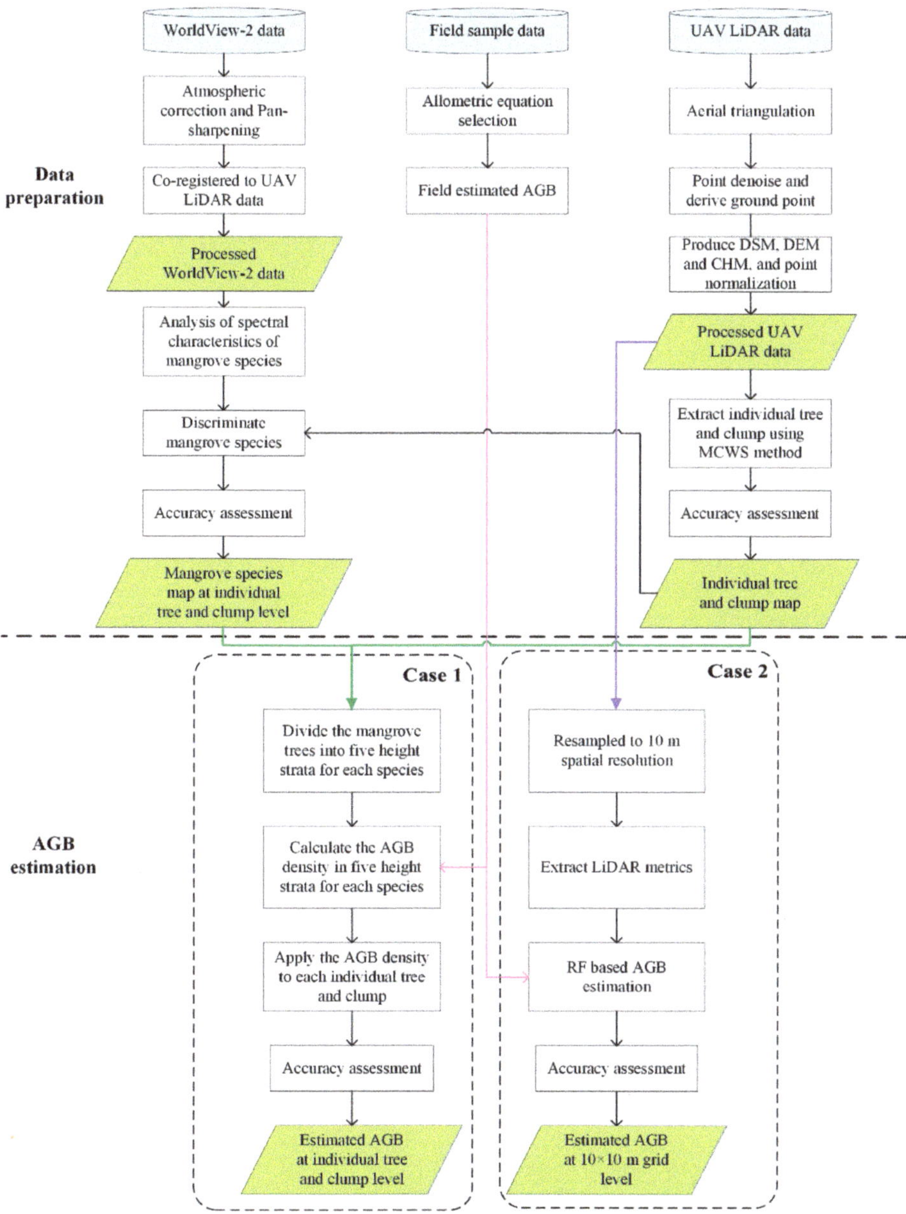

Figure 2. Workflow of estimating mangrove aboveground biomass (AGB). Case 1 denotes the mangrove AGB estimation at individual tree level, namely the individual tree-based inference method. Case 2 denotes the common model-based AGB estimation at grid level, namely the grid-based random forest model method, which is used as a benchmark in this study.

2.3.2. UAV LiDAR Data and Mangrove Canopy Height Model Production

The UAV LiDAR data were collected in March 2018 using a Velodyne VLP-16 puck sensor (Velodyne LiDAR, San Jose, CA, USA) with a laser wavelength of 903 nm. The sensor can emit 300,000 points per second with an accuracy of 3 cm. The average point density was 94 points/m^2. The flight elevation was 52 m with a speed of 5 m/s. An RTK-GPS was simultaneously used to obtain the base station position of the UAV LiDAR system with centimeter-level accuracy.

The raw LiDAR data were first computed using aerial triangulation algorithm to obtain accurate position of each LiDAR point cloud. Then, the noise and outlier points of the LiDAR data were filtered using LiDAR360 software (GreenValley, Beijing, China). Subsequently, the point clouds were classified as non-ground and ground points. The ground points were used to produce the digital elevation model (DEM) using the triangulated irregular net (TIN) interpolation method. Both the ground points and non-ground points were utilized to generate the digital surface model (DSM) using the inverse distance weighted (IDW) interpolation method. The canopy height model (CHM) was produced through subtracting the DEM from the DSM [5]. Finally, the point clouds were normalized using the DEM subtraction method [26] and masked by the mangrove extent, which was produced from the WorldView-2 imagery.

2.4. Mangrove Species Classification and Individual Tree Detection

The mangrove classification was conducted in the R language platform using random forest (RF) algorithm. RF, proposed by Leo Breiman (2001) [46], is an integrated machine learning algorithm based on decision trees. The bootstrap re-sampling is used to construct the decision tree model by randomly sampling from the original sample set. The final result is obtained by voting on the prediction of multiple decision trees. The RF algorithm can be used to classify remote sensing images with complex spatial distribution or to identify multiple data of different statistical distributions and scales.

From the spectral curves of different mangrove species in Figure 3 and the 0.5 m pan-sharpened WorldView-2 imagery, we found that *S. ovata*, *S. apetala*, and *S. alba* communities and the mixed community of *S. ovata* and *A. corniculatum* have similar appearances, textures, and spectral features overall, which means they are difficult to separate. In addition, *S. ovata*, *S. apetala*, and *S. alba* belong to the same genus of Sonneratia. Therefore, these three species were merged into a class of *Sonneratia* spp. *H. littoralis* and *K. candel* were hard to discriminate because of their small coverage areas. *C. tagal* was usually mixed with *L. racemosa* in the study area. *A. corniculatum* and *X. granatum* were companion species growing with dominant species, so it was difficult to identify them. Thus, the mangrove species in the study area were categorized into six types: *B. sexangula* (BS), *E. agallocha* (EA), *Sonneratia* spp. (SS, including *S. alba*, *S. apetala*, and *S. ovata*), *R. apiculata* (RA), *H. tiliaceus* (HT), and *L. racemosa* (LR).

With the development of remote sensing technology, individual tree extraction based on very high-density point clouds is becoming possible [19,26,47–54]. Using the marker-controlled watershed segmentation (MCWS) algorithm [55], Yin and Wang (2019) well detected and delineated individual mangrove tree [26]. Their result indicated that the spatial resolution of CHM should be finer than one-quarter of the crown diameter (CD) in order to correctly delineate crown boundaries and characterize the crown shapes. In the field survey, we found that the crown diameters of a large amount of *L. racemosa* and low *B. sexangula* ranged from 1.0 to 1.5 m. Therefore, the individual tree was detected based on a 0.25 m spatial resolution CHM, which obtained from the point cloud data of UAV LiDAR.

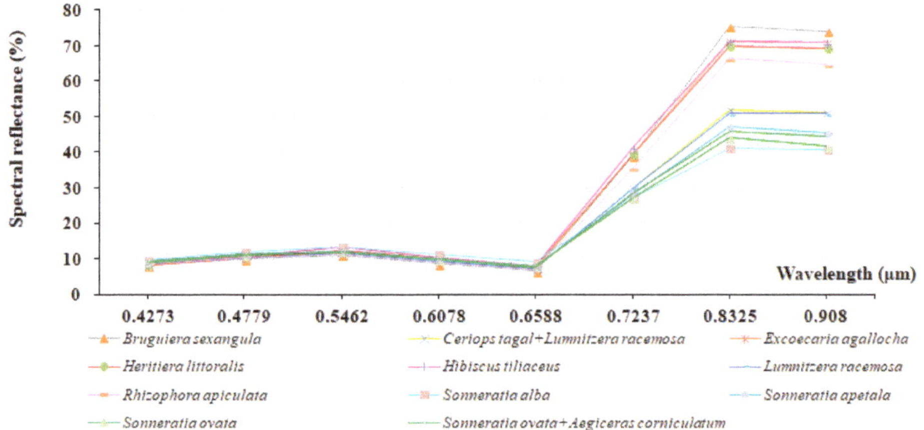

Figure 3. Spectral curves of different mangrove species.

After mangrove individual tree detection, a total of 6800 samples of the six mangrove species types were randomly collected based on our field survey with a uniform distribution in the study area. These samples all lay in homogeneous and monoculture communities. Of the 6800 patches, 70%, namely, 4760, were used as training samples and the other 30%, namely, 2040, were used as independent validation samples.

For the RF classification model, the parameter ntree was set to 1000 and mtry was set to the default value $\sqrt{mtry}$ (*mtry* denotes the number of predictor variables input). We have tested the values of 500, 600, 700, 800, 1000, 1500, and 2000 for the *ntree* parameter, and found that when *ntree* reached 1000 with *mtry* using $\sqrt{mtry}$, the error become convergent. For the detail tuning process of the *ntree* parameter, refer to the study of Pham and Brabyn (2017) [18]. Because the default value was also widely used in previous studies [56], we directly utilized this default setting.

2.5. AGB Estimation

Based on the mangrove species classification map at individual tree level and the obtained tree heights and crown diameters from the UAV LiDAR data, we could divide the mangrove trees into five strata for each species. The nature break method in ArcGIS (ESRI, Redlands, CA, USA) was used for the stratification. Because we have measured 3832 trees and obtained their heights, crown cover, and species, we first classified these trees into the above five strata for each species. Subsequently, we summed the AGB and the crown cover area according to the six mangrove species and the five tree height levels (Equation (1)). We divided the summed AGB by the corresponding summed area to obtain an AGB density for each species at each height level. Finally, these stratum AGB density values were applied to each individual tree and clump based on its species and height level to produce a mangrove AGB map for this study area (case 1, Figure 2). The proposed AGB estimation method was named individual tree-based inference method.

$$\overline{\text{AGB}^{s-h}} = \frac{\sum_i^n \text{AGB}_i^{s-h}}{\sum_i^n \text{Area}_i^{s-h}} \tag{1}$$

where $\overline{\text{AGB}^{s-h}}$ denotes the AGB density of s species at h height level, AGB_i^{s-h} and Area_i^{s-h} respectively denote the AGB and the area of the ith tree, which belongs to s species and h height level.

In this study, we also estimated the mangrove AGB using the common model-based method at grid level (case 2, Figure 2), which was used as a benchmark in this study. To construct the inversion model, the random forest algorithm was employed due to the good performance of RF in biomass prediction [18,46]. The size of the grid is 10×10 m, which is the same to the minimum size of field plots. Prior to fitting model, we extract 52 LiDAR metrics for each field sample plot and each 10×10 m grid. We first linked the field estimated AGB with these LiDAR metrics and selected 11 optimal metrics (presented in Table 3). Subsequently, we constructed a prediction model based on the 11 LiDAR metrics. For the RF regression model, *ntree* was set to 1000 and *mtry* was set to the default value $\sqrt{mtry}$. Finally, we applied this model to all grids and obtained the mangrove AGB map at grid level. This benchmark method was named the grid-based random forest mode method.

Table 3. List of the selected light detection and ranging (LiDAR) metrics.

LiDAR Metrics	Explanation
$CC_{1.3}$	Canopy cover above 1.3 m.
HSD	Standard deviation of heights.
D01	The number of canopy return points in the 1th slice relative to the total points. There are 12 density metrics in this study from 0 to 24 m with an interval of 2 m
HVAR	Variance of heights.
HIQ	Interquartile distance of percentile height.
H05	The 5th percentile of height.
H10	The 10th percentile of height.
H80	The 80th percentile of height.
H90	The 90th percentile of height.
H95	The 95th percentile of height.
CTHK	Canopy thickness.

2.6. Accuracy Assessment

2.6.1. Validation Mangrove Classification

The producer accuracy, user accuracy, and overall accuracy were employed to assess mangrove species classification accuracy based on a confusion matrix using the 2040 independent validation samples.

2.6.2. Validation Biomass Result

The accuracy of the biomass prediction was assessed using the coefficient of determination (R^2), and root mean square error (RMSE) between field estimated and predicted values.

For finer resolution AGB estimation, we used the 45-field estimated AGBs as reference data to assess the predicted AGBs. For the grid AGB estimation, namely, the benchmark case, we employed the 10-fold cross-validation method to validate the predicted AGBs. This cross-validation method is also based on the 45-field estimated AGBs.

2.6.3. Individual Tree Detection

The actual position of the trees was determined by visual interpretation of the UAV LiDAR point clouds. The average point spacing of the UAV LiDAR point clouds is 0.16 m. So, in most cases, individual tree or clump could be identified by inspecting the point clouds from different angles [57–59]. When compare the visual interpretation trees and clumps with the machine segmentation trees and clumps, we could obtain accuracy from spatial matching aspect. Because the visual analysis and discrimination of individual tree and clump cannot avoid arbitrary and subjective issues. We also compared the number of actual mangrove trees within a field plot with the number of machine segmentation trees in the field plot. Therefore, the accuracy of individual tree detection was assessed by above two aspects: spatial position and number.

For the spatial position aspect, we overlaid the LiDAR point clouds and the segmentation layer to identify errors. The position of each segmented tree was established as the location of the highest pixel in the set of cloud points or pixels that make up the crown [58]. The matching procedure pairs the closest observed tree location within a given radius for any segmented tree and then eliminates all the non-minimal pairs with the same observed tree [58]. The search radius in this study was set to 2 m. The mismatch types included omission (OM) and commission (CM) cases. The two mismatch cases were added up to compose the total number of mismatch (TNM, Equation (2)). Then, the total accuracy rate of segmentation (TAR) could be obtained by a ratio of the number of right segmented trees to the total number of segmented trees (TNS, Equation (3)) [58,60].

$$TNM = OM + CM \tag{2}$$

$$TAR = (TNS - TNM)/TNS \times 100 \tag{3}$$

where, TNM denotes the total number of mismatches, TAR denotes the total accuracy rate of tree segmentation, TNS denotes the total number of segmented trees, OM denotes omission cases, and CM denotes commission cases.

For the number aspect, the number of trees (N_i) surveyed in a field plot was used as observed value. The number of individual trees (n_i) derived from the individual tree segmentation in the same field sample plot was used as predicted value. Then, the ratio of the difference between the two values was used to evaluate the number accuracy. The calculation formula is as follows:

$$D_i = (n_i - N_i)/N_i \tag{4}$$

$$DA = \left(\sum M_{|Di| \leq 0.4}/N \right) \times 100 \tag{5}$$

where D_i is deviation degree, DA is the accuracy of the summary metric, M is the number of plots that have different segmentation results, and N is the total number of plots. $D_i > 0$ denotes over-segmentation and $Di < 0$ denotes under-segmentation. When $|D_i| > 0.4$, the deviation of segmentation is large and the accuracy is poor. When $0.3 < |D_i| \leq 0.4$, the accuracy of segmentation is acceptable. When $0.2 < |D_i| \leq 0.3$, the accuracy of segmentation is moderate. When $0.1 < |D_i| \leq 0.2$, the segmentation accuracy is good. When $|D_i| \leq 0.1$, the segmentation accuracy is very good and individual tree can be accurately identified.

3. Results

3.1. Individual Mangrove Tree Extraction

The mismatching statistic of visual interpretation and segmentation results showed that the total accurate rate of segmentation position in 45 sample plots was 87.43% (Table 4).

Table 4. Accuracy assessment of individual tree segmentation in spatial position aspect. TA denotes the total accuracy of individual segmentation.

Type	Number of Mismatches			Number of Segmentations	TA
	Commission	Omission	Sum		
Number	221	47	268	2132	87.43%
Ratio (%)	10.37	2.2	12.57		

Table 5 shows the accuracy of the number of individual tree segmentation. The accuracy of all tree species except *A. corniculatum* tree was 57.78% and that of all tree species was 51.11% (Table 5). Because *A. corniculatum* trees in the study area were small, appearing as lower wood and clustering, this was not conducive to the segmentation of individual tree based on crown.

Table 5. Accuracy assessment of individual tree segmentation in number aspect. OS: over-segmentation, US: under-segmentation.

Scenario	Item	$\lvert D_i \rvert \leq 0.1$	$0.1 < \lvert D_i \rvert \leq 0.2$	$0.2 < \lvert D_i \rvert \leq 0.3$	$0.3 < \lvert D_i \rvert \leq 0.4$	$\lvert D_i \rvert > 0.4$	Total	DA
All species except *A. corniculatum*	No. plots	5	7	5	9	19	45	57.78
	Percentage (%)	11.11	15.56	11.11	20.00	42.22	100.00	
	No. OS plots	1	1	0	0	4	6	
	No. US plots	4	6	5	9	15	39	
All species	No. plots	4	7	4	8	22	45	51.11
	Percentage (%)	8.89	15.56	8.89	17.78	48.89	100.00	
	No. OS plots	1	1	0	0	3	5	
	No. US plots	3	6	4	8	19	40	

There were 5 or 6 cases with over-segmentation and deviations greater than 0.4 among 45 samples, of which more than 75% had an average DBH value larger than 20 cm. The number of samples with under-segmentation and deviations greater than 0.4 in 45 samples was 15–19, and in about 80% of these samples, the average DBH value was less than 10 cm. This means that when DBH < 10 cm, the individual tree segmentation was more likely to have under-segmentation and a large deviation value.

3.2. Finer Resolution Mangrove Classification

Table 6 shows the classification accuracy of mangrove species at individual tree level. Table 7 describes the characteristics of tree patches in different mangrove species in the whole study area. Figure 4 presents the thematic map of the mangrove species.

Table 6. Accuracy of mangrove species classification at individual tree level.

		Predicted							Producer Accuracy (%)	User Accuracy (%)
		BS	EA	HT	LR	RA	SS	Sum		
Observed	BS	470	14	1	3	18	4	510	92.16	82.89
	EA	37	158	15	3	13	0	226	69.91	74.53
	HT	6	26	219	2	4	5	262	83.59	90.12
	LR	3	7	1	321	11	11	354	90.68	93.31
	RA	51	7	6	10	165	10	249	66.27	74.66
	SS	0	0	1	5	10	423	439	96.36	93.38
	Sum	567	212	243	344	221	453	2040	Overall accuracy = 86.08%	

BS: *B. sexangula*; EA: *E. agallocha*; HT: *H. tiliaceus*; LR: *L. racemosa*; RA: *R. apiculata*; SS: *Sonneratia* spp.

Table 7. Characteristics of tree patch in different mangrove species in the whole study area.

Species	Mean Tree Height (m)	Patch Number	Mean Area of Patches (m²)	Total Area (ha)	Percentage of Total Area (%)
BS	8.00	117,413	6.26	73.53	35.01
EA	5.03	76,348	5.21	39.81	18.96
SS	7.84	42,738	6.03	25.78	12.28
RA	6.90	41,450	6.24	25.88	12.32
HT	5.85	40,334	6.19	24.95	11.88
LR	3.78	50,467	3.97	20.05	9.55
Sum	6.43	368,750	5.69	209.99	100.00

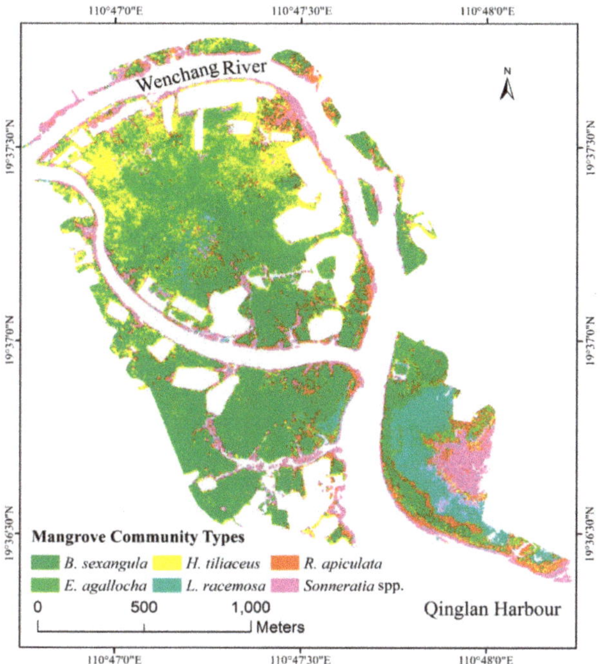

Figure 4. Classification and distribution map of mangroves in the study area.

The overall classification accuracy was 86.08% with the kappa coefficient of 0.83. The majority of mangrove species could be well discriminated with the producer and user accuracies higher than 80%, except for *E. agallocha* and *R. apiculata*. The accuracies of the two species ranged from 66.27% to 74.66%.

Table 7 portrays that *B. sexangula* and *Sonneratia* spp. were the two highest mangrove species (mean height 8.00 and 7.84 m, respectively) in the study area followed by *R. apiculata* (mean height 6.90 m). The lowest tree species is *L. racemosa*. Table 7 also indicates that *B. sexangula* and *E. agallocha* covered 35.01% and 18.96% of the study area, respectively, ranking first and second. The other four species each covered approximately 10% of the study area.

3.3. AGB Estimates and Prediction

The resultant AGB density values for each species at five height levels were presented at Table 8. By applying these AGB density values to individual tree and clump, which have the species and height attributes derived from the UAV LiDAR and WorldView-2 data, the AGB of each patch and the mangrove AGB map of the study area at individual tree level could be obtained. The RF model based on 10 × 10 m grids and LiDAR metrics was used as benchmark and to predict the AGB. Figure 5 delineates the field estimated AGBs versus the predicted AGBs with the accuracy metrics.

Table 8. AGB density of different mangrove species in five height levels (Height: m; AGB: Mg ha^{-1}). The division of five height levels for each species is based on all this type of trees using the nature break method.

		Level 1	Level 2	Level 3	Level 4	Level 5
BS	Height	<5.54	5.54–7.48	7.48–8.91	8.91–10.94	10.94–19.21
	AGB	27.03	77.99	112.32	261.87	363.65
EA	Height	<3.61	3.61–4.77	4.77–5.89	5.89–7.51	7.51–15.77
	AGB	16.62	36.53	45.25	58.53	71.04
HT	Height	<4.77	4.77–6.34	6.34–8.17	8.17–10.82	10.82–19.53
	AGB	30.93	48.92	78.49	95.27	174.47
LR	Height	<3.25	3.25–4.33	4.33–5.25	5.25–6.07	6.07–8.81
	AGB	13.19	16.76	23.27	30.58	38.08
RA	Height	<5.64	5.64–7.46	7.44–9.46	9.46–12.15	12.15–20.20
	AGB	61.47	110.60	191.81	245.74	270.73
SS	Height	<5.92	5.92–8.85	8.85–11.33	11.33–14.20	14.20–21.85
	AGB	50.14	145.83	465.34	654.25	747.58

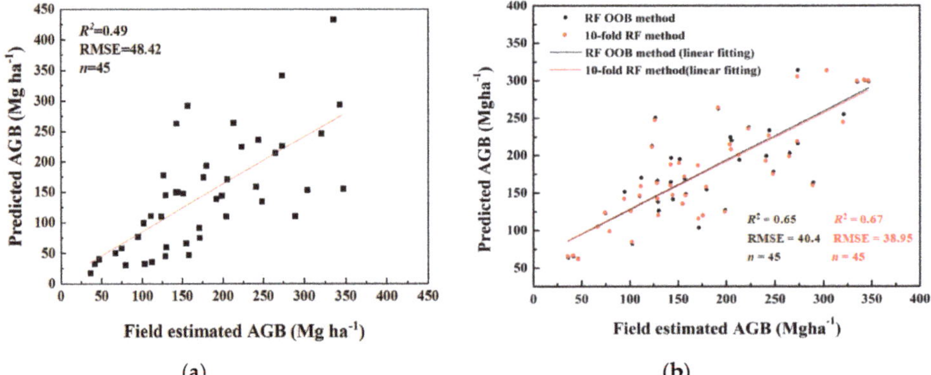

Figure 5. Accuracy comparison of AGB between the field estimated and predicted values. (**a**): Case 1, individual tree-based inference method. (**b**): Case 2, grid-based RF model method.

The R^2 of the individual tree-based inference method is 0.49 with an RMSE of 48.42 Mg ha^{-1}. While, the benchmark, the grid-based RF model method, produced a higher R^2 of 0.67 and a lower RMSE of 38.95 Mg ha^{-1}. Compared with field-estimated AGB, the maximum deviation value of AGB derived from the individual tree-based inference method was 194.01 Mg·ha^{-1}, and the minimum deviation was 1.93 Mg·ha^{-1}. The AGB density of the whole study area calculated from the individual tree-based inference method was 119.34 Mg·ha^{-1}, which was lower than the AGB density produced from the grid-based RF model method (148.97 Mg·ha^{-1}). The results show that it is feasible to use the proposed individual tree-based inference method for mangrove AGB estimation in this relatively complex mangrove forest.

3.4. Spatial Distribution of Mangrove AGB

The comparison of the resultant AGB maps derived from the individual tree-based inference method and the grid-based RF model method are presented in Figure 6. According to above accuracy assessment of the two methods, the spatial distribution of the mangrove AGB in the whole study area can be better represented by the grid-based RF model method based on LiDAR metrics than the individual tree-based inference method. While, the AGB map derived from the latter method was able to portray the AGB distribution at individual tree level for each mangrove species.

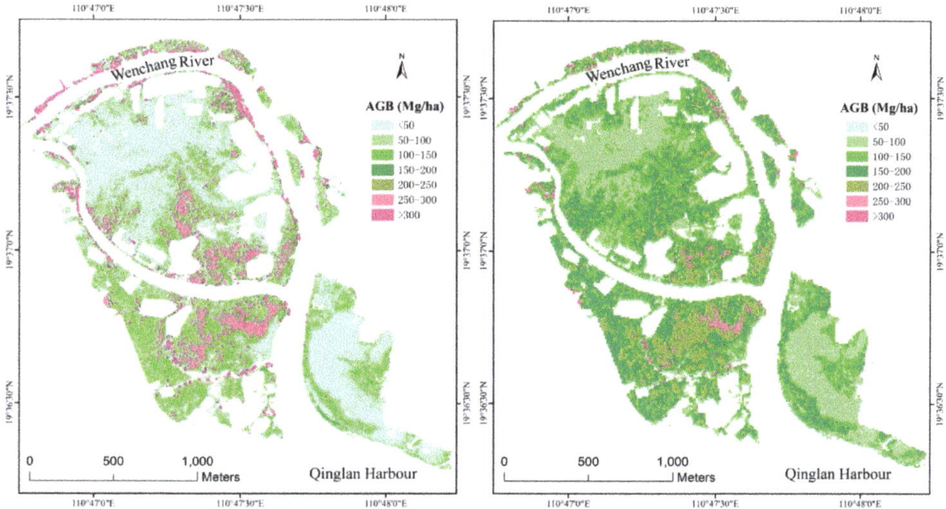

Figure 6. AGB distribution maps of the study area produced from individual tree-based inference method (**left**) and the grid-based random forest (RF) model method (**right**).

The two mangrove AGB maps both show that the mangrove AGB hot spots lie in the southwest and middle of the study area, while the AGB cold spots lie in the southeast and north of the study area (Figure 6). There are also several conspicuous differences between these two maps. First, the low and middle AGB areas of the left map is overall lower than the low and middle AGB areas of the right map with one level difference. Second, the very high AGB areas (>300 Mg ha^{-1}) are more obvious in the left map and cover larger area than those in the right map. Third, the AGB values in the left map are more heterogeneous than the AGB values in the right map.

Figure 7 portrays the AGB maps of different mangrove species, which were produced by the individual tree-based inference method. The AGB of *B. sexangula* showed obvious concentrated distribution characteristics in space with a range of 112.32 to 363.65 Mg ha^{-1}. *E. agallocha* spread out the study area with middle AGB (36.53–58.53 Mg ha^{-1}) accounting for the largest percentage. The AGB of *L. racemosa* was no more than 40 Mg ha^{-1} due to its small individuals and represented disperse distribution. *Sonneratia* spp. had the largest AGB value in the study area, which was mainly distributed in fringe areas to water.

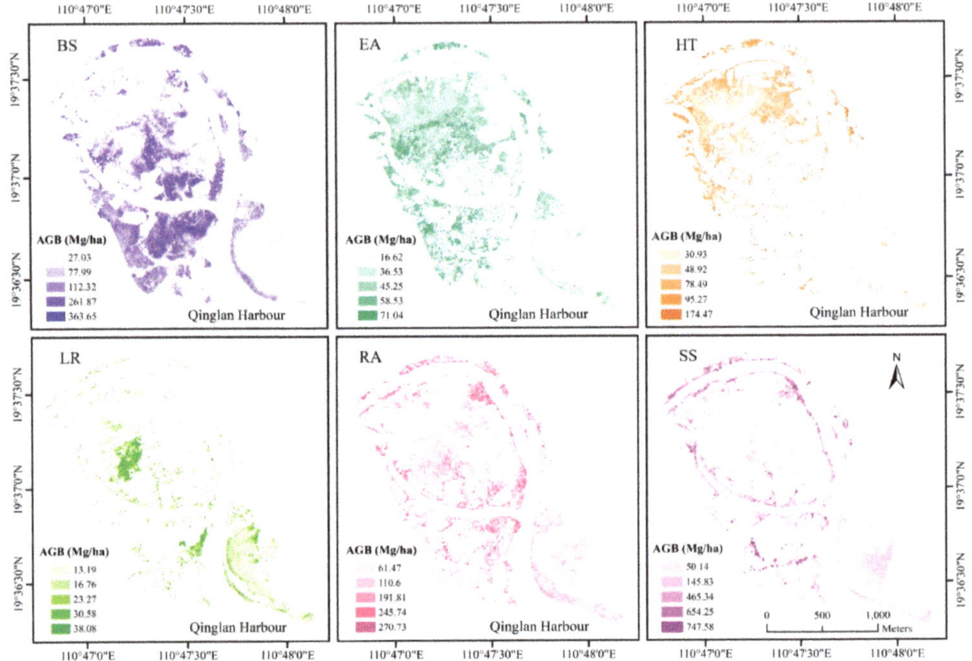

Figure 7. AGB distribution maps in different mangrove species determined from the individual tree-based inference method. BS: *B. sexangula*; EA: *E. agallocha*; HT: *H. tiliaceus*; LR: *L. racemosa*; RA: *R. apiculata*; SS: *Sonneratia* spp.

4. Discussion

4.1. Effect of Mangrove DBH on Individual Tree Segmentation

To explore the relationship between the precision of individual-tree segmentation and the size of the mangroves, the measured DBHs in 45 samples were graded in two-centimeter intervals, and the maximum, minimum, and average DBH values in each sample were determined. Then, the Pearson correlation analysis between the DBH classification and deviation of individual-tree segmentation was completed (Table A1, Appendix A).

The relationship between the DBH and the deviation was as follows: (1) The relationship between DBH and the deviation degree of individual-tree segmentation is relatively complex. Taking 14 cm as the boundary, when DBH < 14 cm, the deviation degree was negatively correlated with DBH, i.e., the deviation degree increased with the decrease in DBH. When DBH >14 cm, the deviation degree was positively correlated with DBH, and the deviation increased with the increase in DBH. This indicates that the accuracy of individual-tree segmentation decreased with the increase in DBH. When DBH < 8 cm, it was significantly negatively correlated with the DBH and the deviation, and when DBH > 30 cm, the correlation coefficient reached the maximum (0.799), it means that the accuracy of individual-tree segmentation decreased considerably. The variation of the correlation coefficient also shows that the accuracy of the segmentation number was relatively ideal within 8 cm ≤ DBH ≤ 30 cm. (2) The deviation degree has a moderate positive correlation with the maximum DBH and average DBH, but has little correlation with the minimum DBH. This means that the maximum and average DBHs are more likely to affect the accuracy of individual-tree segmentation.

4.2. Comparison of the Individual Tree-Based Inference Method and the Grid-Based RF Model Method

Concerning to AGB estimation accuracy, the grid-based RF model method was relatively higher than the individual tree-based inference method. This may be due to the fact that the former had taken fully into account and selected a large number of vegetation indicators reflecting mangrove biomass information.

In terms of mangrove AGB mapping, the grid-based RF model method can be well compared with the field AGB due to the same size of resample grids and field plots. However, whether this grid size is a more accurate AGB mapping method requires more tests and comparison of different sizes of resample grid. The individual tree-based inference method can well reflect the spatial distribution of AGB at different heights for mangrove species, and it is also more beneficial to managers to find problems. From the workflow of this study (Case 1, Figure 2) and the method Section 2.5, we could find that the proposed inference method is simple and it is easy to repeat this method in other mangrove forests or fields.

The results of the detailed comparison between the two AGB estimation methods are shown in Table 9. Although the individual tree-based inference method is considered to provide finer individual-tree information, but the current CHM-based individual tree segmentation still has insurmountable defects. For example, many lower and smaller trees are obscured by the upper tree crown and cannot be effectively identified. In addition, clump mangrove species, such as *L. racemosa*, cannot be well distinguished, too. These limitations inevitably reduce the accuracy of the individual-tree segmentation and then pull down the overall estimation accuracy of the AGB.

Table 9. Comparison of the individual tree-based inference method and the grid-based RF model method.

	Individual Tree-Based Inference Method	Grid-Based RF Model Method
Merits	(1) The AGB is extrapolated on the basis of species, height stratification. The precise tree height can be obtained from LiDAR point cloud data. (2) Finer estimation and mapping of biomass can be conducted on the tree scale, which is more conducive to accurate and differentiated management. (3) Producing large amount of data and detail expression, so it is suitable for application in management or management decision-making in small or specific areas.	(1) Using the RF selection of LiDAR indexes, the relationship of AGB = f (LiDAR indexes) is constructed, and the equal area extrapolation of sample-plot AGB is conducted. (2) The method is relatively mature, and it is the popular method for AGB estimation and mapping at present. (3) The amount of data produced is moderate, the operation speed is fast, and it is suitable for application in government decision-making at the regional level.
Demerits	(1) The extrapolation of AGB lacks a strictly mathematical model. And using tree height stratification to extrapolate AGB, it implies a premise that there is an internal relationship between tree height and AGB, but how this relationship is still not clear. (2) Owing to the survival competition of cluster mangroves, the individuals may tend to have taller tree height, but not necessarily have larger DBH. The AGB estimated by species-specific allometric models based on DBH is also not necessarily suitable for the AGB estimation based on height stratification. (3) Because the individual tree segmentation is mainly based on the CHM layer, the AGB of the lower wood layer under the CHM layer may be ignored.	(1) The AGB extrapolation method is affected by the distribution details of gaps or non-trees in sample plots. (2) AGB estimation based on LiDAR indices focus on the external structural characteristics of plants, such as tree height, and crown width, while neglecting the inherent characteristics of plants, such as wood density, which affects the accuracy of AGB estimation. (3) This extrapolation is not conducive to the fine expression of AGB at the mangrove plant type and individual tree level.
Commonground	The initial AGB values for extrapolation depend on the estimation of species-specific allometric models in the field survey.	

4.3. AGB Comparison with Mangroves in Other Areas

The mangroves in the study area had an AGB value of 82.088–209.520 Mg·ha^{-1}. Compared with the mangrove AGB in other parts of the world, the AGB of *B. sexangular*-dominated forest is 204.933 Mg·ha^{-1}, lower than the value of 279.00 Mg·ha^{-1} in Indonesia [61]. The AGB of *E. agallocha*-dominated forest is 82.088 Mg·ha^{-1}, higher than the value of 14.93 Mg·ha^{-1} for pure *E. agallocha forest* in India (Central Sundarbans) [62]. The AGB of *Sonneratia* spp.-dominated forest is 277.201 Mg·ha^{-1}, higher than the value of 169.61 Mg·ha^{-1} in India (Western Sundarbans) [62], the value of 22.30 Mg·ha^{-1} in China (Guangdong, Futian) [63] and the value of 189.37 Mg·ha^{-1} in China (Guangdong, Leizhou) [64], but lower than the value of 281.20 Mg·ha^{-1} in Thailand (Southern Ranong) [65]. The AGB of *R. apiculata*-dominated forest is 209.52 Mg·ha^{-1}, much lower than those of both pure and mixed forests found in tropical regions (214.00 Mg·ha^{-1} in India (Andaman Island) [66], 216.00 Mg·ha^{-1} in Thailand (Chumphon Sawi Bay) [67], 298.50 Mg·ha^{-1} in Thailand (Ranong Southern) [65], 295.50–350.30 Mg·ha^{-1} in Vietnam (Mekong delta) [68], 460.00 Mg·ha^{-1} in Malaysia (Matang) [69], and 356.80 Mg·ha^{-1} in Indonesia (Halmahera) [70]).

Overall, the AGB of this study essentially aligns with the latitude law of AGB distribution in mangroves, i.e., AGB increases with the decrease in latitude.

5. Conclusions

In this study, we combined the advantages of WorldView-2, UAV LiDAR, and field survey data and proposed a novel method to estimate mangrove AGB at individual tree scale, i.e., individual tree-based inference method, and compared it with the benchmark grid-based RF model method. Although the AGB estimation accuracy of this new method is less than that of the grid-based RF inversion method (R^2 of 0.49 vs. 0.67 and RMSE of 48.42 Mg ha^{-1} vs. 38.95 Mg ha^{-1}), the individual tree-based inference method still has some merits. The individual tree-based inference method can show the spatial distribution details of mangrove AGB in different mangrove species, which is more beneficial to the fine management of mangroves. Considering that the mangrove forest in the study area is complex, the results of the newly proposed method are relatively satisfying.

There are many uncertainties in mangrove AGB estimation due to sampling type, spatial resolution of remote sensing data, local topography, biophysical conditions, forest structure, wood density, tree species, size of trees, selection of allometric models, and technical factors related to data processing. Traditional field surveys are still the most important base for establishing a reliable relationship between biomass and remote sensing variables. With the support of a variety of remote sensing data, more targeted field survey methods, such as tree species, habitat, forest age and wood density, will be further tried.

Author Contributions: Conceptualization, P.Q., D.W. and X.Z.; Methodology, P.Q. and D.W.; Software, P.Q. and D.W.; Validation, P.Q. and D.W.; Formal analysis, P.Q. and D.W.; Investigation, P.Q., D.W., X.Y., G.X. and Z.Z.; Resources, P.Q.; Data curation, P.Q.; Writing—original draft preparation, P.Q.; Writing—review & editing, P.Q., D.W., X.Z. and X.Y.; Visualization, P.Q. and D.W.; Supervision, D.W.; Project administration, P.Q. and S.X.; Funding acquisition, P.Q. and S.X.

Funding: National Natural Science Foundation of China: 41361090, 41877411. This research received no external funding.

Acknowledgments: Authors are thankful to the Administration Station of Qinglan Harbor Mangrove Nature Reserve for providing maps of the Nature Reserve. This study is supported by the National Natural Science Foundation of China (No. 41361090, 41877411).

Conflicts of Interest: The authors declare no conflict of interest.

Appendix A

Table A1. Pearson correlation analysis of diameter at breast height (DBH) classification and deviation of individual-tree segmentation measured in two-centimeter intervals.

	DBH_1	DBH_2	DBH_3	DBH_4	DBH_5	DBH_6	DBH_7	DBH_8	DBH_9	DBH_{10}	DBH_{11}	DBH_{12}	DBH_{13}	DBH_{14}	DBH_{15}	DBH_{16}	DBH_{17}	DBH_{18}	DBH_{19}	DBH_{20}	DBH_{21}	Max.	Min.	Mean
DBH_1	1																							
DBH_2	0.094	1																						
DBH_3	0.078	0.173	1																					
DBH_4	0.157	−0.244	0.258	1																				
DBH_5	−0.002	−.202	0.258	0.834 **	1																			
DBH_6	0.243	−0.213	0.146	0.804 **	0.814 **	1																		
DBH_7	0.292	−0.374 *	−0.154	0.611 **	0.541 **	0.676 **	1																	
DBH_8	0.084	−0.276	−0.275	0.262	0.226	0.469 **	0.604 **	1																
DBH_9	0.086	−0.248	−0.208	0.323 *	0.304 *	0.523 **	0.576 **	0.535 **	1															
DBH_{10}	−0.044	−0.064	−0.306 *	−0.045	0.072	0.152	0.338 *	0.568 **	0.489 **	1														
DBH_{11}	−0.034	−0.097	−0.124	0.197	0.289	0.400 **	0.475 **	0.315 *	0.608 **	0.347 *	1													
DBH_{12}	−0.056	0.084	−0.284	−0.145	−0.085	0.074	0.124	0.241	0.509 **	0.508 **	0.617 **	1												
DBH_{13}	−0.074	−0.164	−0.268	−0.266	−0.131	−0.158	0.003	0.117	0.184	0.147	0.295 *	0.141	1											
DBH_{14}	0.087	−0.016	−0.249	−0.265	−0.154	−0.083	−0.052	0.328 *	0.381 **	0.442 **	0.275	0.390 **	0.624 **	1										
DBH_{15}	−0.094	0.075	−0.211	−0.085	−0.049	−0.023	0.008	−0.037	0.217	0.088	0.485 **	0.531 **	0.609 **	0.345 *	1									
DBH_{16}	−0.081	−0.147	−0.197	−0.229	−0.115	−0.149	−0.020	−0.026	0.019	−0.043	0.286	0.146	0.835 **	0.393 **	0.659 **	1								
DBH_{17}	0.087	−0.045	−0.053	−0.142	−0.028	0.018	−0.004	−0.036	0.036	−0.040	0.334 *	0.106	0.823 **	0.347 *	0.623 **	0.831 **	1							
DBH_{18}	0.008	−0.098	−0.188	−0.147	−0.060	0.026	0.039	0.181	0.551 **	0.234	0.405 **	0.357 *	0.687 **	0.791 **	0.402 **	0.376 *	0.464 **	1						
DBH_{19}	−0.056	−0.105	−0.119	−0.081	−0.040	−0.008	0.056	0.221	0.294 *	0.109	0.200	0.145	0.499 **	0.543 **	0.096	0.139	0.218	0.764 **	1					
DBH_{20}	−0.066	−0.103	−0.058	−0.058	0.042	0.062	0.059	0.102	−0.031	0.079	0.298 *	−0.016	0.170	0.117	0.056	0.408 **	0.159	−0.058	−0.044	1				
DBH_{21}	−0.039	−0.079	−0.097	−0.134	−0.066	−0.031	−0.017	0.102	0.383 **	0.132	0.269	0.201	0.522 **	0.628 **	0.133	0.176	0.262	0.866 **	0.892 **	−0.033	1			
Max.	0.126	−0.294	−0.320 *	0.130	0.207	0.393 **	0.465 **	0.574 **	0.528 **	0.462 **	0.622 **	0.519 **	0.518 **	0.532 **	0.469 **	0.463 **	0.422 **	0.467 **	0.382 **	0.379 *	0.307 *	1		
Min.	−0.306 *	−0.120	−0.162	0.031	−0.001	−0.090	0.140	−0.012	−0.155	−0.076	0.006	−0.110	0.001	−0.241	0.156	0.092	−0.031	−0.251	−0.215	0.038	−0.265	−0.159	1	
Mean	−0.005	−0.250	−0.518 **	−0.151	−0.037	0.040	0.260	0.372 *	0.412 **	0.488 **	0.487 **	0.565 **	0.692 **	0.628 **	0.660 **	0.634 **	0.483 **	0.493 **	0.296 *	0.306 *	0.309 *	0.791 **	0.142	1
DA	−0.211	−0.356 *	−0.388 **	−0.116	−0.024	−0.002	0.302 *	0.309 *	0.371 *	0.289	0.360 *	0.226	0.799 **	0.394 **	0.569 **	0.637 **	0.603 **	0.522 **	0.395 **	0.054	0.391 **	0.556 **	0.086	0.695 **

DBH_1, DBH_2, DBH_{20}, DBH_{21} denote DBH < 4 cm, 4 ≤ DBH < 6cm, 40 ≤ DBH < 42 cm, DBH ≥ 42 cm, respectively; Max. means maximum DBH, Min. means minimum DBH; * Significant at 5%; ** Significant at 1%.

References

1. Saenger, P. *Mangrove Ecology, Silviculture and Conservation*; Kluwer Academic: Dordrecht, The Nederlands, 2003.
2. Constanza, R.; d'Arge, R.; de Groot, R.; Farber, S.; Grasso, M.; Hannon, B.; Limberg, K.; Naeem, S.; O'Neill, R.V.; Paruelo, J.; et al. The value of the world's ecosystem service and natural capital. *Ecol. Econ.* **1997**, *25*, 3–15. [CrossRef]
3. Donato, D.C.; Kauffman, J.B.; Murdiyarso, D.; Kurnianto, S.; Stidham, M.; Kanninen, M. Mangroves among the most carbon–rich forests in the tropics. *Nat. Geosci.* **2011**, *4*, 293–297. [CrossRef]
4. Barbier, E.B.; Hacker, S.D.; Kennedy, C.; Koch, E.W.; Stier, A.C.; Silliman, B.R. The value of estuarine and coastal ecosystem services. *Ecol. Monogr.* **2011**, *81*, 169–193. [CrossRef]
5. Hickey, S.M.; Callow, N.J.; Phinn, S.; Lovelock, C.E.; Duarte, C.M. Spatial complexities in aboveground carbon stocks of a semi–arid mangrove community: A remote sensing height–biomass–carbon approach. *Estuar. Coast. Shelf Sci.* **2018**, *200*, 194–201. [CrossRef]
6. Giri, C.; Ochieng, E.; Tieszen, L.L.; Zhu, Z.; Singh, A.; Loveland, T.; Masek, J.; Duke, N. Status and distribution of mangrove forests of the world using earth observation satellite data. *Glob. Ecol. Biogeogr.* **2011**, *20*, 154–159. [CrossRef]
7. Hamilton, S.E.; Friess, D.A. Global carbon stocks and potential emissions due to mangrove deforestation from 2000 to 2012. *Nat. Clim. Chang.* **2018**, *8*, 240–244. [CrossRef]
8. Fehrmann, L.; Kleinn, C. General considerations about the use of allometric equations for biomass estimation on the example of Norway spruce in central Europe. *Ecol. Manag.* **2006**, *236*, 412–421. [CrossRef]
9. Komiyama, A.; Ong, J.E.; Poungparn, S. Allometry, biomass, and productivity of mangrove forests: A review. *Aquat. Bot.* **2008**, *8*, 128–137. [CrossRef]
10. Ketterings, Q.M.; Coe, R.; van Noordwijk, M.; Ambagu, Y.; Palm, C.A. Reducing uncertainty in use of allometric biomass equations for predicting above–ground tree biomass in mixed secondary forests. *Ecol. Manag.* **2001**, *146*, 199–202. [CrossRef]
11. Overman, J.P.M.; Witte, H.J.L.; Saldarriaga, J.G. Evaluation of regression models for above–ground biomass determination in Amazon rainforest. *J. Trop. Ecol.* **1994**, *10*, 207–218. [CrossRef]
12. Proisy, C.; Couteron, P.; Fromard, F. Predicting and mapping mangrove biomass from canopy grain analysis using Fourier–based textural ordination of IKONOS images. *Remote Sens. Environ.* **2007**, *109*, 379–392. [CrossRef]
13. Jachowski, N.R.A.; Quak, M.S.Y.; Friess, D.A.; Duangnamon, D.; Webb, E.; Ziegler, A.D. Mangrove biomass estimation in Southwest Thailand using machine learning. *Appl. Geogr.* **2013**, *45*, 311–321. [CrossRef]
14. Hamdan, O.; Aziz, H.K.; Hasmadi, I.M. L–band ALOS PALSAR for biomass estimation of Matang Mangroves, Malaysia. *Remote Sens. Environ.* **2014**, *155*, 69–78. [CrossRef]
15. Hartoko, A.; Chayaningrum, S.; Febrianti, D.A.; Ariyanto, D. Carbon biomass algorithms development for mangrove vegetation in Kemujan, Parang Island Karimunjawa National Park and Demak Coastal Area—Indonesia. *Procedia Environ. Sci.* **2015**, *23*, 39–47. [CrossRef]
16. Aslan, A.; Rahman, A.F.; Warren, M.W.; Robeson, S.M. Mapping spatial distribution and biomass of coastal wetland vegetation in Indonesian Papua by combining active and passive remotely sensed data. *Remote Sens. Environ.* **2016**, *183*, 65–81. [CrossRef]
17. Castillo, J.A.A.; Apan, A.A.; Maraseni, T.N.; Salmo, S.G., III. Estimation and mapping of above–ground biomass of mangrove forests and their replacement land uses in the Philippines using Sentinel imagery. *ISPRS J. Photogramm. Remote Sens.* **2017**, *134*, 70–85. [CrossRef]
18. Pham, L.T.H.; Brabyn, L. Monitoring mangrove biomass change in Vietnam using SPOT images and an object–based approach combined with machine learning algorithms. *ISPRS J. Photogramm. Remote Sens.* **2017**, *128*, 86–97. [CrossRef]
19. Dalponte, M.; Frizzera, L.; Ørka, H.O.; Gobakken, T.; Næsset, E.; Gianelle, D. Predicting stem diameters and aboveground biomass of individual trees using remote sensing data. *Ecol. Indic.* **2018**, *85*, 367–376. [CrossRef]
20. Owers, C.J.; Rogers, K.; Woodroffe, C.D. Terrestrial laser scanning to quantify above–ground biomass of structurally complex coastal wetland vegetation. *Estuar. Coast. Shelf Sci.* **2018**, *204*, 164–176. [CrossRef]

21. Demir, N. Using UAVs for detection of trees from digital surface models. *J. For. Res.* **2018**, *29*, 813–821. [CrossRef]

22. Kestur, R.; Angural, A.; Bashir, B.; Omkar, S.N.; Anand, G.; Meenavathi, M.B. Tree Crown Detection, Delineation and Counting in UAV Remote Sensed Images: A Neural Network Based Spectral–Spatial Method. *J. Indian Soc. Remote Sens.* **2018**, *46*, 991–1004. [CrossRef]

23. Disney, M.; Burt, A.; Calders, K.; Schaaf, C.; Stovall, A. Innovations in ground and airborne technologies as reference and for training and validation: Terrestrial laser scanning (TLS). *Surv. Geophys.* **2019**. [CrossRef]

24. Kędra, K.; Barbeito, I.; Dassot, M.; Vallet, P.; Gazda, A. Single–image photogrammetry for deriving tree architectural traits in mature forest stands: A comparison with terrestrial laser scanning. *Ann. For. Sci.* **2019**, *76*, 5. [CrossRef]

25. Carrijo, J.V.N.; de Freitas Ferreira, A.B.; Ferreira, M.C.; de Aguiar, M.C.; Miguel, E.P.; Matricardi, E.A.T.; Rezende, A.V. The growth and production modeling of individual trees of Eucalyptus urophylla plantations. *J. For. Res.* **2019**. [CrossRef]

26. Yin, D.; Wang, L. Individual mangrove tree measurement using UAV–based LiDAR data: Possibilities and challenges. *Remote Sens. Environ.* **2019**, *223*, 34–49. [CrossRef]

27. Shao, Z.F.; Zhang, L.J.; Wang, L. Stacked sparse autoencoder modeling using the synergy of airborne LiDAR and satellite optical and SAR data to map forest above-ground biomass. *IEEE J. Sel. Top. Appl. Earth Obs. Remote Sens.* **2017**, *10*, 5569–5582. [CrossRef]

28. Baral, S. *Mapping Carbon Stock Using High Resolution Satellite Images in Sub–Tropical Forest of Nepal*; University of Twente University of Faculty of Geo-Information and Earth Observation (ITC): Enschede, The Netherlands, 2011.

29. Maharjan, S. *Estimation and Mapping above Ground Woody Carbon Stocks Using LiDAR Data and Digital Camera Imagery in the Hilly Forests of Gorkha, Nepal*; University of Twente University of Faculty of Geo-Information and Earth Observation (ITC): Enschede, The Netherlands, 2012.

30. Karna, Y.K.; Hussin, Y.A.; Gilani, H.; Bronsveld, M.C.; Murthy, M.S.R.; Qamer, F.M.; Karky, B.S.; Bhattarai, T.; Aigong, X.; Baniya, C.B. Integration of WorldView–2 and airborne LiDAR data for tree species level carbon stock mapping in Kayar Khola watershed, Nepal. *Int. J. Appl. Earth Obs. Geoinf.* **2015**, *38*, 280–291. [CrossRef]

31. Zhang, L.; Shao, Z.; Liu, J.; Cheng, Q. Deep learning based retrieval of forest aboveground biomass from combined lidar and landsat 8 data. *Remote Sens.* **2019**, *11*, 1459. [CrossRef]

32. Nandy, S.; Ghosh, S.; Kushwaha, S.P.S.; Kumar, A.S. Remote sensing–based forest biomass assessment in northwest Himalayan landscape. In *Remote Sensing of Northwest Himalayan Ecosystems*; Navalgund, R.R., Kumar, A.S., Nandy, S., Eds.; Springer: Singapore, 2019.

33. Meyer, V.; Saatchi, S.; Ferraz, A.; Xu, L.; Duque, A.; García, M.; Chave, J. Forest degradation and biomass loss along the Chocó region of Colombia. *Carbon Balance Manag.* **2019**, *14*, 2. [CrossRef]

34. Kellner, J.R.; Armston, J.; Birrer, M.; Cushman, K.C.; Duncanson, L.; Eck, C.; Falleger, C.; Imbach, B.; Král, K.; Krůček, M.; et al. New Opportunities for Forest Remote Sensing Through Ultra High Density Drone LiDAR. *Surv. Geophys.* **2019**. [CrossRef]

35. Hamilton, S.E.; Castellanos–Galindo, G.A.; Millones–Mayer, M.; Chen, M. Remote sensing of mangrove forests: Current techniques and existing databases. In *Threats to Mangrove Forests*; Makowski, C., Finkl, C.W., Eds.; Coastal Research Library: Boca Raton, FL, USA, 2018; Volume 25, pp. 497–520.

36. Fragoso–Campón, L.; Quirós, E.; Mora, J.; Gallego, J.A.G.; Durán–Barroso, P. Overstory–understory land cover mapping at the watershed scale: Accuracy enhancement by multitemporal remote sensing analysis and LiDAR. *Environ. Sci. Pollut. Res.* **2019**. [CrossRef] [PubMed]

37. Cintron, G.; Schaeffer–Novelli, S.Y. Methods for studying Mangrove structure. In *The Mangrove Ecosystems: Research Methods*; Snedaker, S.C., Snedaker, J., Eds.; UNESCO: Paris, France, 1984; pp. 91–113.

38. Clough, B.F.; Scott, K. Allometric relationships for estimating above–ground biomass in six mangrove species. *For. Ecol. Manag.* **1989**, *27*, 117–127. [CrossRef]

39. Hossain, M.; Siddique, M.R.H.; Saha, S.; Abdullah, S.M.R. Allometric models for biomass, nutrients and carbon stock in Excoecaria agallocha of the Sundarbans, Bangladesh. *Wetl. Ecol. Manag.* **2015**, *23*, 765–774. [CrossRef]

40. Ong, J.E.; Gong, W.K.; Wong, C.H. Allometry and partitioning of the mangrove, Rhizophora apiculata. *For. Ecol. Manag.* **2004**, *188*, 395–408. [CrossRef]

41. Fromard, F.; Puig, H.; Mougin, E.; Marty, G.; Betoulle, J.L.; Cadamuro, L. Structure of above–ground biomass and dynamics of mangrove ecosystems: New data from French Guiana. *Oecologia* **1998**, *115*, 39–53. [CrossRef]

42. Tam, N.F.Y.; Wong, Y.S.; Lan, C.Y.; Chen, G.Z. Community structure and standing crop biomass of a mangrove forest in Futian Nature Reserve, Shenzhen, China. *Hydrobiologia* **1995**, *295*, 193–201. [CrossRef]

43. Kusmana, C.; Hidayat, T.; Tiryana, T.; Rusdiana, O. Allometric models for above-and below–ground biomass of Sonneratia spp. *Glob. Ecol. Conserv.* **2018**, *15*, e00417. [CrossRef]

44. Komiyama, A.; Poungparn, S.; Kato, S. Common allometric equations for estimating the tree weight of mangroves. *J. Trop. Ecol.* **2005**, *21*, 471–477. [CrossRef]

45. Wang, D.; Wan, B.; Qiu, P.; Su, Y.; Guo, Q.; Wang, R.; Sun, F.; Wu, X. Evaluating the Performance of Sentinel-2, Landsat 8 and Pléiades-1 in Mapping Mangrove Extent and Species. *Remote Sens.* **2018**, *10*, 1468. [CrossRef]

46. Breiman, L. Random Forests. *Mach. Learn.* **2001**, *45*, 5–32. [CrossRef]

47. Leboeuf, A.; Beaudoin, A.; Fournier, R.A.; Guindon, L.; Luther, J.E.; Lambert, M.-C. A shadow fraction method for mapping biomass of northern boreal black spruce forests using QuickBird imagery. *Remote Sens. Environ.* **2007**, *110*, 488–500. [CrossRef]

48. Holmgren, J.; Persson, A.; Söderman, U. Species identification of individual trees by combining high resolution LiDAR data with multi–spectral images. *Int. J. Remote Sens.* **2008**, *29*, 1537–1552. [CrossRef]

49. Othmani, A.; Piboule, A.; Krebs, M.; Stolz, C.; Voon, L.L.Y. Towards automated and operational forest inventories with T–LiDAR. In Proceedings of the 11th International Conference on LiDAR Applications for Assessing Forest Ecosystems (SilviLaser 2011), Hobart, Australia, 16–20 October 2011.

50. Sousa, A.M.O.; Gonçalves, A.C.; Mesquita, P.; da Silva, J.R.M. Biomass estimation with high resolution satellite images: A case study of Quercus rotundifolia. *ISPRS J. Photogramm. Remote Sens.* **2015**, *101*, 69–79. [CrossRef]

51. Aihua Li Glenn, N.F.; Olsoy, P.J.; Mitchell, J.J.; Shrestha, R. Aboveground biomass estimates of sagebrush using terrestrial and airborne LiDAR data in a dryland ecosystem. *Agric. For. Meteorol.* **2015**, *213*, 138–147.

52. Dalponte, M.; Coomes, D.A. Tree–centric mapping of forest carbon density from airborne laser scanning and hyperspectral data. *Methods Ecol. Evol.* **2016**, *7*, 1236–1245. [CrossRef]

53. Stovall, A.E.L.; Anderson–Teixeira, K.J.; Shugart, H.H. Assessing terrestrial laser scanning for developing non–destructive biomass allometry. *For. Ecol. Manag.* **2018**, *427*, 217–229. [CrossRef]

54. Bazezew, M.N.; Hussin, Y.A.; Kloosterman, E.H. Integrating airborne LiDAR and terrestrial laser scanner forest parameters for accurate above–ground biomass/carbon estimation in Ayer Hitam tropical forest, Malaysia. *Int. J. Appl. Earth Obs. Geoinf.* **2018**, *73*, 638–652. [CrossRef]

55. Wang, L.; Gong, P.; Biging, G.S. Individual tree–crown delineation and treetop detection in high–spatial–resolution aerial imagery. *Photogramm. Eng. Remote Sens.* **2004**, *70*, 351–357. [CrossRef]

56. Duro, D.C.; Franklin, S.E.; Dubé, M.G. A comparison of pixel-based and object-based image analysis with selected machine learning algorithms for the classification of agricultural landscapes using SPOT-5 HRG imagery. *Remote Sens. Environ.* **2012**, *118*, 259–272. [CrossRef]

57. Yin, D.; Wang, L. How to assess the accuracy of the individual tree-based forest inventory derived from remotely sensed data: A review. *Int. J. Remote Sens.* **2016**, *37*, 4521–4553. [CrossRef]

58. Strîmbu, V.F.; Strîmbu, B.M. A graph-based segmentation algorithm for tree crown extraction using airborne LiDAR data. *ISPRS J. Photogramm. Remote Sens.* **2015**, *104*, 30–43. [CrossRef]

59. Zhou, T.; Popescu, S.; Lawing, A.; Eriksson, M.; Strimbu, B.; Bürkner, P. Bayesian and classical machine learning methods: A comparison for tree species classification with LiDAR waveform signatures. *Remote Sens.* **2018**, *10*, 39. [CrossRef]

60. Pouliot, D.A.; King, D.J.; Bell, F.W.; Pitt, D.G. Automated tree crown detection and delineation in high-resolution digital camera imagery of coniferous forest regeneration. *Remote Sens. Environ.* **2002**, *82*, 322–334. [CrossRef]

61. Kusmana, C.; Sabiham, S.; Abe, K.; Watanabe, H. An estimation of above ground tree biomass of a mangrove forest in East Sumatra, Indonesia. *Tropics.* **1992**, *1*, 243–257. [CrossRef]

62. Mitra, A.; Zaman, S. Abiotic Variables of the Marine and Estuarine Ecosystems. In *Basics of Marine and Estuarine Ecology*; Mitra, A., Zaman, S., Eds.; Springer: New Delhi, India, 2016; pp. 182–184.

63. Zan, Q.J.; Wang, Y.J.; Liao, B.W.; Zheng, D.Z. Biomass and net productivity of Sonneratia apetala, S. caseolaris mangrove man-made forest. *J. Wuhan Bot. Res.* **2001**, *19*, 391–396.

64. Han, W.D.; Gao, X.M.; Teunissen, E. Study on Sonneratia apetala productivity in restored forests in Leizhou Peninsula, China. *J. For. Res.* **2001**, *12*, 229–234.

65. Komiyama, A.; Ogino, K.; Aksomkoae, S.; Sabhasri, S. Root biomass of a mangrove forest in southern Thailand 1: Estimation by the trench method and the zonal structure of root biomass. *J. Trop. Ecol.* **1987**, *3*, 97–108. [CrossRef]

66. Mall, L.P.; Singh, V.P.; Garge, A. Study of biomass, litter fall, litter decomposition and soil respiration in monogeneric mangrove and mixed mangrove forests of Andaman Islands. *Trop. Ecol.* **1991**, *32*, 144–152.

67. Alongi, D.M.; Dixon, P. Mangrove primary production and above and below–Ground biomass in Sawi Bay, southern Thailand. *Phuket Mar. Biol. Cent. Spec. Publ.* **2000**, *22*, 31–38.

68. Phan, S.M.; Nguyen, H.T.T.; Nguyen, T.K.; Lovelock, C. Modelling above ground biomass accumulation of mangrove plantations in Vietnam. *For. Ecol. Manag.* **2019**, *432*, 376–386. [CrossRef]

69. Putz, F.; Chan, H.T. Tree growth, dynamics, and productivity in a mature mangrove forest in Malaysia. *For. Ecol. Manag.* **1986**, *17*, 211–230. [CrossRef]

70. Komiyama, A.; Moriya, H.; Prawiroatmodjo, S.; Toma, T.; Ogino, K. Forest primary productivity. In *Biological System of Mangrove*; Ogino, K., Chihara, M., Eds.; Ehime University: Matsuyama, Japan, 1988; pp. 97–117.

Article

Nondestructive Estimation of the Above-Ground Biomass of Multiple Tree Species in Boreal Forests of China Using Terrestrial Laser Scanning

Shilin Chen [1,2], Zhongke Feng [1,2,*], Panpan Chen [1,2], Tauheed Ullah Khan [3] and Yining Lian [1,2]

[1] Precision Forestry Key Laboratory of Beijing, Beijing Forestry University, Beijing 100083, China; chenshilin@bjfu.edu.cn (S.C.); chenpanpan@bjfu.edu.cn (P.C.); lyining@bjfu.edu.cn (Y.L.)
[2] Mapping and 3S Technology Center, Beijing Forestry University, Beijing 100083, China
[3] School of Nature Conservation, Beijing Forestry University, Beijing 100083, China; tauheed@bjfu.edu.cn
*   Correspondence: zhongkefeng@bjfu.edu.cn

Received: 15 September 2019; Accepted: 19 October 2019; Published: 23 October 2019

Abstract: Above-ground biomass (AGB) plays a pivotal role in assessing a forest's resource dynamics, ecological value, carbon storage, and climate change effects. The traditional methods of AGB measurement are destructive, time consuming and laborious, and an efficient, relatively accurate and non-destructive AGB measurement method will provide an effective supplement for biomass calculation. Based on the real biophysical and morphological structures of trees, this paper adopted a non-destructive method based on terrestrial laser scanning (TLS) point cloud data to estimate the AGBs of multiple common tree species in boreal forests of China, and the effects of differences in bark roughness and trunk curvature on the estimation of the diameter at breast height (DBH) from TLS data were quantitatively analyzed. We optimized the quantitative structure model (QSM) algorithm based on 100 trees of multiple tree species, and then used it to estimate the volume of trees directly from the tree model reconstructed from point cloud data, and to calculate the AGBs of trees by using specific basic wood density values. Our results showed that the total DBH and tree height from the TLS data showed a good consistency with the measured data, since the bias, root mean square error (RMSE) and determination coefficient (R^2) of the total DBH were −0.8 cm, 1.2 cm and 0.97, respectively. At the same time, the bias, RMSE and determination coefficient of the tree height were −0.4 m, 1.3 m and 0.90, respectively. The differences of bark roughness and trunk curvature had a small effect on DBH estimation from point cloud data. The AGB estimates from the TLS data showed strong agreement with the reference values, with the RMSE, coefficient of variation of root mean square error (CV(RMSE)), and concordance correlation coefficient (CCC) values of 17.4 kg, 13.6% and 0.97, respectively, indicating that this non-destructive method can accurately estimate tree AGBs and effectively calibrate new allometric biomass models. We believe that the results of this study will benefit forest managers in formulating management measures and accurately calculating the economic and ecological benefits of forests, and should promote the use of non-destructive methods to measure AGB of trees in China.

Keywords: terrestrial laser scanning; above-ground biomass; nondestructive method; DBH; bark roughness

---

1. Introduction

Forest biomass is an important indicator of forest productivity, carbon storage and forest carbon sequestration capacity, and it has been widely investigated by the scientific community [1–3]. As a developing country, China has taken measures to increase forest biomass and carbon storage by limiting deforestation and afforestation, and positively supports and implements the mechanism of

Reducing emissions from deforestation and forest degradation in developing countries (REDD+). Accurate assessment of forest biomass plays a pivotal role in afforestation management planning, forest resource monitoring, the assessment of the ecological value of forests, climate change impacts and policy formulation for forest harvesting, conservation and management [4,5]. The assessment of forest biomass includes the estimation of both above-ground biomass (AGB) and underground biomass. However, the underground biomass is not only difficult to quantify, but it is relatively small to the AGB [6]. Therefore, the estimation of AGB has always been the main focus in biomass research. AGB calculations rely on tree structure parameters, such as diameter at breast height (DBH), tree height, crown radius, etc., form which the AGB can be calculated using allometric biomass models, which can be very effective when applied to tree species and productivity ranges with reliable calibration data. Conventional methods for AGB measurement, which involve cutting down trees and then drying them for weighing, are destructive, time-consuming, expensive and laborious, and are consequently rarely adopted [7,8]. Moreover, the conventional methods can be used only for a small area, as their accuracy could be compromised when used to estimate the AGB of a forest spanning over a larger region [2,8–11].

The use of advanced technologies in forestry, especially remote sensing technology, provides an alternative tool to estimate the AGB with ease and high precision [10,12]. Satellite remote sensing technology provides distinct advantages for the assessment and mapping of large-scale and multi-temporal forest biomass and carbon stocks [13], but it is not applicable or uncertain for forest AGB assessment at the plot and tree level. Recently, the light detection and ranging (LiDAR) technology was developed and advanced rapidly with its special utilization in forest inventory. Primarily, the LiDAR include airborne laser scanning (ALS), terrestrial laser scanning (TLS) and mobile laser scanning (MLS). MLS mounted on vehicles, which is an efficient and effective way to obtain 3D point cloud data in urban forests or forest areas on flat terrain. It relies on GNSS (Global Navigation Satellite System) signals for positioning and the coordinate calculation of points [14,15]. This technology has the ability to generate high spatial resolution and accurate three-dimensional (3D) point cloud data. Consequently, it has been widely applied in forestry surveys to acquire basic tree parameters [16], as well as estimate AGB and carbon storage [17,18]. ALS can produce large-scale 3D point cloud data in a short time, from which tree height, DBH, canopy height and density metrics can be obtained, and then the AGB of trees can be evaluated. The accuracy of AGB estimates by this technology is not higher than that of conventional methods [17], but it is nevertheless higher than that of satellite remote sensing and UAV (unmanned aerial vehicle) aerial photogrammetry [19,20]. However, this method of assessing the biomass by ALS is prone to problems, including large estimation uncertainties, large costs, and limited information [21,22]. The system's performance is compromised in forest areas with weak GNSS signals or large variations of topography [16,23]. TLS can generate detailed and accurate parameter information of the 3D structure of trees by calculating the time difference between the emission and return of laser pulses and analyzing the energy of the returned laser pulses, which is not affected by GNSS signals and offers opportunities for a consistent and robust framework to support AGB estimates [3,24].

Terrestrial laser scanning (TLS) has shown great potential for accurately assessing forest biomass with greater precision than inferred from the nationwide allometric biomass models [8,25]. Yao et al. [26] used high-precision TLS data to obtain accurate tree structure parameters, and calculated the biomass of New England forest stands using allometric biomass models of specific tree species, demonstrating the accuracy and effectiveness of measuring forest AGB using non-destructive methods. Seidel et al. [27] measured the DBH of individual trees from the TLS data, and predicted the biomass of trees via a regional allometric biomass model. The mean absolute error and the mean relative error were 12.9 kg and 16.4%, respectively, which significantly reduced fieldwork efforts in dense forests when compared to traditional diameter tallying by calipers or tapes. However, this non-destructive measurement method still relies on an allometric biomass model established by empirical relationships extracted from a sample of trees to evaluate the AGB. Fundamentally, the essence of this method is still to estimate the biomass of trees based on limited structure parameters. A different approach has been developed to reconstruct the complete 3D structure of trees from TLS data rather than several tree structure parameters. The quantitative structure models (QSM) developed by Raumonen et al. [28] and improved by Raumonen et al. [25] and Calders et al. [2] were used to reconstruct the morphological structure of individual trees from TLS data. The volume of the reconstructed model, including the trunk and branches, can be measured from a single tree model constructed by a least squares, cylinder fitting algorithm [28]. The estimated tree volume is converted into the AGB by multiplying by the basic wood density values of a specific tree species. This method estimates the tree AGB based on a real biological morphological structure model of a specific tree species, which is completely different from the allometric biomass model, which only depends on a limited number of tree structure parameters [8,29].

Some study results have shown the feasibility and effectiveness of the QSM method for the estimation of forest AGB. Raumonen et al. [25] used the QSM algorithm to reconstruct the tree 3D structure models of oak and eucalyptus, and then calculated the AGB of oak and eucalyptus using basic wood density values. Compared with destructively harvested biomass, the calculated biomass of oak was overestimated by 15.3% to 18.8%, and the average relative absolute error of eucalyptus biomass was about 28.5%. Calders et al. [2] used this algorithm to estimate the biomass of 65 eucalyptus trees in tropical areas, with a coefficient of variation of root mean square error (CV(RMSE)), and concordance correlation coefficient (CCC) of 16.1% and 0.98, respectively, showing a high biomass estimation accuracy. At the same time, Tanago et al. [8] validated the applicability of the QSM algorithm to the estimation of the biomass of large trees under complex conditions in tropical regions. The CV (RMSE) and CCC of the estimated total AGB of trees were 28.37% and 0.95% respectively. Although these studies all used the QSM algorithm to estimate the AGB of trees, only a few species were involved, and the QSM algorithm had an impact on the reconstruction results of trees with different biophysical and morphological structures, which directly affected the final estimation accuracy of biomass [25].

In this study, we used a non-destructive method (QSM) to estimate the AGB of trees from TLS point cloud data based on their true morphological structure. We hypothesized that different tree species have different bark roughness and trunk curvature, which could affect the estimation of DBH from point cloud data. To our knowledge, very limited work has been done to investigate the effects of bark roughness and trunk curvature on DBH estimates. Therefore, the main objectives of this study included (1) a quantitative analysis of the influence of TLS data on DBH estimation of tree species with different bark roughness and trunk curvature, and (2) to optimize the QSM algorithm based on 100 trees of 10 different species. (3) We estimated the AGB of 10 tree species with different biophysical and morphological structures using a non-destructive method, and the estimation accuracy of AGB of trees was evaluated by comparing with the results of the regional allometric biomass model of specific tree species from specific areas.

2. Materials and Methods

2.1. Study Area

The study area was located in Beijing province (39.43°–41.05° N, 115.42°–117.50° E) (Figure 1). The long-term annual averages of minimum and maximum daily air temperatures in Beijing are 9 °C and 19 °C, respectively, with an altitude range of 20–1500 m. Ten rectangular plots, including one with dimensions of 32 m × 32 m (low stem density) and nine with dimensions of 16 m × 16 m, were used to obtain experimental data in the study; of those, 2, 2, 3 and 3 were respectively located in Chaoyang District, Huairou District, Changping District and Fangshan District. The topography of the plots was characterized by gentle slopes of less than 45 degrees. Data acquisition was carried out in coniferous, broad-leaved and mixed arbor plantations during the period from March to May of 2018. The main tree species were gingko (*Ginkgo biloba*), saliz matsudana (*Salix matsudana*), Chinese scholartree (*Sophora japonica*), Chinese pine (*Pinus tabulaeformis*), Chinese catalpa (*Catalpa bungee*), white wax *(Fraxinus pennsylvanica)*, Chinese white poplar *(Populus tomentosa)*, locust *(Robinia pseudoacacia)*, metasequoia *(Metasequoia glyptostroboides)* and China savin *(Juniperus chinensis)*, which are common trees in the boreal forests of china. The trees specimens had different bark roughness and trunk curvature. There were low shrubs in the study area, all of which were less than 1.2 m in height, that had no effect on the acquisition of DBH from the point cloud data of trunks.

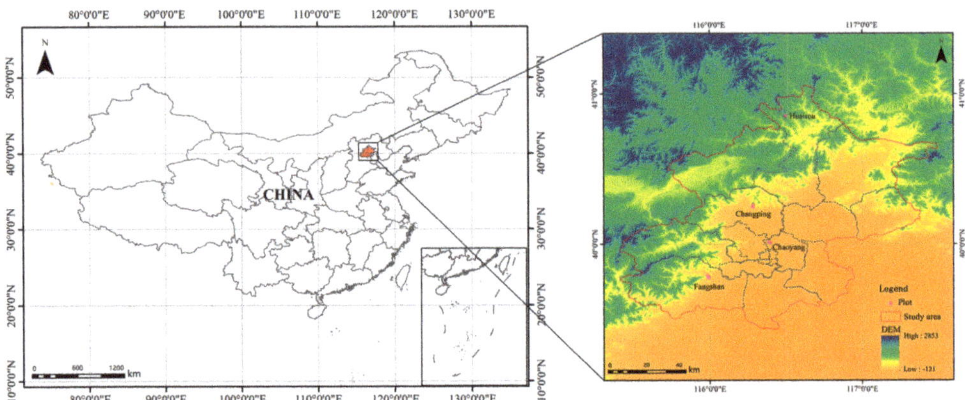

Figure 1. Location of the study area. (The pink dots depict the location of the plots).

2.2. Data Acquisition

2.2.1. Field Data Collection

Field surveys were conducted to collect data, including tree species, DBH and tree height information. The data was collected in ten different sampling plots using traditional forest inventory methods. An experienced taxonomist accompanied the survey team to identify the tree species in the field. The allometric growth biomass models used in this study all require DBH greater than 5 cm. Therefore, the DBH of each tree of more than 5 cm was manually measured using a diameter tape with millimeter accuracy at DBH height (1.3 m vertical above the ground from the base of the tree). A total station with centimeter-scale ranging accuracy developed by the South Surveying and Mapping Technology CO., LTD (Guangzhou, China) was used to accurately measure the tree height in the plot, which was achieved by the principle of triangulation. The structure parameters of the ten forest types based on field forest inventory are summarized in Table 1. The growth model of a specific tree species was used to estimate the tree height in cases where the treetop was not visible due to occlusion by adjacent trees (a total of 13 tree heights were calculated by the tree growth model) [30–32]. The DBH and tree height data of 322 trees obtained in the field were used as references, which were provided in Table S1 of Supplementary Materials.

Table 1. Structural characteristics of ten forest types based on field forest inventory data.

Plot	Dominant Species	Number of Trees	DBH (cm)				Tree Height (m)			
			Mean	SD	Min	Max	Mean	SD	Min	Max
1	*Ginkgo biloba*	78	17.4	1.6	13.5	21.5	11.2	1.4	7.9	14.0
2	*Salix matsudana*	30	22.9	5.5	12.3	40.6	16.3	2.6	8.4	20.6
3	*Sophora japonica*	33	14.1	3.5	8.8	21.5	13.4	2.5	6.9	16.6
4	*Pinus tabulaeformis*	28	17.5	2.3	12.3	23.1	9.2	1.5	5.4	11.5
5	*Catalpa bungei*	22	21.7	2.4	16.2	25.1	16.3	1.3	12.7	18.9
6	*Fraxinus pennsylvanica*	29	18.4	2.0	14.5	22.8	10.0	0.8	8.7	12.3

Table 1. *Cont.*

Plot	Dominant Species	Number of Trees	DBH (cm)				Tree Height (m)			
			Mean	SD	Min	Max	Mean	SD	Min	Max
7	*Populus tomentosa*	24	21.1	2.1	17.7	26.2	20.4	2.0	15.3	23.8
8	*Robinia pseudoacacia*	25	18.9	3.8	12.8	27.2	15.7	2.3	10.5	18.6
9	*Metasequoia glyptostroboides*	31	14.6	2.1	8.9	18.6	11.6	1.3	9.4	15.4
10	*Juniperus chinensis*	22	23.1	3.5	14.9	28.5	11.3	0.8	9.7	13.4

SD stands for standard deviation; Min stands for the minimum value of DBH; Max stands for the maximum value of DBH.

2.2.2. TLS Data Collection

The TLS data were collected in the spring of 2018 using a FARO Focus S 150 terrestrial laser scanner produced by FARO Technologies Company (FL, America), a phase-based scanner with a field of view of 360° horizontally and approximately 300° vertically, and a minimum horizontal and vertical step size of 0.009° (approximately 40,000 laser pulses for a full hemispherical scan). The maximum rate of the laser scanner data acquisition was 976,000 points per second, and an acquisition rate of 244,000 points per second was used in this study. The scanner employed a continuous wave of 1500 nm to measure distances with a range of up to 153 m. The laser scanning level of the scanner is one level. The system's distance error is less than 1 mm within a distance of 25 m, which helped to acquire highly accurate data for the surveyed forest sample plots. More detailed information about the device is provided in Table 2. The laser scanner device with an embedded microcomputer can store and preprocess the point cloud data after completion of data collection. The laser scanner had corresponding FARO Scene desktop software (FARO Technologies, Inc., version 7.1.0, www.faro.com) for post-processing of point cloud data. The FARO Scene software was used to assess the quality of the point cloud data and filter out "ghost points" and discrete points. Finally, by utilizing the functions of "Clear Sky" and "Clear Contour," we obtained more reliable 3D spatial data for further modelling.

Table 2. Technical specifications of the FARO Focus S 150 instrument.

Parameter	Value
Laser measurement principle	Phase-based
Data Acquisition Speed	9.76×10^5 points/sec
Maximum Range	150 m
Laser Power	20 mW
Beam Divergence Angle	0.3 mrad
Scanner Line Speed	2880 rpm
Angular Resolution	±0.009°
Battery Life	4.5 h
Total Weight	4.2 kg

In order to reduce the effects of tree occlusion and terrestrial vegetation, we used a multi-scan (MS) approach to obtain better point cloud coverage for the 16 m × 16 m plot, setting scanning positions at four corners of the plot and the center to perform four consecutive scans of the trees in a clockwise sequence. In order to obtain similar point cloud coverage, we scanned the 32 m × 32 m plots according to the layout principle of 16 m × 16 m plots, and scanned a total of 13 positions. Six highly reflective target spheres were placed throughout the plot for registration of point cloud data at different scanning sites using Scene software, and the accuracy of registration was within 1 cm. In addition, the acquisition of all the plot scanning data was conducted at the same scan resolution to eliminate the effects of different resolution on the modeling results, and carried out under windless conditions, which helps to avoid inconsistencies in the spatial position of the same branches at different scanning locations.

2.3. Processing of the Point Cloud Data

2.3.1. Filtering

Filtering of the acquired point cloud data is the basis for accurate and precise modeling. In addition, reducing the size of point cloud data helps to save memory and reduces computation time [25,28]. First, point cloud data outside the plot area was removed. In order to obtain data for the complete canopy of the trees located at the edges of the plot, we set a 5 m buffer zone along the boundary of each plot. Any data beyond this range was excluded from further modeling. Subsequently, point cloud data that did not contain information on tree attributes in the plot was manually removed by visual inspection. Secondly, the ground points and other unrelated understory point clouds were also filtered out. We used the open source software CloudCompare (https://www.danielgm.net/cc/) to automatically filter the ground points. In addition, the height of the understory shrub vegetation in the plots was less than 1.5 m, so the shrub point cloud data in the plot was removed. Third, in order to make the volume of the reconstructed trunks more similar to the volume of the real trees, the noise and outliers were removed from the data, since these could result in the fitting of wrong cylinder models and overestimating the trunk volume [28,33]. A sphere with a radius of 0.3 cm was used to filter noise points in cloud data of plots while removing outliers. Finally, visual inspection of the point cloud data of the trees in the plot was done to manually remove the points from adjacent crowns or stems, if present [34].

2.3.2. Extracting of Individual Trees

In order to achieve accurate modeling of trees and acquisition of their structural parameters, individual trees were extracted from the acquired point cloud data of each plot. The approaches typically used to extract individual trees include fully manually, semi-automatic and automated methods [25,35–37]. Manual extraction of individual trees from large point clouds is inefficient and time-consuming, and it is difficult to accurately segment branches of trees in a dense forest. Therefore, a bottom-up automatic extraction algorithm of individual trees based on biological theory and metabolic ecology theory was used in this study.

The segmentation algorithm of individual trees used in the study, called comparative shortest-path (CSP), was developed by Tao et al. [37]. It is a bottom-up method based on the 3D structure of point cloud data. The extraction of individual trees from the plot was mainly composed of three parts. First, ground points were filtered and normalized. Since the plot terrain was not strictly a two-dimensional plane, ground points were separated before individual tree extraction in order to eliminate the influence of topographic fluctuation on the z-value of the point cloud data. The filtered ground points were generated using a digital elevation model (DEM) of 0.3 m resolution via Kriging interpolation, and the minimum z value of the point cloud data in each grid was used as the real ground to normalize the point cloud data in a plot area. Then, the density-based spatial clustering of applications with a noise algorithm (DBSCAN) was used for automatic identification of tree trunks in the plot, which required the definition of the minimum number of cluster points or the radius of neighboring points to determine whether a point belonged to the real trunk surface [37,38]. The point cloud data of a single trunk in the plot was segmented automatically. A 10 cm thick point cloud slice from the trunk (1.3 m vertical above the ground from the base of the tree) was used to calculate the DBH value as a seed point for the tree. Finally, the seed points and the CSP algorithm based on the metabolic ecology theory were used to segment the canopy point clouds of the trees. According to the distribution of trees in the forest, the canopy point cloud was mainly divided in three ways. For the forest plots where the sparsely distributed canopies did not intersect, the point cloud from the trunk was extended to the points of branches according to the shortest 3D Euclidean geometric distance to complete the segmentation of canopy point cloud data. For adjacent trees with the same size and canopy intersection, the shortest path distance from the point to the trunk was calculated based on biological theory, and the points were assigned to the closest respective target trunks. The remaining points were sequentially calculated one by one to determine the point cloud data of each tree. For the canopy intersection of trees with large differences in size, the distance from the point to the trunk was converted and reduced according to a power-law relationship between the branch length and the branch radius in according to metabolic ecology theory. Then, the converted distances were compared to determine the shortest distance, which yielded the target tree to which the point belonged. The target tree of each point in the point cloud was determined successively until the whole point cloud data was segmented. The conversion formula based on the power-law relationship [37] was as follows:

$$D_{i \to Trunk}^{T} = D_{i \to Trunk} / DBH^{2/3} \tag{1}$$

where $D_{i \to Trunk}^{T}$ is the distance from the i-th point to the base of the trunk after transformation according to the power-law relationship; $D_{i \to Trunk}$ is the biological distance from the i-th point to the base of the trunk; $DBH^{2/3}$ is the fixed scale. More detailed information on the algorithm can be found in Tao et al. [37].

We used the segmentation algorithm to extract individual trees from 10 plots and visually inspected the results (Figure 2). The point cloud data of a single tree after segmentation was further checked one by one, and those trees that were not correctly classified were re-segmented manually, and abnormal points that did not belong to a specific tree were removed to generate more accurate point cloud data of single trees. The accuracy of single-tree segmentation directly affects the modeling results of the QSM algorithm, and thus has an impact on the estimation of aboveground biomass [3,28].

Figure 2. Individual trees extracted automatically from the point cloud data of the plot.

2.3.3. Acquisition of Tree Structure Parameters from TLS Data

The basic parameters of tree height and DBH were obtained from the point cloud data of the plots. The point cloud slices of 10 cm thickness (1.25 m–1.35 m) were intercepted from the filtered point cloud data, which after eliminating the effect of topographic fluctuation, and the DBH values of each tree were calculated using the least squares circle fitting method, which was considered to be the most accurate DBH estimation method compared with the least square cylinder fitting and the circle Hough transformation [3,39]. In dense forests it is difficult to measure the height of trees, because their tops are often occluded by adjacent trees. In order to obtain accurate and reliable tree height information, the top of each tree should be accurately detected by the scanner, and a sufficient number of tree vertex points should be obtained, which is about 1–2 cm in point space at the top of the tree [40]. The scanning positions of the 10 plots in this study were set up to ensure the detection of tree tops. The point cloud data of individual trees was used to calculate the tree height, which was defined as:

$$h_{tree} = Z_{max} \tag{2}$$

where h_{tree} is the calculated height of a tree, while Z_{max} is the maximum value of the Z coordinate in a single vertex of the tree point cloud data.

The DBH and tree height values obtained from point cloud data were also compared with those obtained from field measurements. Analysis of variance (ANOVA) was used to test whether the differences in DBH and tree height estimates were statistically significant at the 95% level of significance, and to analyze DBH and tree height, respectively [41]. The linear regression models between estimates and reference values are also used to show the differences between estimates and reference values. Precision indicators including bias, root mean squared error (RMSE), relative bias and relative RMSE (RMSE%) were used to evaluate the precision of DBH and tree height obtained from point cloud data.

2.4. Model Reconstruction and Algorithmic Optimization

2.4.1. QSM Reconstruction Method for Individual Trees

The QSM method can realize the rapid modeling of point cloud data of individual trees and accurately calculate the modelled tree volume from the reconstructed structure, including the volume of branches and the trunk. The method mainly achieved accurate tree structure reconstruction and acquisition of important structural parameters through the following eight steps. (1) The filtered point cloud data of individual tree is used for tree model reconstruction in QSM algorithm, and the corresponding cover sets are generated for the point cloud on the tree surface. (2) The neighbor-relation between the coverage sets is determined according to the size of the cover sets. (3) According to the size of the cover sets in conjunction with their eigenvalues and vectors, the local geometric features of the tree, the local directions of the trunk and branches, and the angle between the trunk and branches are determined. (4) The base of the trunk and branches are identified based on the characteristics of cover sets and the component parts of the tree are extracted by "growing". (5) The generated cover sets of the tree surface are segmented into the corresponding real trunk and branches. (6) A series of cylinders with different radii, lengths and directions are used to conduct least-squares piecewise fitting for the point cloud data segmented in the preceding step, and the structure of each part of the tree is reconstructed. (7) The gap between the fitted cylinders are found and supplemented with cylinders to complete the cylinder model of the whole tree. (8) The volume and length of the trunk and branches are calculated automatically from the reconstructed model. The method is described and validated in detail by Raumonen et al. [28]. There are, mainly, seven parameters in the process of tree reconstruction, as follows:

$$tree\_qsm\ (d,\ r,\ n,\ d_1,\ r_1,\ n_1,\ l) \tag{3}$$

where d and d_1 represent the minimum distance between the two centers of the cover sets; r and r_1 are the radii of the balls used to generate the cover sets; n and n_1 are the minimum thresholds for the number of points in a given ball; l is the ratio of the length to the radius of the fitting cylinder.

The first three parameters (d, r and n) were used to generate the first cover sets of a single tree, which primarily filters out the noise in the point cloud data for individual trees caused by small branches and leaves. The parameters d_1, r_1 and n_1 were used to generate a finer cover, mainly to determine the neighbor-relation of the cover sets and the local characteristics of the tree. Parameter l controls the magnitude and is used to fit the cylinder. The smaller the cylinder, the more detailed the local characteristics of the tree. Figure 3 shows the point cloud data for a single specimen of *Ginkgo biloba* and the 3D structure reconstructed using the QSM method.

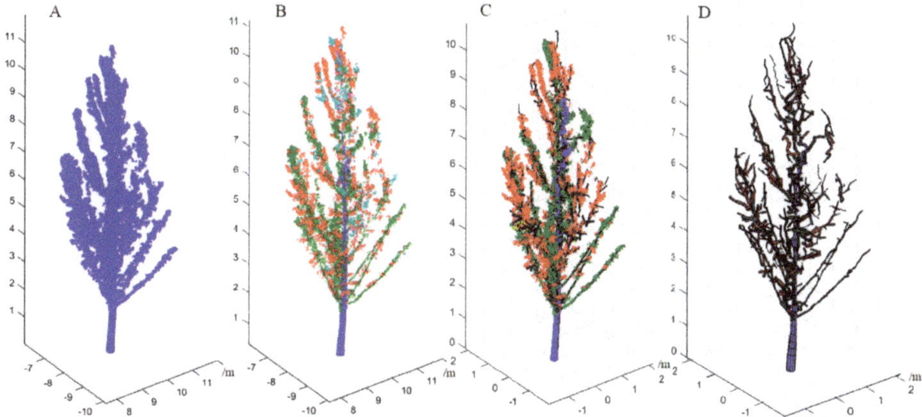

Figure 3. 3D structural model of a single *Ginkgo biloba* tree reconstructed from point cloud data using the QSM (quantitative structure model) algorithm ((**A**) original point cloud data for the trees; (**B**) the original point clouds of the tree are classified, with blue dots representing the trunk, green dots representing the branches and dots in other colors representing the rest of the tree; (**C**) cylinders plotted over the point cloud data; (**D**) reconstructed 3D cylinder model of the tree).

2.4.2. Optimization and Sensitivity Analysis of the QSM Algorithm

The algorithm has two important parameters, d and l, which were used to generate the reconstructed model by defining the covering sets of different sizes and the ratio of the length and radius of the cylinder, which has an important impact on the reconstruction results [3,8]. Therefore, we used different parameter values to test ten tree species with different ecological structures to determine the optimal parameters for each tree species and the sensitivity of the algorithm to different tree species with different ecological structures. Ten trees of each species were randomly selected to optimize the QSM algorithm and determine the optimal model parameters of different tree species. All the remaining data were used for the independent evaluation and precision evaluation of the estimated AGB.

We determined the optimal parameter d from different test values (2.0, 3.0, 4.0, 5.0, 6.0 and 7.0) and the optimal parameter l from different test values (3, 4, 5 and 6), which gave the most accurate biomass estimation according to Tanago et al. [8] and Raumonen et al. [28]. The other parameters r, n, d_1, r_1 and n_1, used the default values $d + 2$, 3, 6, 7 and 1 cm, respectively. Firstly, the average values of the individual tree volume of the selected tree species were obtained from the 10 model realizations using different parameter values for d, after which the biomass calculated by multiplying with the basic wood density of specific tree species was compared with the biomass calculated using the regional allometric model of the same tree species. The biomass of ten trees of the tree species was calculated respectively according to this process, and the RMSE was calculated from the estimated biomass value and the reference biomass value. The optimal parameter d was determined from the corresponding parameter value with the smallest RMSE. Then, once the optimal parameter d was determined, the average value of the individual tree volume was calculated again from the 10 model realizations using different parameter values l. The biomass of the tree was calculated by multiplying its average volume by its density, and the calculated tree biomass was compared with the biomass calculated using the regional allometric model. The biomass of ten trees of the tree species was calculated respectively according to this process, and the RMSE was calculated from the estimated biomass value and the reference biomass value. The optimal parameter l was determined using the minimum RMSE. By this time, the optimal parameters d and l of a tree species were determined. The same operation was performed for the remaining tree species in turn to determine the optimal QSM algorithm for different tree species. Finally, the optimal parameters d and l of ten species were determined respectively, and the optimized QSM algorithm was used to estimate the individual tree volume of the remaining data sets and calculate the individual tree biomass via the same process as described above.

We analyzed the sensitivity of the QSM algorithm using the modeling result for different parameter values of d and l. Sensitivity analysis mainly includes two aspects: the sensitivity analysis of the QSM algorithm for different parameters of the same tree species, and the sensitivity analysis of QSM the algorithm for the same parameters of different tree species. The QSM algorithm for tree reconstruction was implemented in MATLAB software (The MathWorks, Inc., Version Matlab 2018a, Natick, MA, USA) using a Windows 10 64-bit operating system (Microsoft Corporation, Redmond, WA, USA).

2.5. Estimation of AGB

In the estimation of AGB, the differences in tree biomass caused by the errors in the measurement operation, the allometric model and the instrument itself were considered to be minimal and negligible [8,42]. Furthermore, the stem contributed about 70%–80% of the total AGB [43]. We calculated from the data released by the forestry ministerial standard of the People's Republic of China that the biomass of the branches accounted for about 20%–30% of the total AGB. By contrast, the leaves accounted for only 10% of the total biomass, and we consequently did not consider them when assessing the AGB via the QSM method.

2.5.1. AGB Estimation from Allometric Biomass Models

The biomasses of individual trees of ten species were estimated using the regional biomass models specific for each tree species. Those biomass equations were based on variables, such as specific tree species, DBH and tree height, which were collected during the field inventory. The biomass models of the ten tree species used in this study, which evaluate the total AGB, including the biomass of the trunks, branches and leaves, were obtained from published literature and national standards or references (see Table 3). The table contains the information on the number of modelled tree species, the DBH range, tree height range, biomass model parameters, tree species density and coefficient of determination (R^2). The modeling data for the tree species, including *Robinia pseudoacacia*, *Salix matsudana*, *Juniperus chinensis*, *Sophora japonica* and *Fraxinus pennsylvanica*, was not available in detail; therefore, only some information obtained from references is shown. The individual tree biomass obtained from these biomass models was used as a reference value and compared with the biomass calculated from the reconstructed QSM models.

Table 3. Information of the allometric biomass models of ten tree species.

Model Source	Species	Number	DBH Range (cm)	Tree Height (m)	Allometric Biomass Model AGB$_{est}$=	Wood Density(ρ) (g/cm^3)	R^2
SFAC [44]	*Pinus tabulaeformis*	149	5.0–32.9	1.6–20.1	$0.067765D^{2.18050}H^{0.43610}$	0.424	0.9492
Zeng [45]	*Populus tomentosa*	602	5.0–48.9	2.4–31.1	$0.06304D^{2.2460}H^{0.3588}$	0.452	0.9506
Xiaver et al. [46]; Zhuang et al. [47]	*Metasequoia glyptostroboides*	10	8.4–27.5	15.4 (mean)	$0.05488(D^2H)^{0.8583}$	0.284	0.9970
Liu et al. [48]	*Ginkgo biloba*	13	10.0–27.2	11.1–14.5	$e^{(-1.18+2.62\ln D-0.79\ln H)}$	0.455	0.9810
Zhang [49]	*Catalpa bungei*	3	10.6–17.8	7.9–10.2	$0.053(D^2H)^{0.895}+0.018D^{0.037}$	0.520	0.9310
Zhou et al. [50]	*Robinia pseudoacacia*	-	5.0–30.0	-	$0.0261(D^2H)^{0.9613}+0.1012(D^2H)^{0.659}+0.0057(D^2H)^{0.8455}$	0.678	0.9600
Zhou et al. [50]	*Salix matsudana*	-	5.0–38.0	-	$0.075(D^2H)^{0.8210}+0.0110(D^2H)^{0.9430}+0.0130(D^2H)^{0.7520}$	0.506	0.9210
SFAC [51]	*Juniperus chinensis*	-	3.3–33.0	-	$0.2479D^{2.0333}$	0.597	-
Wang [52]	*Sophora japonica*	-	-	-	$0.03D^2H+0.714$	0.636	-
Li et al. [53]	*Fraxinus pennsylvanica*	-	-	-	$0.0495502(D^2H)^{0.952453}$	0.569	-

D stands for DBH; H stands for tree height.

2.5.2. AGB Estimation from TLS-QSM

We used point cloud data of individual trees to fit the topological structure of whole trees using cylinders via the QSM algorithm, calculated the volume of each cylinder part, and then managed to calculate the volume of the trunk and branches of each tree. For given parameters d and l, the random coverage set generated by the QSM algorithm showed small differences in the reconstruction results. Therefore, the volume of the individual tree was defined as the average volume of 10 reconstructed models. The AGB of individual trees was obtained by multiplying the volume of a specific tree (branches and trunks) by the basic wood density of the corresponding tree species. The basic wood density values of specific tree species were obtained from the Global Wood Density Database [54,55] or the literature. We calculated the RMSE, RMSE% and concordance correlation coefficient (CCC) of the biomass of the QSM model relative to the reference data to evaluate the QSM algorithm. The RMSE and RMSE% are defined in the following equations:

$$RMSE = \sqrt{\frac{\sum_1^n \left(AGB^i_{est} - AGB^i_{ref}\right)^2}{n}} \tag{4}$$

$$RMSE\% = \frac{RMSE}{\overline{AGB_{ref}}} \times 100\% \tag{5}$$

where AGB^i_{est} is the i-th estimation value of AGB, AGB^i_{ref} is the ith reference value of AGB, $\overline{AGB_{ref}}$ is the mean of the reference AGB and n is the number of trees.

3. Results

3.1. DBH and Tree Height

A total of 322 trees belonging to ten different species were successfully extracted without commission or omission errors. The estimated DBHs were compared with the reference DBHs, which were measured using a diameter tape (Figure 4). Our estimates showed that most of the DBHs of the ten tree species obtained from the TLS data were below the 1:1 dashed line, indicating that the DBH estimated from point cloud data was smaller than the one measured in the field (Figure 4a). The R^2 of the linear regression model describing the agreement of LiDAR DBH data with the measured DBH values was 0.97, and its slope was also 0.97. A detailed analysis of the effects of different bark roughness and trunk curvature on DBH estimation for each tree species is provided in Appendix A. The underestimation of DBH from TLS data did not change significantly with the increase of DBH. As shown in Figure 4b, most of the DBH residuals of the ten tree species were above the $y = 0$ line, which was consistent with the result that the DBH values from TLS data in Figure 4a were smaller than the values measured in the field. There was no significant difference in the residuals of DBH among different tree species, most of which were between −0.5 cm and 2.0 cm.

A comparison of tree heights from TLS and field measurements, with the tree height measured by the total station as a reference is shown in Figure 5. Our estimates showed that the determination coefficient (R^2) of the linear regression model describing the agreement between the LiDAR tree height data and measured tree height data was 0.90, and its slope was 0.99. The tree height of *Populus tomentosa* estimated using the TLS data was significantly higher than that measured in situ. The estimated tree heights of *Robinia pseudoacacia* and *Catalpa bungei* were consistent with the measured values and were evenly distributed on both sides of the 1:1 line. The estimated height of other tree species was mostly lower than the measured values. A detailed analysis of the effects of different tree species on tree height estimation is provided in Appendix A. Figure 5b illustrates that the residual values of most tree heights were more uniformly distributed on both sides of the $y = 0$ line, and most of the residual values were between −1 m and 2 m. The tree heights obtained from point cloud data were largely underestimated compared with the heights measured in situ. No significant difference was observed in the distribution range of residual values with the increase of tree height.

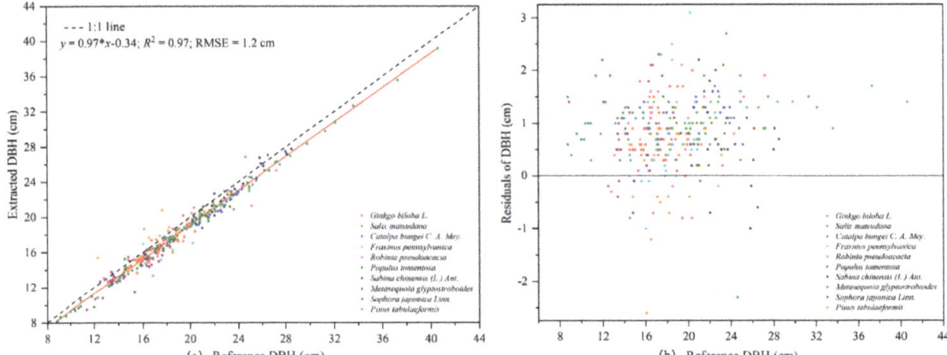

Figure 4. Comparison of diameters at breast height (DBHs) from terrestrial laser scanning (TLS) data and field measurements (reference DBH). (**a**) Comparison of DBHs and references from different tree species; (**b**) residuals of DBHs from different tree species.

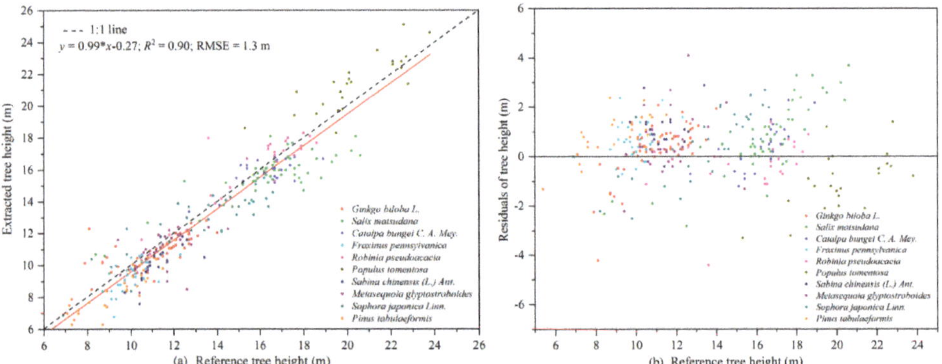

Figure 5. Comparison of tree height from TLS data and field measurements (reference tree height). (**a**) Comparison of tree height and references from different tree species; (**b**) residuals of tree height from different tree species.

The accuracy of DBH and tree height estimation using TLS data is shown in Table 4. It can be seen that the total bias and RMSE of the DBH calculated via the least square circle algorithm were −0.8 cm and 1.2 cm, respectively. Among the 322 selected trees with reference DBH values ranging from 8.8 cm to 40.6 cm, trees with a DBH bias of less than 1.5 cm accounted for 89.1% of the total. The bias and RMSE of tree heights were −0.4 m and 1.3 m, respectively. The bias of tree height from TLS data was within 1.5 m and accounted for 80.8% of the total. A more detailed analysis of tree height and DBH of different tree species is provided in Appendix A.

Table 4. The accuracy of the DBH and tree height estimates utilizing TLS data.

	Bias	**Bias%**	**RMSE**	**RMSE%**
DBH (cm)	−0.8	−4.3	1.2	6.1
Tree height (m)	−0.4	−3.2	1.3	9.7

RMSE stands for root mean square error; RMSE% represents relative root mean square error.

The analysis of variance showed that there was no statistically significant difference between the estimated DBH and tree height for all trees and the field measurements ($\alpha = 0.05$; Table 5). The results of variance analysis of DBH and tree height supported the hypothesis of equality, with p-values of 0.921 and 0.056, respectively.

Table 5. The results of ANOVA analysis for DBH and tree height estimations for all trees with degrees of freedom (DF), mean squared error (MS), F-value and p-value.

Parameter	DF	MS	F-Value	p-Value
DBH	1	1.5	0.010	0.921
Tree height	1	54.3	3.672	0.056

3.2. Optimization Parameters of the QSM Algorithm

The tree structures of different tree species were reconstructed with different parameters of d and l values, and the optimal parameters of the reconstructed model were obtained as shown in Table 6. The optimal value for parameter d was 2 or 3 cm for other tree species except for *Salix matsudana*, *Fraxinus pennsylvanica* and *Pinus tabulaeformis*. The minimum relative RMSE of *Populus tomentosa* was 7.1%, while the maximum relative RMSE of *Pinus tabulaeformis* was 25.6%. The optimal value of the parameter l for different tree species was evenly distributed between 3, 4 and 5, and there was no obvious difference between different forest types. The optimal l value of the algorithm had a significant effect on the reconstruction results of *Ginkgo biloba*, and the relative RMSE increased from 21.5% to 14.9%. For other tree species, the optimal l value had no significant effect on the accuracy of the reconstruction results. Therefore, the optimal values of these parameters for the different tree species were used to reconstruct the model of the remaining data.

Table 6. Optimal values of the model parameters d and l for ten tree species.

Species	d/cm	$d_{RMSE\%}$	l	$l_{RMSE\%}$
Ginkgo biloba	2	21.5	3	14.9
Salix matsudana	6	8.6	3	8.0
Catalpa bungei	3	8.0	5	8.5
Fraxinus pennsylvanica	5	15.8	4	17.1
Robinia pseudoacacia	2	7.5	5	8.1
Populus tomentosa	2	7.1	4	7.0
Juniperus chinensis	2	14.3	3	17.0
Metasequoia glyptostroboides	3	12.0	5	12.7
Sophora japonica	2	17.3	4	19.0
Pinus tabulaeformis	4	25.6	4	24.7

3.3. Estimation of Above-Ground Biomass

When the QSM algorithm was used to reconstruct tree structure, nine individual trees failed to model, including two *Ginkgo biloba*, one *Salix matsudana*, one *Robinia pseudoacacia*, two *Metasequoia glyptostroboides* and three *Pinus tabulaeformis*. Consequently, the data of these nine trees were eliminated from modeling. A comparison of the AGB of each tree from the TLS-QSM model and the basic wood density with the corresponding reference AGB is shown in Figure 6. The comparison between the AGB calculated by TLS-QSM and the AGB from the regional allometric model indicated that the total RMSE and CV (RMSE) of the AGB obtained via TLS-QSM were 17.4 kg and 13.6%, respectively. The correlation between the AGB values from TLS-QSM and the allometric model, expressed by a linear regression model, was 0.95. The TLS-QSM approach and regional allometric model were highly consistent in regard to the 95% confidence interval level (CCC = 0.97), with no major systematic deviation from the 1:1 line (slope of 1.02). Hence, there was no tendency to overestimate or underestimate the AGBs of larger trees. Figure 6b shows the distribution of the AGB residuals from different tree species. Compared with the regional allometric models, the TLS-QSM method for calculating AGB with basic density has neither systematically tended to overestimate nor underestimate the AGB of a particular tree species for different tree species.

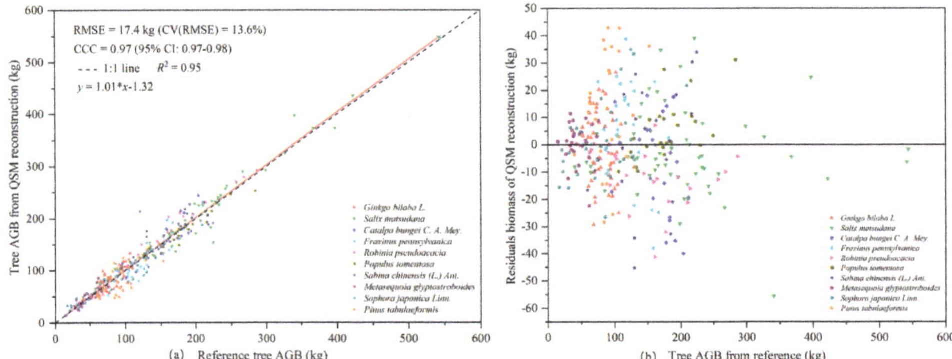

Figure 6. Comparison of the tree above ground biomass (AGB) obtained through quantitative structure model (QSM) reconstruction and the AGB derived from allometric models. (**a**) Comparison of tree AGB and references from ten different tree species; (**b**) residuals of tree AGB from ten different tree species.

Figure 7 depicts the average modeled AGB from ten different QSMs for each tree and compares them to the reference AGB. The QSM algorithm showed significant differences in modeling accuracy and stability for tree species with different morphological and topological structural characteristics. The best AGB modeling results were achieved for *Salix matsudana* and *Populus tomentosa*, with RMSE, CV (RMSE) and CCC values of 16.2 kg and 11.6 kg, 7.7% and 6.5% and 0.99 and 0.96, respectively (Figure 7). When modeling *Juniperus chinensis* and *Pinus tabulaeformis*, we found that the branches of some of the trees were disturbed and destroyed by people. Therefore, the AGB-calculated by the model was quite different from that of the regional allometric model, and we did not conduct a more detailed analysis of those values. *Ginkgo biloba* showed large deviations and large uncertainties compared to other tree species. Our results showed that small trees showed lesser uncertainties and lesser deviations from the reference than large trees. The linear regression slopes of the models for *Catalpa bungei* and *Robinia pseudoacacia* were 1.17 and 1.05, respectively, indicating that the TLS-QSM approach slightly overestimated the AGBs of larger trees. By contrast, the TLS-QSM approach slightly underestimated the AGBs of the large *Populus tomentosa* and *Sophora japonica*, and the slope of their linear regression models was 0.84.

The analysis of variance showed that there was no statistically significant difference between the AGB estimates and reference values for different tree species ($\alpha = 0.05$; Table 7). The results of variance analysis supported the hypothesis of equality. When the same analysis was performed for ten different tree species using the assumptions of equal AGB means, *p*-values were 0.895, 0.811, 0.542, 0.651, 0.559, 0.685, 0.725, 0.881, 0.787 and 0.117 for *Salix matsudana, Ginkgo biloba, Catalpa bungei, Fraxinus pennsylvanica, Robinia pseudoacacia, Populus tomentosa, Juniperus chinensis, Metasequoia glyptostroboides, Sophora japonica* and *Pinus tabulaeformis*, respectively.

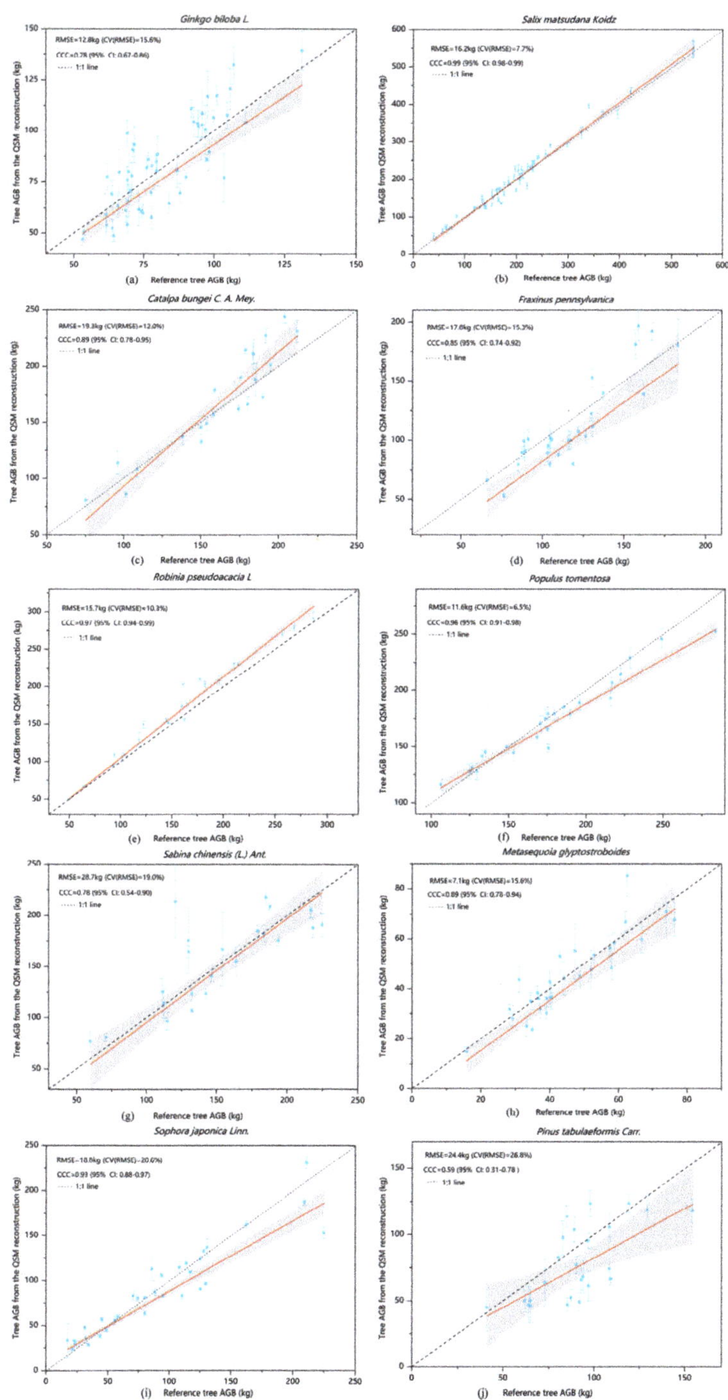

Figure 7. Scatterplot of biomass estimates of different tree species according to the TLS-QSM reconstruction model and the allometric model. (**a–j**) represent ten different tree species. The black dotted line depicts the 1:1 line. The red line represents the fitted linear regression model of the QSM-reconstructed biomass and the biomass according to the allometric model. The grey band depicts the 95% confidence interval of the regression. Error bars are the standard deviation of the 10 QSM model realizations per tree.

Table 7. The results of ANOVA analysis for AGB estimations for different tree species with degrees of freedom (DF), mean squared errors (MSs), F-values and p-values.

Species	DF	MS	F-Value	p-Value
Salix matsudana	1	203.1	0.018	0.895
Ginkgo biloba	1	21.9	0.058	0.811
Catalpa bungei	1	550.2	0.378	0.542
Fraxinus pennsylvanica	1	236.6	0.207	0.651
Robinia pseudoacacia	1	1578.7	0.346	0.559
Populus tomentosa	1	281.9	0.167	0.685
Juniperus chinensis	1	252.3	0.125	0.725
Metasequoia glyptostroboides	1	5.3	0.022	0.881
Sophora japonica	1	168.3	0.073	0.787
Pinus tabulaeformis	1	1758.2	2.549	0.117

3.4. Sensitivity Analysis of the QSM Algorithm

We mainly analyzed the sensitivity and stability of the QSM algorithm according to the RMSE of tree AGBs generated using different values for the parameters d and l, and the standard deviation produced by the given parameter values in the QSM algorithm for ten modeling times per tree. Tables 8 and 9 respectively, show the differences of AGB for different tree species, calculated using different values for the parameters d and l in the QSM algorithm. The different values of d had a great influence on the volume of the trunks and branches reconstructed by the QSM algorithm, and consequently a great impact on the evaluation of AGB (Table 8). Different tree species had large errors in the reconstruction results for the same value of d, which indicated that the algorithm was impacted by the different morphological structures of different tree species. We also found that different parameter values had no significant impact on the reconstructed trunk volume, but had a great impact on the volume of the reconstructed branches. Table 9 describes the effects of different values of the parameter l on the tree AGB reconstructed by the QSM algorithm. It can be seen that different values of l had no significant impact on the reconstructed AGB of specific trees, which indicates that the parameter l of the QSM algorithm has low sensitivity and high stability for the reconstruction results.

Table 8. Sensitivity of the QSM algorithm to the parameter d.

Species	RMSE (d = 2)/kg	RMSE (d = 3)/kg	RMSE (d = 4)/kg	RMSE (d = 5)/kg	RMSE (d = 6)/kg	RMSE (d = 7)/kg
Ginkgo biloba	17.6	22.5	29.5	41.1	62.3	75.2
Salix matsudana	23.9	21.4	23.9	18.7	18.1	19.3
Catalpa bungei	14.9	12.9	22.0	18.3	25.0	27.4
Fraxinus pennsylvanica	23.2	19.9	22.6	18.4	23.5	20.2
Robinia pseudoacacia	11.4	17.3	23.0	22.4	28.2	28.6
Populus tomentosa	12.7	17.8	21.1	30.3	40.0	54.3
Juniperus chinensis	21.6	42.3	60.8	77.7	91.1	85.0
Metasequoia glyptostroboides	9.8	5.5	6.3	10.9	13.9	18.1
Sophora japonica	15.8	19.2	25.6	30.5	38.1	42.2
Pinus tabulaeformis	31.1	28.3	23.3	30.6	36.6	45.7

Table 9. Sensitivity of the QSM algorithm to the parameter *l*.

Species	RMSE ($l = 3$)/kg	RMSE ($l = 4$)/kg	RMSE ($l = 5$)/kg	RMSE ($l = 6$)/kg
Ginkgo biloba	12.2	12.7	12.8	14.9
Salix matsudana	16.8	17.4	22.3	22.8
Catalpa bungei	14.5	14.3	13.7	15.0
Fraxinus pennsylvanica	21.2	19.9	24.0	24.8
Robinia pseudoacacia	13.2	13.9	11.9	12.3
Populus tomentosa	13.2	12.4	13.9	13.7
Juniperus chinensis	25.7	30.2	32.5	32.1
Metasequoia glyptostroboides	7.6	6.7	5.8	6.0
Sophora japonica	17.5	17.4	19.1	19.2
Pinus tabulaeformis	24.8	22.5	24.3	24.5

4. Discussion

TLS is a powerful and effective tool for obtaining 3D point cloud data of the morphological structures of individual trees, and then extracting a variety of geometric and statistical parameters. The least squares circle fitting algorithm was used to fit point clouds at DBHs of trees with different bark roughness and trunk curvature. The results showed that different bark roughness and trunk curvature had a minor influence on the evaluation of DBH in laser point clouds [56,57]. The comparison of DBH from TLS data and field measurements shows a high consistency. The precision of the DBH estimation in our study (RMSE = 1.2 cm) was higher than that reported by Calders et al. [2] and Cabo et al. [58]. The slope of the fitting line between the estimated DBH and the measured data was 0.97, which indicated that the DBH of trees was slightly underestimated on the whole. In addition to the influence of bark roughness, the main reasons for the smaller DBH estimation were the shape of the trunk and the low density and uneven distribution of point cloud data [40,56]. The precision of the tree height estimation (RMSE = 1.3 m) in our study was lower than that reported by Cabo et al. [58]. That was likely due to the higher trunk density in our plots, which caused occlusion of tree tops, leading to the modelling error of the tree height-DBH model. Obtaining tree height from point cloud data has an obvious advantage over other instruments for measuring tree height since it does not rely on the artificial selection of tree vertices. This method is more advantageous for measuring tree height in dense forests where the tops of trees cannot be seen [59,60].

A QSM algorithm optimized for different tree species based on their biophysical characteristics can provide reliable and highly accurate estimates of AGB. The optimal value for the parameter *d* was 2 or 3 cm for all investigated tree species except *Salix matsudana*, *Fraxinus pennsylvanica* and *Pinus tabulaeformis*, which is consistent with the results of Calders et al. [2] and Tanago et al. [8]. The optimal value for the parameter *l* for different tree species was always 3, 4 or 5, which was consistent with the results of Raumonen et al. [28]. The optimum value for the parameters *d* and *l* for different trees species were obviously different, which was mainly influenced by the different biophysical and morphological structures of the trees. Different parameters of the QSM algorithm have a great impact on the final model reconstruction results. The method we used to reconstruct the model from TLS data does not require prior assumptions about tree structure, nor does it rely on limited tree structure parameters. Calders et al. [2] used the QSM algorithm to reconstruct the tree structure of *Eucalyptus leucoxylon*, *E. microcarpa* and *E. tricarpa*, and then evaluated the tree AGB. The results of the AGBs calculated using their TLS-QSM method were highly consistent with the reference values, with the CV (RMSE) and CCC being 16.1% and 0.98, respectively. The accuracy of the total AGB estimated using the optimized QSM algorithm in our study was higher than that reported by Calders et al. [2], and our CV (RMSE) and CCC were 13.6% and 0.97 respectively. The accuracy of the AGB estimated in our study was significantly higher than that of an estimate for 29 tropical rainforest trees species during the foliage period (CV(RMSE) = 28.37%, R^2 = 0.90, CCC = 0.95) [8]. The modelled volumes of different tree species calculated using the QSM algorithm were quite different. In addition to the influence of the biophysical structure of trees, an important reason is the difference in the stem density in the sample plots populated by different species. Among of 322 trees, nine failed to model when performing the 3D reconstruction of tree structure. Possible reasons include a lack of point cloud data in some trees and problems in the algorithm itself. Because different point cloud densities will affect the results of QSM modeling, we used the same scanning resolution and set the same scanning mode and the same point cloud data processing method in ten sample plots to ensure that the single trees used for QSM model reconstruction had similar point cloud densities, which reduces the impact of different point cloud density on the modelling results. The point cloud data of *Ginkgo biloba* were acquired in the period of leaf growth, and the acquired data contained the point cloud data of leaves, which led to discrepancies in the cylinders fitted in the modeling of branches, and further. led to a large standard deviation in the biomass of individual modelled trees (Figure 7a). By contrast, the point cloud data of *Salix matsudana* and *Populus tomentosa* were less affected by leaves and the biophysical morphologies of the trees were relatively simple, so that the modelled results showed high consistency with the reference biomasses (Figure 7b,f). When the QSM algorithm was used to repeatedly model the same individual tree a total of 10 times, the modelled volume of the trunk was very similar, whereas the modelled volume of the branches varied greatly, up to several times. An important reason was that the point cloud that newly grows out of the leaves caused the fitted branches to be bulky. This method can not only calculate the volume and biomass of trees, but also monitor the annual natural growth of tree volume, as well as changes of branches and trunks [28]. Kaasalainen et al. [61] used this method to monitor the same tree in successive years and quantitatively evaluate its growth of tree volume.

There are many factors influencing the calculated AGBs of trees, including allometric biomass models. The allometric biomass models for a specific area have high accuracy in assessing the biomass of individual trees. Therefore, all allometric biomass models used in this study were specific for the area and tree species under investigation. In this study, the single variable (DBH) AGB prediction model was only adopted for *Juniperus chinensis*, and the two variable (DBH and tree height) AGB prediction model, which has higher accuracy of biomass assessment [62], was used for all the other tree species. There are three main reasons for the inconsistency of the biomass calculated by the QSM algorithm using basic wood density and regional allometric biomass models. First, the model data of the regional allometric biomass equations of different tree species are inconsistent, which lead to differences of model accuracy and inevitable systematic errors of the model itself, all of which had an impact on the final biomass results for specific trees. Secondly, there is sensitivity and instability of the QSM algorithm in the process of volume reconstruction of individual trees and the influence of the differences of basic wood density on the calculation of the AGB of trees. Thirdly, we did not consider the biomass of leaves in this study, because we found that it accounts for only a small part (about 10%) of the total AGB, but it also directly affected the calculated AGBs of trees. In addition, dead branches, lack of branches, incomplete extraction of a tree's point cloud and other factors also lead to differences between the calculated AGBs of trees reconstructed using the QSM algorithm and the biomasses calculated by the model.

We optimized the reconstruction process of the QSM algorithm based on parameters d and l, which have the most influence on the reconstruction results [2]. The results showed that the optimum values of parameter d from most of the ten tree species was 2 or 3 cm, which s consistent with the results of Raumonen et al. [28,63]. For different tree species, the optimal QSM algorithm had obvious differences. At the same time, the influence of the parameter l on the modeling results was not very significant. This provides an important reference for researchers who use the QSM algorithm for other research. The QSM algorithm should further implement the automatic determination of the optimal parameters of different tree species in the process of structural reconstruction and the automatic output of the average values of multiple modeling realization as the final result to improve the accuracy of wood volume and biomass assessment.

Future research will focus on more accurate estimation of tree volume and biomass from point cloud data and how to use non-destructive approaches to replace the destructive felling of trees to create and correct new allometric biomass models for specific tree species, especially for large trees.

5. Conclusions

We quantitatively analyzed the effects of different bark roughness and trunk curvature on DBH estimation, and the results indicated that neither parameter had a severe impact. The TLS-QSM method optimized in our study can be used to accurately evaluate the AGB of trees with different morphological and topological structures from 3D reconstructed data, which compared with the reference AGB from specific allometric biomass models. However, since we did not harvest trees in this study, the AGB estimated from TLS data was not compared to the actual value of biomass. The results of this study show that our optimized approach can provide a potential possibility for the development and calibration of allometric biomass models, especially for large trees and precious tree species that are not usually harvested and measured. Using our optimized QSM algorithm, we can continuously monitor different trees to evaluate their growth, health, economic value and ecological benefits. In addition, we can also analyze more potential information in a given forest structure based on the model reconstructed by our optimized algorithm, including the height under the first branch, the height of the tree crown, tree crown shape, crown volume and so on.

Supplementary Materials: The following are available online at http://www.mdpi.com/1999-4907/10/11/936/s1, Table S1: Original Data.

Author Contributions: Conceptualization, S.C. and Z.F.; methodology, S.C.; validation, S.C. and P.C.; formal analysis, S.C. and T.U.K.; investigation, S.C., P.C. and Y.L.; writing—original draft preparation, S.C.; writing—review and editing, S.C. and T.U.K.; proof reading, T.U.K. and P.C.

Funding: This research was funded by the Fundamental Research Funds for the Central Universities (grant number 2015ZCQ-LX-01), the National Natural Science Foundation of China (grant number U1710123).

Acknowledgments: The authors wish to say thanks to J.S., Y.F., and Z.Z., for helping with collecting data.

Conflicts of Interest: The authors declare no conflict of interest.

Appendix A

The least squares circle fitting algorithm was used to fit the DBHs of 10 cm thick slices of trees with different bark roughness and trunk curves. The estimated DBH from TLS point cloud data was compared with the measured DBHs, as shown in Figure A1. The estimated DBHs and reference DBHs of different tree species showed a good linear fit. Bark roughness of ten tree species was measured using a method similar to that of Sioma et al. [64]. Finally, we classified *Salix matsudana, Robinia pseudoacacia, Metasequoia glyptostroboides* and *Pinus tabulaeformis* into the rough bark group; *Catalpa bungei, Fraxinus pennsylvanica, Sabina chinensis* and *Sophora japonica* into the moderately rough bark group; and *Ginkgo biloba* and *Populus tomentosa* into the smooth bark group. Analysis of variance (ANOVA) was used to test whether the differences between estimates and reference values of DBH for different tree species were statistically significant at the 95% level of significance. As can be seen from Figure A1 and Table A1, the DBHs of the ten tree species from LiDAR point cloud data were underestimated to different extents compared with the measured DBH. The DBH of the same tree species was not overestimated with the increase of DBH. The RMSEs of *Ginkgo biloba* in the smooth bark group and *Juniperus chinensis* in the moderately rough bark group were the smallest, both of which were 0.9 cm. The RMSE of the moderately rough bark group was mostly 1.2 cm. The assessment accuracy of *Metasequoia glyptostroboides, Robinia pseudoacacia* and *Pinus tabulaeformis* in the rough bark group was the lowest, with RMSE values of 1.4, 1.3, and 1.3 cm, respectively. Therefore, we could infer that the assessment accuracy of the DBH decreased gradually with the increase of bark roughness, which indicated that DBH assessment from laser point cloud data would be affected by bark roughness. However, the accuracy of DBH evaluation of different species with different bark roughness was not very different.

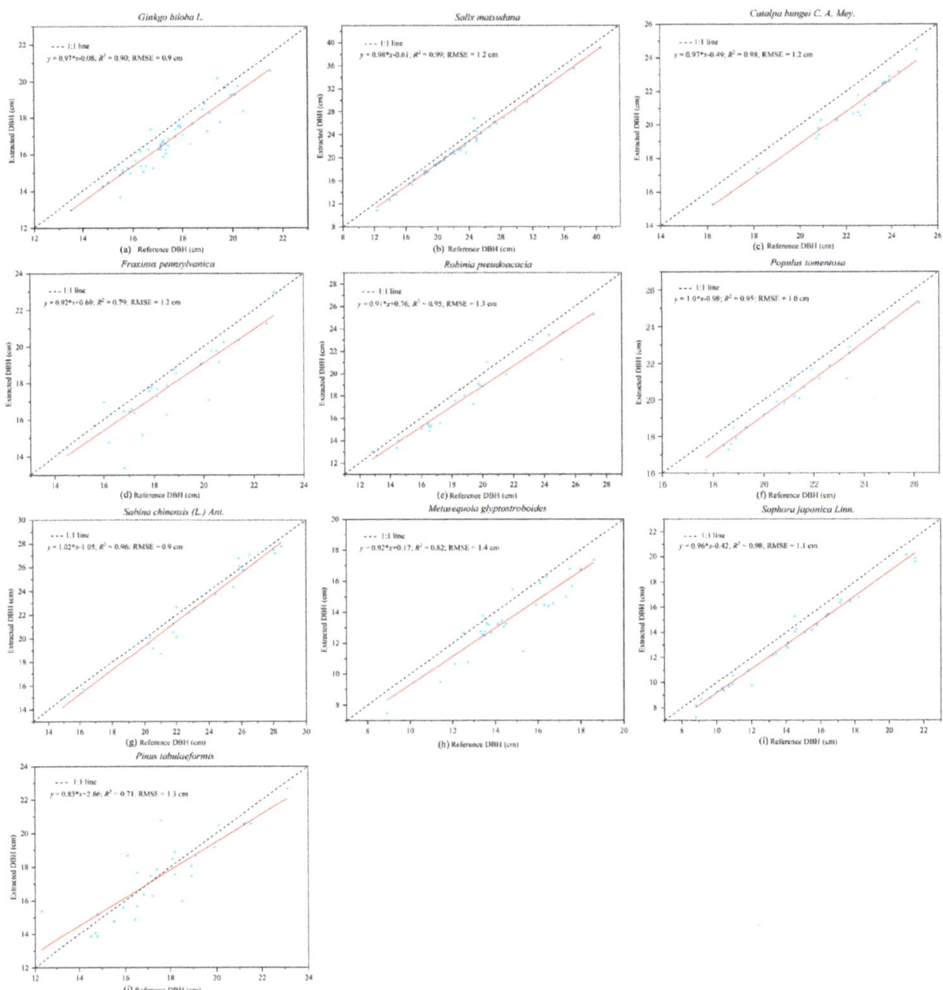

Figure A1. Comparison of DBH calculated for tree species with different bark roughness and trunk curvature with measured DBH (**a–j** stands for the different tree species).

The accuracy of the DBH estimation for ten different tree species utilizing TLS data is shown in Table A1. It can be seen that all deviations are less than zero, which indicates that all the calculated DBH values were underestimated compared with the measured DBH. *Catalpa bungei* had the largest bias of −1.2 cm and a relative bias of −5.4%. The bias of *Pinus tabulaeformis* was the smallest, at −0.1 cm, and the relative bias was −0.4%. The RMSE of DBH values of the ten tree species ranged from 0.9 cm to 1.4 cm, which indicates that the differences between species were not very large. The RMSE values of *Metasequoia glyptostroboides*, *Robinia pseudoacacia* and *Pinus tabulaeformis* were large, at 1.4, 1.3 and 1.3 cm, respectively, and the corresponding relative RMSE values were 9.6%, 7.0% and 7.4%. The RMSE values of *Ginkgo biloba* and *Juniperus chinensis* were the smallest, both at 0.9 cm, and the relative RMSE values were 4.9%, and 3.9% respectively, which was consistent with the preceding results.

Table A1. The accuracy of the DBH estimations for ten different tree species utilizing TLS data.

Species	Bias (cm)	Bias%	RMSE (cm)	RMSE%
Salix matsudana	−1.1	−4.6	1.2	5.3
Ginkgo biloba	−0.7	−3.9	0.9	4.9
Catalpa bungei	−1.2	−5.4	1.2	5.6
Fraxinus pennsylvanica	−0.7	−4.0	1.2	6.6
Robinia pseudoacacia	−1.0	−5.3	1.3	7.0
Populus tomentosa	−0.8	−3.9	1.0	4.5
Juniperus chinensis	−0.5	−2.2	0.9	3.9
Metasequoia glyptostroboides	−1.1	−7.3	1.4	9.6
Sophora japonica	−0.9	−6.7	1.1	7.6
Pinus tabulaeformis	−0.1	−0.4	1.3	7.4

The analysis of variance showed that there was no statistically significant difference between the DBH estimates and reference values of other tree species except *Ginkgo biloba* (α = 0.05; Table A2). The ANOVA results of the DBH of *Ginkgo biloba* trees rejected the assumption of equal DBH mean (p-value = 0.037 < 0.05), and the results of ANOVA of other tree species supported this hypothesis. When the same analysis was performed for other tree species using the assumptions of equal DBH means, p-values were 0.337, 0.081, 0.185, 0.372, 0.198, 0.608, 0.052, 0.278 and 0.941 for *Salix matsudana*, *Catalpa bungei*, *Fraxinus pennsylvanica*, *Robinia pseudoacacia*, *Populus tomentosa*, *Juniperus chinensis*, *Metasequoia glyptostroboides*, *Sophora japonica* and *Pinus tabulaeformis*, respectively.

Table A2. The results of ANOVA analysis for DBH estimations for different tree species with degrees of freedom (DF), mean squared errors (MSs), F-values and p-values.

Species	DF	MS	F-Value	p-Value
Salix matsudana	1	28.8	0.932	0.337
Ginkgo biloba	1	12.4	4.475	0.037
Catalpa bungei	1	14.6	3.203	0.081
Fraxinus pennsylvanica	1	8.3	1.800	0.185
Robinia pseudoacacia	1	11.6	0.814	0.372
Populus tomentosa	1	8.4	1.706	0.198
Juniperus chinensis	1	3.6	0.268	0.608
Metasequoia glyptostroboides	1	16.9	3.931	0.052
Sophora japonica	1	13.6	1.199	0.278
Pinus tabulaeformis	1	0	0.005	0.941

A comparison of tree heights calculated from laser point clouds with measured tree heights for different tree species is shown in Figure A2. We can see that there are obvious differences in tree height comparisons among different tree species. The tree height of all species except *Populus tomentosa* was underestimated to different extents. The reference tree height was obtained using the total station, and the tree height was calculated using the growth model of the specific tree species in the specific area where the tree top was not visible. Therefore, the measured tree height in the field was close to that measured in a destructive way. The slopes of linear fitting for *Ginkgo biloba*, *Salix matsudana*, *Catalpa bungei*, *Sophora japonica* and *Pinus tabulaeformis* were all less than 0.70, whereby the heights of large trees were overestimated and those of small trees were underestimated. A likely reason for this is the occlusion of small trees by nearby large trees, especially in forest types with high tree density, which leads to the mistake of dividing a portion of the large trees into small trees, so that the height of the small trees is overestimated. In the case of large trees, due to the occlusion by lower trees and branches preventing the acquisition of sufficient point cloud data in the top part of the tree, there is a subsequent underestimation of tree height. As shown in Figure A2 and Table A3, there was no significant difference in the tree height estimated using TLS point cloud data between coniferous and broad-leaved forests. The estimation accuracies of the tree heights of *Salix matsudana* and *Populus tomentosa* were the lowest, with an RMSE of 1.6 m in both cases, and relative the RMSEs of 9.8%, and 7.7%, respectively. The tree height estimation accuracies of *Catalpa bungei* and *Fraxinus pennsylvanica* were the highest, with RMSE 1.0 m in both cases, and relative RMSEs of 6.3%, and 9.9%, respectively.

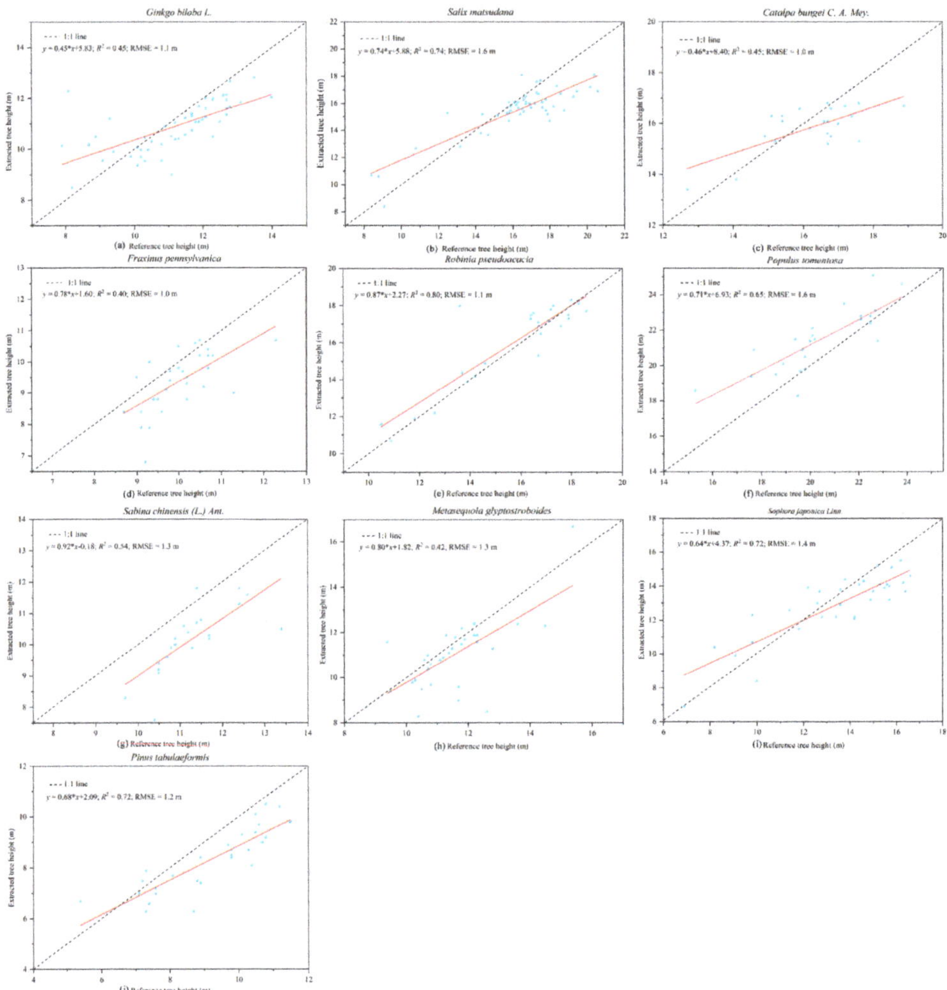

Figure A2. Comparison of tree heights calculated from LiDAR point cloud data of different tree species with measured tree heights (**a**–**j** stand for different tree species).

Table A3. The accuracy of the tree height estimation for ten different tree species utilizing TLS data.

Species	Bias (m)	Bias%	RMSE (m)	RMSE%
Salix matsudana	−0.8	−4.9	1.6	9.8
Ginkgo biloba	−0.3	−2.7	1.1	9.8
Catalpa bungei	−0.4	−2.7	1.0	6.3
Fraxinus pennsylvanica	−0.7	−6.5	1.0	9.9
Robinia pseudoacacia	0.3	1.8	1.1	6.9
Populus tomentosa	1.1	5.2	1.6	7.7
Juniperus chinensis	−1.1	−10.0	1.3	11.7
Metasequoia glyptostroboides	−0.5	−4.7	1.3	11.3
Sophora japonica	−0.5	−3.9	1.4	10.7
Pinus tabulaeformis	−0.9	−9.6	1.2	13.0

References

1. Whitehead, D. Forests as carbon sinks—benefits and consequences. *Tree Physiol.* **2011**, *31*, 893–902. [CrossRef] [PubMed]
2. Calders, K.; Newnham, G.; Burt, A.; Murphy, S.; Raumonen, P.; Herold, M.; Culvenor, D.; Avitabile, V.; Disney, M.; Armston, J.; et al. Nondestructive estimates of above-ground biomass using terrestrial laser scanning. *Methods Ecol. Evol.* **2015**, *6*, 198–208. [CrossRef]
3. Calders, K.; Burt, A.; Newnham, G.; Disney, M.; Murphy, S.; Raumonen, P.; Herold, M.; Culvenor, D.; Armston, J.; Avitabile, V.; et al. Reducing uncertainties in above-ground biomass estimates using terrestrial laser scanning. In Proceedings of the 2015 Silvilaser, La Grande Motte, France, 28–30 September 2015; pp. 197–199.
4. Sarker, L.R.; Nichol, J.E. Improved forest biomass estimates using ALOS AVNIR-2 texture indices. *Remote Sens. Environ.* **2011**, *115*, 968–977. [CrossRef]
5. Gleason, C.J.; Im, J. Forest biomass estimation from airborne LiDAR data using machine learning approaches. *Remote Sens. Environ.* **2012**, *125*, 80–91. [CrossRef]
6. Disney, M.; Boni Vicari, M.; Burt, A.; Calders, K.; Lewis, S.L.; Raumonen, P.; Wilkes, P. Weighing trees with lasers: Advances, challenges and opportunities. *Interface Focus* **2018**, *8*, 20170048. [CrossRef]
7. Clark, D.B.; Kellner, J.R. Tropical forest biomass estimation and the fallacy of misplaced concreteness. *J. Veg. Sci.* **2012**, *23*, 1191–1196. [CrossRef]
8. Gonzalez de Tanago, J.; Lau, A.; Bartholomeus, H.; Herold, M.; Avitabile, V.; Raumonen, P.; Martius, C.; Goodman, R.C.; Disney, M.; Manuri, S.; et al. Estimation of above-ground biomass of large tropical trees with terrestrial LiDAR. *Methods Ecol. Evol.* **2018**, *9*, 223–234. [CrossRef]
9. Prado-Junior, J.A.; Schiavini, I.; Vale, V.S.; Arantes, C.S.; van der Sande, M.T.; Lohbeck, M.; Poorter, L. Conservative species drive biomass productivity in tropical dry forests. *J. Ecol.* **2016**, *104*, 817–827. [CrossRef]
10. Gourlet-Fleury, S.; Rossi, V.; Rejou-Mechain, M.; Freycon, V.; Fayolle, A.; Saint-André, L.; Cornu, G.; Gérard, J.; Sarrailh, J.-M.; Flores, O.; et al. Environmental filtering of dense-wooded species controls above-ground biomass stored in African moist forests. *J. Ecol.* **2011**, *99*, 981–990. [CrossRef]
11. Calders, K.; Newnham, G.; Herold, M.; Murphy, S.; Culvenor, D.; Raumonen, P.; Burt, A.; Armston, J.; Avitabile, V.; Disney, M. Estimating above ground biomass from terrestrial laser scanning in Australian Eucalypt Open Forest. In Proceedings of the SilviLaser 2013, Beijing, China, 9–11 October 2013.
12. Muscarella, R.; Kolyaie, S.; Morton, D.C.; Zimmerman, J.K.; Uriarte, M. Effects of topography on tropical forest structure depend on climate context. *J. Ecol.* **2019**, *00*, 1–15. [CrossRef]
13. Soenen, S.A.; Peddle, D.R.; Hall, R.J.; Coburn, C.A.; Hall, F.G. Estimating aboveground forest biomass from canopy reflectance model inversion in mountainous terrain. *Remote Sens. Environ.* **2010**, *114*, 1325–1337. [CrossRef]
14. Barber, D.; Mills, J.; Smith-Voysey, S. Geometric validation of a ground-based mobile laser scanning system. *ISPRS J. Photogramm. Remote Sens.* **2008**, *63*, 128–141. [CrossRef]
15. Pu, S.; Rutzinger, M.; Vosselman, G.; Oude Elberink, S. Recognizing basic structures from mobile laser scanning data for road inventory studies. *ISPRS J. Photogramm. Remote Sens.* **2011**, *66*, S28–S39. [CrossRef]

16. Chen, S.; Liu, H.; Feng, Z.; Shen, C.; Chen, P. Applicability of personal laser scanning in forestry inventory. *PLoS ONE* **2019**, *14*, e0211392. [CrossRef]

17. Ene, L.; Næsset, E.; Gobakken, T.; Mauya, E.; Bollandsås, O.; Gregoire, T.G.; Ståhl, G.; Zahabu, E. Large-scale estimation of aboveground biomass in miombo woodlands using airborne laser scanning and national forest inventory data. *Remote Sens. Environ.* **2016**, *186*, 626–636. [CrossRef]

18. Næsset, E.; Ørka, H.O.; Solberg, S.; Bollandsås, O.M.; Hansen, E.H.; Mauya, E.; Zahabu, E.; Malimbwi, R.; Chamuya, N.; Olsson, H.; et al. Mapping and estimating forest area and aboveground biomass in miombo woodlands in Tanzania using data from airborne laser scanning, TanDEM-X, RapidEye, and global forest maps: A comparison of estimated precision. *Remote Sens. Environ.* **2016**, *175*, 282–300. [CrossRef]

19. Tymen, B.; Réjou-Méchain, M.; Dalling, J.W.; Fauset, S.; Feldpausch, T.R.; Norden, N.; Phillips, O.L.; Turner, B.L.; Viers, J.; Chave, J. Evidence for arrested succession in a liana-infested Amazonian forest. *J. Ecol.* **2016**, *104*, 149–159. [CrossRef]

20. Edrisi, S.A.; Abhilash, P.C. Sustainable bioenergy production from woody biomass: Prospects and promises. *J. Clean. Prod.* **2015**, *102*, 558–559. [CrossRef]

21. Skowronski, N.S.; Clark, K.L.; Gallagher, M.; Birdsey, R.A.; Hom, J.L. Airborne laser scanner-assisted estimation of aboveground biomass change in a temperate oak–pine forest. *Remote Sens. Environ.* **2014**, *151*, 166–174. [CrossRef]

22. Cao, L.; Gao, S.; Li, P.; Yun, T.; Shen, X.; Ruan, H. Aboveground Biomass Estimation of Individual Trees in a Coastal Planted Forest Using Full-Waveform Airborne Laser Scanning Data. *Remote Sens.* **2016**, *8*, 729. [CrossRef]

23. Kukko, A.; Kaijaluoto, R.; Kaartinen, H.; Lehtola, V.V.; Jaakkola, A.; Hyyppä, J. Graph SLAM correction for single scanner MLS forest data under boreal forest canopy. *ISPRS J. Photogramm. Remote Sens.* **2017**, *132*, 199–209. [CrossRef]

24. Wang, Y.; Pyörälä, J.; Liang, X.; Lehtomäki, M.; Kukko, A.; Yu, X.; Kaartinen, H.; Hyyppä, J. In situ biomass estimation at tree and plot levels: What did data record and what did algorithms derive from terrestrial and aerial point clouds in boreal forest. *Remote Sens. Environ.* **2019**, *232*, 111309. [CrossRef]

25. Raumonen, P.; Casella, E.; Calders, K.; Murphy, S.; Åkerblom, M.; Kaasalainen, M. Massive-Scale Tree Modelling from TLS Data. *ISPRS Ann. Photogramm. Remote Sens. Spatial Inf. Sci.* **2015**, *II-3/W4*, 189–196. [CrossRef]

26. Yao, T.; Yang, X.; Zhao, F.; Wang, Z.; Zhang, Q.; Jupp, D.; Lovell, J.; Culvenor, D.; Newnham, G.; Ni-Meister, W.; et al. Measuring forest structure and biomass in New England forest stands using Echidna ground-based lidar. *Remote Sens. Environ.* **2011**, *115*, 2965–2974. [CrossRef]

27. Seidel, D.; Albert, K.; Ammer, C.; Fehrmann, L.; Kleinn, C. Using terrestrial laser scanning to support biomass estimation in densely stocked young tree plantations. *Int. J. Remote Sens.* **2013**, *34*, 8699–8709. [CrossRef]

28. Raumonen, P.; Kaasalainen, M.; Åkerblom, M.; Kaasalainen, S.; Kaartinen, H.; Vastaranta, M.; Holopainen, M.; Disney, M.; Lewis, P. Fast Automatic Precision Tree Models from Terrestrial Laser Scanner Data. *Remote Sens.* **2013**, *5*, 491–520. [CrossRef]

29. Momo Takoudjou, S.; Ploton, P.; Sonké, B.; Hackenberg, J.; Griffon, S.; de Coligny, F.; Kamdem, N.G.; Libalah, M.; Mofack, G., II; Le Moguédec, G.; et al. Using terrestrial laser scanning data to estimate large tropical trees biomass and calibrate allometric models: A comparison with traditional destructive approach. *Methods Ecol. Evol.* **2018**, *9*, 905–916. [CrossRef]

30. Wu, B.; Yan, Q.; Chi, M.; Zhang, H. Harvest evaluation model and system of fast-growing and high-yield poplar plantation. *Math. Comput. Model.* **2010**, *51*, 1444–1452. [CrossRef]

31. Liu, M.; Feng, Z.; Zhang, Z.; Ma, C.; Wang, M.; Lian, B.-L.; Sun, R.; Zhang, L. Development and evaluation of height diameter at breast models for native Chinese Metasequoia. *PLoS ONE* **2017**, *12*, e0182170. [CrossRef]

32. Gao, X.; Xing, Z. Study on Relation of Diameter at Breast Height and Height of *Sophora japonica* in Liaocheng City. *North. Hortic.* **2010**, *10*, 128–130. (In Chinese)

33. Torresan, C.; Chiavetta, U.; Hackenberg, J. Applying quantitative structure models to plot-based terrestrial laser data to assess dendrometric parameters in dense mixed forests. *For. Syst.* **2018**, *27*, 4. [CrossRef]

34. Srinivasan, S.; Popescu, S.C.; Eriksson, M.; Sheridan, R.D.; Ku, N.-W. Multi-temporal terrestrial laser scanning for modeling tree biomass change. *For. Ecol. Manag.* **2014**, *318*, 304–317. [CrossRef]

35. Li, W.; Guo, Q.; Jakubowski, M.K.; Kelly, M. A New Method for Segmenting Individual Trees from the Lidar Point Cloud. *Photogramm. Eng. Rem. Sens.* **2012**, *78*, 75–84. [CrossRef]

36. Lu, X.; Guo, Q.; Li, W.; Flanagan, J. A bottom-up approach to segment individual deciduous trees using leaf-off lidar point cloud data. *ISPRS J. Photogramm. Remote Sens.* **2014**, *94*, 1–12. [CrossRef]

37. Tao, S.; Wu, F.; Guo, Q.; Wang, Y.; Li, W.; Xue, B.; Hu, X.; Li, P.; Tian, D.; Li, C.; et al. Segmenting tree crowns from terrestrial and mobile LiDAR data by exploring ecological theories. *ISPRS J. Photogramm. Remote Sens.* **2015**, *110*, 66–76. [CrossRef]

38. Wu, J.; Cawse-Nicholson, K.; van Aardt, J. 3D Tree Reconstruction from Simulated Small Footprint Waveform Lidar. *Photogramm. Eng. Rem. Sens.* **2013**, *79*, 1147–1157. [CrossRef]

39. Tansey, K.; Selmes, N.; Anstee, A.; Tate, N.J.; Denniss, A. Estimating tree and stand variables in a Corsican Pine woodland from terrestrial laser scanner data. *Int. J. Remote Sens.* **2009**, *30*, 5195–5209. [CrossRef]

40. Liang, X.; Kankare, V.; Hyyppä, J.; Wang, Y.; Kukko, A.; Haggrén, H.; Yu, X.; Kaartinen, H.; Jaakkola, A.; Guan, F. Terrestrial laser scanning in forest inventories. *ISPRS J. Photogramm. Remote Sens.* **2016**, *115*, 63–77. [CrossRef]

41. Luoma, V.; Saarinen, N.; Wulder, M.A.; White, J.C.; Vastaranta, M.; Holopainen, M.; Hyyppä, J. Assessing Precision in Conventional Field Measurements of Individual Tree Attributes. *Forests* **2017**, *8*, 38. [CrossRef]

42. Berger, A.; Gschwantner, T.; Mcroberts, R.E.; Schadauer, K. Effects of Measurement Errors on Individual Tree Stem Volume Estimates for the Austrian National Forest Inventory. *For. Sci.* **2014**, *60*, 14–24. [CrossRef]

43. Zeng, W.S.; Tang, S.Z. *A New General Allometric Biomass Model*; Nature Publishing Group: London, UK, 2011.

44. State Forestry Administration of China (SFAC). *Tree Biomass Models and Related Parameters to Carbon Accounting for Pinus tabulaeformis*; China Standard Press: Beijing, China, 2015.

45. Zeng, W.-S. Developing Tree Biomass Models for Eight Major Tree Species in China. In *Biomass Volume Estimation and Valorization for Energy*; Jaya, S.T., Ed.; Books on Demand: Pasig City, Philippines, 2017; pp. 3–21. [CrossRef]

46. Xiaver, B. *Allometric Estimation of the Aboveground Biomass and Carbon in Metasequoia glyptostroboide Plantations in Shanghai*; Cranfield University: Bedfordshire, UK, 2009.

47. Zhuang, H.; Becuwe, X.; Xiao, C.; Wang, Y.; Wang, H.; Yin, S.; Liu, C. Allometric Equation-Based Estimation of Biomass Carbon Sequestration in *Metasequoia glyptostroboides* Plantations in Chongming Island, Shanghai. *J. Shanghai Jiaotong Univ.* **2012**, *30*, 48–55. (In Chinese)

48. Liu, K.; Cao, L.; Wang, G.; Cao, F. Biomass allocation patterns and allometric models of *Ginkgo biloba*. *J. Beijing For. Univ.* **2017**, *39*, 12–20. (In Chinese)

49. Zhang, X. Study on the Aboveground Biomass of the Deciduous Trees in Yanqing, Beijin. *For. Sci. Technol.* **2012**, *37*, 1. (In Chinese)

50. Zhou, G.; Yin, G.; Tang, X.; Wang, W. *China's Forest Ecosystem Carbon Storage-Biomass Equation*; Science Press: Beijing, China, 2018; pp. 40–80. (In Chinese)

51. State Forestry Administration of China (SFAC). *Tree Biomass Models and Related Parameters to Carbon Accounting for Cunninghamia Lanceolata*; China Standard Press: Beijing, China, 2014.

52. Wang, X. *Research on Biomass Model of Sophora japonica Linn in Beijing*; Beijing Forestry University: Beijing, China, 2011. (In Chinese)

53. Li, H.; Lei, Y. *Estimation and Evaluation of Forest Biomass Carbon Storage in China*; China Forestry Publishing House: Beijing, China, 2010; pp. 52–58. (In Chinese)

54. Jerome, C.; David, C.; Steven, J.; Lewis, S.L.; Swenson, N.G.; Zanne, A.E. Towards a worldwide wood economics spectrum. *Ecol. Lett.* **2010**, *12*, 351–366.

55. Zanne, A.E.; Lopez-Gonzalez, G.; Coomes, D.A.; Ilic, J.; Jansen, S.; Lewis, S.L.; Miller, R.B.; Swenson, N.G.; Wiemann, M.C.; Chave, J. Data from: Towards a worldwide wood economics spectrum. *Dryad Digit. Repos.* **2009**, *12*, 351–366.

56. Bauwens, S.; Bartholomeus, H.; Calders, K.; Lejeune, P. Forest Inventory with Terrestrial LiDAR: A Comparison of Static and Hand-Held Mobile Laser Scanning. *Forests* **2016**, *7*, 127. [CrossRef]

57. Kałuża, T.; Sojka, M.; Strzeliński, P.; Wróżyński, R. Application of Terrestrial Laser Scanning to Tree Trunk Bark Structure Characteristics Evaluation and Analysis of Their Effect on the Flow Resistance Coefficient. *Water* **2018**, *10*, 753. [CrossRef]

58. Cabo, C.; Ordóñez, C.; Lopez-Sanchez, C.; Armesto, J. Automatic dendrometry: Tree detection, tree height and diameter estimation using terrestrial laser scanning. *Int. J. Appl. Earth Obs.* **2018**, *69*, 164–174. [CrossRef]

59. Liu, G.; Wang, J.; Dong, P.; Chen, Y.; Liu, Z. Estimating Individual Tree Height and Diameter at Breast Height (DBH) from Terrestrial Laser Scanning (TLS) Data at Plot Level. *Forests* **2018**, *9*, 398. [CrossRef]

60. Kankare, V.; Holopainen, M.; Vastaranta, M.; Puttonen, E.; Yu, X.; Hyyppä, J.; Vaaja, M.; Hyyppä, H.; Alho, P. Individual tree biomass estimation using terrestrial laser scanning. *ISPRS J. Photogramm. Remote Sens.* **2013**, *75*, 64–75. [CrossRef]

61. Kaasalainen, S.; Krooks, A.; Liski, J.; Raumonen, P.; Kaartinen, H.; Kaasalainen, M.; Puttonen, E.; Anttila, K.; Mäkipää, R. Change Detection of Tree Biomass with Terrestrial Laser Scanning and Quantitative Structure Modelling. *Remote Sens.* **2014**, *6*, 3906–3922. [CrossRef]

62. Segura, M.; Kanninen, M. Allometric Models for Tree Volume and Total Aboveground Biomass in a Tropical Humid Forest in Costa Rica. *Biotropica* **2005**, *37*, 2–8. [CrossRef]

63. Raumonen, P.; Kaasalainen, S.; Kaasalainen, M.; Kaartinen, H. Approximation of Volume and Branch Size Distribution of Trees from Laser Scanner Data. *Int. Arch. Photogramm. Remote Sens. Spat. Inf. Sci.* **2011**, *38*, W12. [CrossRef]

64. Sioma, A.; Socha, J.; Klamerus-Iwan, A. A New Method for Characterizing Bark Microrelief Using 3D Vision Systems. *Forests* **2018**, *9*, 30. [CrossRef]

 *forests*

Article

Estimating Forest Aboveground Carbon Storage in Hang-Jia-Hu Using Landsat TM/OLI Data and Random Forest Model

Meng Zhang [1,2,3], Huaqiang Du [1,2,3,*], Guomo Zhou [1,2,3], Xuejian Li [1,2,3], Fangjie Mao [1,2,3], Luofan Dong [1,2,3], Junlong Zheng [1,2,3], Hua Liu [1,2,3], Zihao Huang [1,2,3] and Shaobai He [1,2,3]

[1] State Key Laboratory of Subtropical Silviculture, Zhejiang A & F University, Hangzhou 311300, China; 2017103242004@stu.zafu.edu.cn (M.Z.); zhougm@zafu.edu.cn (G.Z.); 2017303661004@stu.zafu.edu.cn (X.L.); maofj@zafu.edu.cn (F.M.); 2017103241008@stu.zafu.edu.cn (L.D.); 2017103241009@stu.zafu.edu.cn (J.Z.); 2018103242003@stu.zafu.edu.cn (H.L.); 2018103241008@stu.zafu.edu.cn (Z.H.); 2018103241005@stu.zafu.edu.cn (S.H.)

[2] Key Laboratory of Carbon Cycling in Forest Ecosystems and Carbon Sequestration of Zhejiang Province, Zhejiang A & F University, Hangzhou 311300, China

[3] School of Environmental and Resources Science, Zhejiang A & F University, Hangzhou 311300, China

*   Correspondence: duhuaqiang@zafu.edu.cn

Received: 27 September 2019; Accepted: 7 November 2019; Published: 9 November 2019

Abstract: Dynamic monitoring of carbon storage in forests resources is important for tracking ecosystem functionalities and climate change impacts. In this study, we used multi-year Landsat data combined with a Random Forest (RF) algorithm to estimate the forest aboveground carbon (AGC) in a forest area in China (Hang-Jia-Hu) and analyzed its spatiotemporal changes during the past two decades. Maximum likelihood classification was applied to make land-use maps. Remote sensing variables, such as the spectral band, vegetation indices, and derived texture features, were extracted from 20 Landsat TM and OLI images over five different years (2000, 2004, 2010, 2015, and 2018). These variables were subsequently selected according to their importance and subsequently used in the RF algorithm to build an estimation model of forest AGC. The results showed the following: (1) Verification of classification results showed maximum likelihood can extract land information effectively. Our land cover classification yielded overall accuracies between 86.86% and 89.47%. (2) Additionally, our RF models showed good performance in predicting forest AGC, with R^2 from 0.65 to 0.73 in the training and testing phase and a RMSE range between 3.18 and 6.66 Mg/ha. RMSEr in the testing phase ranged from 20.27 to 22.27 with a low model error. (3) The estimation results indicated that forest AGC in the past two decades increased with density at 10.14 Mg/ha, 21.63 Mg/ha, 26.39 Mg/ha, 29.25 Mg/ha, and 44.59 Mg/ha in 2000, 2004, 2010, 2015, and 2018. The total forest AGC storage had a growth rate of 285%. (4) Our study showed that, although forest area decreased in the study area during the time period under study, the total forest AGC increased due to an increment in forest AGC density. However, such an effect is overridden in the vicinity of cities by intense urbanization and the loss of forest covers. Our study demonstrated that the combined use of remote sensing data and machine learning techniques can improve our ability to track the forest changes in support of regional natural resource management practices.

Keywords: Landsat dataset; forest AGC estimation; random forest; spatiotemporal evolution

---

1. Introduction

Forests comprise a major part of the terrestrial ecosystems, occupying about 30% of the world's land area, and they are the main contributor to carbon (C) emissions and removal [1–3]. They store more than 80% of forest aboveground carbon (AGC) in terrestrial ecosystems, more than 70% of global soil organic C [4–6] and more than double the amount of C in the atmosphere [7]. As an important part of terrestrial ecosystems, forest ecosystems are a huge global carbon pool and they will likely play a long-term and sustained role in mitigating the impacts of global warming [8–11]. Forest AGC is not only an important indicator reflecting the basic characteristics of forest ecosystems, but also a basis for evaluating forest structural functions and production potentials [12–14]. Accurate quantitative evaluation of forest AGC storages and their spatiotemporal patterns is critical for understanding the mechanisms that control the global terrestrial C cycle [15–20].

The methods of estimating in forest AGC generally include field survey, model simulation, and remote sensing inversion [21–24]. The traditional field survey method has high precision [17], but it is often limited by manpower and material resources, yielding short durations of observation. Additionally, because the field survey only covers limited locations, when used alone, it cannot estimate AGC over large areas. The ecological process models take the biological characteristics and growth mechanisms of vegetation into account, resulting in a higher accuracy of estimation [25]. However, it needs various parameters of vegetation to simulate forest AGC. If the input data are insufficient or missing, it will have a great impact on the prediction results. Remote sensing technology has been commonly used in monitoring forest AGC over broad areas due to its wide coverage of observation, timeliness, and repetitive data availability [26,27]. As spectral characteristics of land cover show great differences [28–32], it is one of the important links in current research to accurately quantify various indicators of forest resources [33,34]. Moreover, studies have found that remote sensing data and its derived bands have good practicability for simulating forest AGC [35–37]; this is especially the case when combined with machine learning algorithms that allow for large scale automated analysis of high dimensional data from satellites [38]. The machine learning approach can derive rich information from remote sensing data as the input data, and continuously optimize the algorithm's performance via empirical learning to make the results more feasible and credible [39–41]. Remotely sensed datasets, combined with machine learning algorithms for intelligent estimation of forest AGC, support more efficient and precise observation and management of forest resources [42–44].

Among the numerous machine learning techniques, random forest (RF) has recently emerged as popular due to its ability to select and rank a large number of predictor variables [45,46] and its reliance on an ensemble of decision trees as a strategy to improve model robustness. RF is unexcelled in accuracy among the current algorithms and it can run efficiently on large data bases. It can not only handle thousands of input variables without variable deletion, but can also give estimates of what variables are important in the model. Khatami et al. [47] classified images used a surveillance classification algorithm based on remote sensing data to classify images, and the results showed that the RF algorithm is superior to the traditional decision tree algorithm. Bargiel et al. [48] included phenology factors in the classification model and revealed that RF based on the phenology can be better to identify the crop types. Chen et al. [49] used climatic factors to simulate carbon dioxide flux and the results showed that the RF model had R^2 values of 0.96 and 0.85 at the training and testing phases.

In this study, we focused on Hang-Jia-Hu, a region with rapid economic development in the northwestern part of Zhejiang, China. This region has a forest area of nearly 6.06 million hectares and LUCC caused by urban expansion due to socio-economic factors has great impact on forest resources. Therefore, the timely and efficient estimation of forest AGC in Hang-Jia-Hu has significant importance to the rational allocation of forest resources. Dynamic monitoring of forest AGC in Hang-Jia-Hu was carried out based on the Landsat time series remote sensing data combined with the RF algorithm. The importance of variables in the models and spatiotemporal evolution of forest AGC in the past two decades were analyzed.

2. Materials and Methods

2.1. Study Area

Hang-Jia-Hu is located in the northwestern part of Zhejiang Province, China, ranging from 118°50′15′′ E to 121°19′6′′ E and from 29°42′52′′ N to 31°11′53′′ N (Figure 1). Its climate is subtropical monsoon and the annual average temperature is 15–18 °C, with an average annual precipitation of about 1100 mm. The 18.1 million hectares study area administratively covered the entire Huzhou City and Jiaxing City and the northeastern part of Hangzhou City. Most forest in this place is distributed in the southwest of the study area with the main forest types being broad-leafed forest (BLF), coniferous forest (CNF), and bamboo forest (BMF).

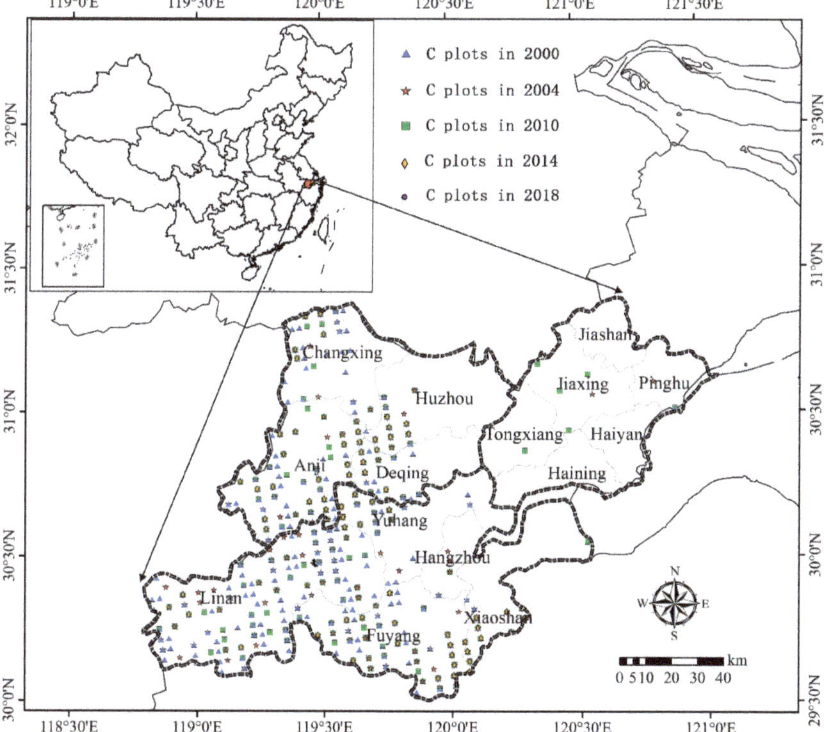

Figure 1. Study area and forest aboveground carbon (AGC) plots of different years.

2.2. Datasets and Processing

2.2.1. Processing Landsat Times Series Products

30-m multispectral data of Landsat5 TM (2000, 2004, 2010) and Landsat8 OLI (2015, 2018) were downloaded from the United States Geological Survey (USGS). We selected cloud-free images from the years with ground observations for four scenes that cover our study area (Table 1). Few scenes image contain a high amount of cloud, but the region covering the study area is cloudless. Additionally, due to the poor image quality from 2014, we used the images in 2015 to correspond with field data in 2014.

Satellite remote sensed data are easily influenced by water vapor, aerosol, bidirectional reflection, and data transmission, which will result in serious fluctuations of time series data and influence the desired effect in data analysis [50,51]. Therefore, this study applied the Fast line-of-Sight Atmospheric Analysis of Spectral Hypercubes (FLAASH) [52–54] to eliminate such atmospheric interference in each image. A digital elevation model (DEM) [55,56] was used to make terrain corrections, as terrain factors may affect the brightness values of original imagery.

Table 1. Acquisition date and cloud coverage (C) (%) of the Landsat datasets.

WRS2 row/path	2000		2004		2010		2015		2018	
	TM 5	C	TM 5	C	TM 5	C	OLI 8	C	OLI 8	C
118039	06/06/2000	6.12	19/07/2004	14.2	17/07/2009	0.07	03/08/2015	4.68	15/01/2018	1.28
119038	31/07/2000	0.16	23/05/2004	0.03	24/05/2010	0.00	22/05/2015	15.4	23/02/2018	0.99
119039	17/09/2000	0.03	14/10/2004	0.00	24/05/2010	0.00	13/10/2015	0.63	28/04/2018	12.4
120039	10/10/2000	0.02	08/12/2004	0.01	19/08/2010	11.2	06/02/2015	19.9	19/04/2018	0.05

In addition to the original image bands, we calculated vegetation indices and texture variables to use as input data to the forest AGC model. The vegetation indices [57–59] included in this study included: NDVI (Normalized Difference Vegetation Index), SAVI (Soil Adjusted Vegetation Index), SR (Simple Ratio Index), DVI (Difference Vegetation Index), and EVI (Enhanced Vegetation Index). Additionally, the texture variables based on the gray-level co-occurrence matrix (GLCM) [60,61] included Mean, Variance, Homogeneity, Contrast, Dissimilarity, Entropy, Angular second moment, and Correlation [27,58,59,62–64] with different windows (3×3, 5×5, 7×7, 9×9, and 11×11) [27,65,66]. There were 251 total variables derived with the details that are shown in Table 2.

Table 2. Information of remote sensing variables.

Type	Name	Calculation Model	Abbreviation	Remarks
Original Band	Band1	band1 (band2*)	B1	Suitable Landsat5 TM (2000, 2004, 2010) and Landsat8 OLI (2015, 2018) data with*
	Band2	band2 (band3*)	B2	
	Band3	band3 (band4*)	B3	
	Band4	band4 (band5*)	B4	
	Band5	band5 (band6*)	B5	
	Band6	band6 (band7*)	B6	
Vegetation Index	NDVI	$(B4 - B3)/(B4+B3)$	NDVI	L take value for 0.5 [57,67]
	SAVI	$(B4 - B3)(1 + L)/(B4 + B3 + L)$	SAVI	
	EVI	$2.5\,(B4 - B3)/(B4+6B3 - 7.5B1+1)$	EVI	
	SR	$B5/B4$	SR	
	DVI	$B5 - B4$	DVI	
Texture	Mean	$\sum_{i=0}^{N-1}\sum_{j=0}^{N-1} iP(i,j)$	ME	$P(i,j) = V(i,j) / \sum_{i=0}^{N-1}\sum_{j=0}^{N-1} V(i,j)$
	Variance	$\sum_{i=0}^{N-1}\sum_{j=0}^{N-1} (i - mean)^2 P(i,j)$	VA	$V(i,j)$ is the ith row of the jth column in the Nth moving window; $u_x = \sum_{j=0}^{N-1} j \sum_{i=0}^{N-1} P(i,j)$
	Homogeneity	$\sum_{i=0}^{N-1}\sum_{j=0}^{N-1} \frac{P(i,j)}{1+(i-j)^2}$	HO	$u_y = \sum_{i=0}^{N-1} i \sum_{j=0}^{N-1} P(i,j)$
	Contrast	$\sum_{\|i-j\|=0}^{N-1} \|i - j\|^2 \left\{ \sum_{i=1}^{N}\sum_{j=1}^{N} P(i,j) \right\}$	CON	$\sigma_x = \sum_{j=0}^{N-1} (j - u_i)^2 \sum_{i=0}^{N-1} P(i,j)$
	Dissimilarity	$\sum_{\|i-j\|=0}^{N-1} \|i - j\| \left\{ \sum_{i=1}^{N}\sum_{j=1}^{N} P(i,j) \right\}$	DI	$\sigma_y = \sum_{i=0}^{N-1} (i - u_j)^2 \sum_{j=0}^{N-1} P(i,j)$
	Entropy	$-\sum_{i=0}^{N-1}\sum_{j=0}^{N-1} P(i,j)\log(P(i,j))$	EN	
	Angular second moment	$\sum_{i=0}^{N-1}\sum_{j=0}^{N-1} P(i,j)^2$	SE	
	Correlation	$\frac{\sum_{i=0}^{N-1}\sum_{j=0}^{N-1} P(i,j)^2 - \mu_x\mu_y}{\sigma_x\sigma_y}$	COR	

2.2.2. Processing Observed Data

Forest AGC plots and classification verification plots in 2000, 2004, 2010, and 2014 are all derived from the data of National Forest Inventory (NFI) [20,27] in Zhejiang province. The investigation method of NFI was systematic sampling, which is usually evenly placed at the intersection of kilometer grids of 1:50,000 topographic maps. Each plot size is 28.5 m × 28.5 m. Forest AGC data that were used in 2018 were derived from a field survey in July, 2019. Each plot of 30 m × 30 m was placed to cover a homogenous area of bamboo, inside which the number of trees and average diameter at breast height (DBH) were measured. After calculating the bamboo biomass of each plot by using the growth model, bamboo AGC was calculated by using the conversion coefficients between biomass and carbon stocks. Classification verification data in 2018 were manually and evenly selected from the result of unsupervised classified data and visual interpretation.

Land use in this paper was classified into six types: urban, water, cultivated land (CTL), BLF, CNF, and BMF based on the data of the NFI and field survey [26]. Maximum likelihood [68,69] was applied to make the land use classification map. The classification training samples in this paper are uniformly and evenly selected by the visual interpretation based on the land use spectral reflectance characteristics. The training samples of each land use type were consistent (400 pixels) and they could cover the study area. The total number of classification verification samples in 2000, 2004, 2010, 2014, and 2018 were 171, 175, 191, 156, and 240, respectively, and the numbers of forest verification samples were 120, 116, 137, 106, and 160. Table 3 details forest AGC plots used to construct the AGC estimation model from 2000 to 2018.

Table 3. Summary of the forest AGC plots of different years.

Year	Sample Size			Unit: Mg/ha			
	Training	Testing	Total	Min	Max	Mean	SD
2000	160	69	229	1.28	32.11	10.17	6.02
2004	138	59	197	4.92	38.84	20.28	6.66
2010	123	52	175	5.33	43.48	22.89	9.13
2014	-	102	102	2.18	37.49	17.98	7.97
2018	-	42	42	14.08	40.88	27.26	5.93

2.3. Random Forest and AGC Estimation

Breiman first made formal definition of the RF in 2001, which is a bagging of uncorrelated CART trees learned with randomized node optimization [70–72]. First, the algorithm generates N bootstrap samples for the training dataset. Second, a regression tree model is built for each bootstrap sample. Finally, the predicted results are obtained by averaging the predictions from all individual regressions [20]. RF has three important parameters: (1) the number of random regression trees (Ntree); (2) number of variables to be randomly sampled at each node in a tree (Mtry), used to search for the variable that best partitions samples in the training data set and the default number is 1/3 of input variables [73]; and, (3) the minimum number of terminal nodes (Nodesize) where the default value is 5 in regression analysis [8,49].

Forest AGC samples of different years were divided into two groups to optimize the RF model in this study (Table 3), with one group of data accounting for 70% samples represented training samples of the RF model and the remaining 30% represented test samples to test the model accuracy. Training data were also used to evaluate the importance of input variables to forest AGC and select input variables for constructing the RF model. Forest in this paper includes three types and each of them has corresponding forest AGC plots in 2000, 2004, and 2010. In 2014 and 2018, the AGC plots only have the forest type of BMF. Therefore, we decided to use data in 2000, 2004, and 2010 to construct individual AGC estimation models by using the forest non-stratification method [65] and applied the corresponding model to simulate forest AGC in 2000, 2004 and 2010. Subsequently, these three models are applied to predict forest AGC in 2015 and 2018 and the BMF AGC plots in 2014 and 2018 are used for verification. The estimation result in 2015 and 2018 with the highest accuracy in the verification phase will be chosen as the final simulation result. In this paper, we used "RandomForest" package in the R statistical software [74,75] to construct the forest AGC model. During the process of applying the model established, we put the entire image data into the model and run it, and use Python to read the model result into image format, and then perform spatial analysis.

2.4. Accuracy Assessment

The determination coefficient (R^2), root mean square error (*RMSE*), and relative RMSE (*RMSEr*) [27,65] were used to evaluate the accuracy of the RF model. A higher accuracy is indicated by the higher values of R^2 and lower values of *RMSE* and *RMSEr*. R^2, *RMSE*, and *RMSEr* were calculated, as follows:

$$R^2 = 1 - \sum\nolimits_{i=1}^{n} (p_i - o_i)^2 / \sum\nolimits_{i=1}^{n} (o_i - \overline{o_i})^2 \qquad (1)$$

$$RMSE = \sqrt{\frac{1}{n}\sum\nolimits_{i=1}^{n} (p_i - o_i)^2} \qquad (2)$$

$$RMSEr = \frac{RMSE}{\overline{y}} \times 100 \qquad (3)$$

In Equations (1)–(3), R^2, *RMSE*, and *RMSEr* are the indexes to measure model accuracy; p_i and o_i are the value of forest AGC samples in the predicted and observed phase in the year of 2000, 2004, and 2010; $\overline{o_i}$ is the average value of observed plots; $\overline{y}$ is the average value of forest AGC in testing samples; and, i is the number of samples.

3. Results

3.1. Land Use Classification

Figure 2 presents the classification results using maximum likelihood methods.

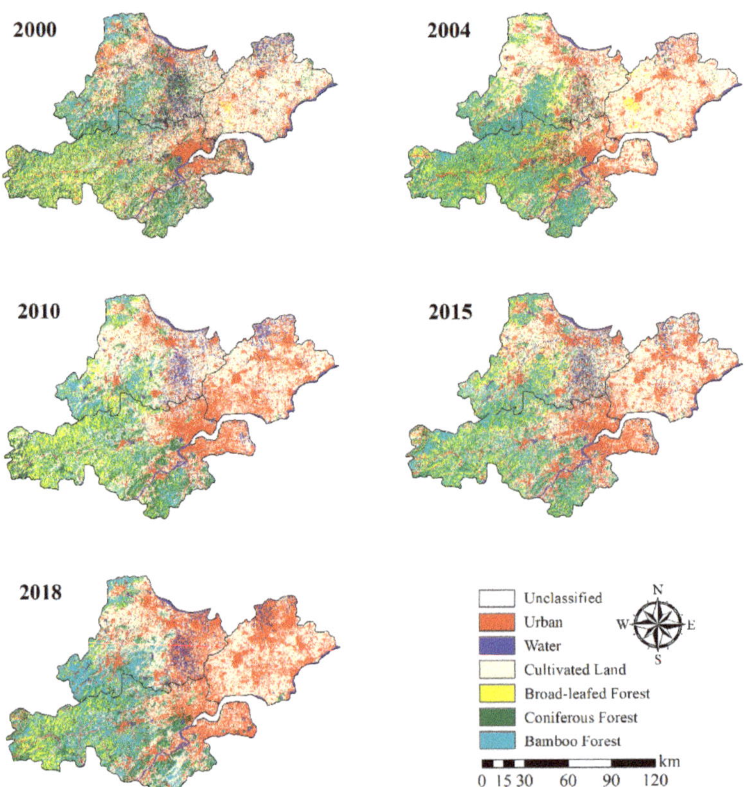

Figure 2. Land use maps from 2000 to 2018.

Table 4 shows the overall accuracy of classification by using confusion matrix. It illustrates that the overall classification accuracies in different years were above 86.86%, and the highest was 89.47% with a kappa coefficient above 0.84 and a high of 0.87. The classification accuracies of the BLF, BMF, and CNF were between 86.67% and 89.62%. Classification verification results show a high level of precision. Based on this high precision classification result, we extracted BLF, BMF and CNF separately and use it to mask the forest AGC estimation results.

Table 4. Overall accuracy and kappa coefficient of land use classification.

Year	Overall Accuracy (%)		Forest Extraction Accuracy (%)			
	Accuracy (%)	Kappa Coefficient	BLF	BMF	CNF	Forest
2000	89.47	0.87	87.71	81.25	91.30	86.67
2004	86.86	0.84	90.00	86.54	88.87	88.89
2010	88.48	0.85	89.15	84.85	90.91	88.57
2015	89.10	0.87	93.75	82.14	91.30	89.62
2018	87.90	0.86	92.50	85.00	90.00	89.17

3.2. RF Model Construction

3.2.1. Parameters Optimization of RF

Training data were used to input into RF to traverse all of the variables values and finally obtain the optimal parameters. Figure 3 shows the results. Figure 3a presents Mtry, which is used to determine the minimum amounts of variables in each tree to construct RF. As can be seen from Figure 3a, when Mtry is a certain value, the model error is the smallest, which is the minimum Mtry value that is required. Ntree in Figure 3b shows that model error tends to be stable when the Mtry is big enough. Additionally, we used a 10-fold cross-validation to observe the effect of the number of variables on the model error. Results in Figure 3c represent the variation trend of average model errors in this process. It shows that the variables in the model are not as good as possible. When the model traverses all possibilities, the model error will tend to be stable until a certain amount, when the value of the model error reaches a minimum. Table 5 lists specific settings for different parameter values of different models.

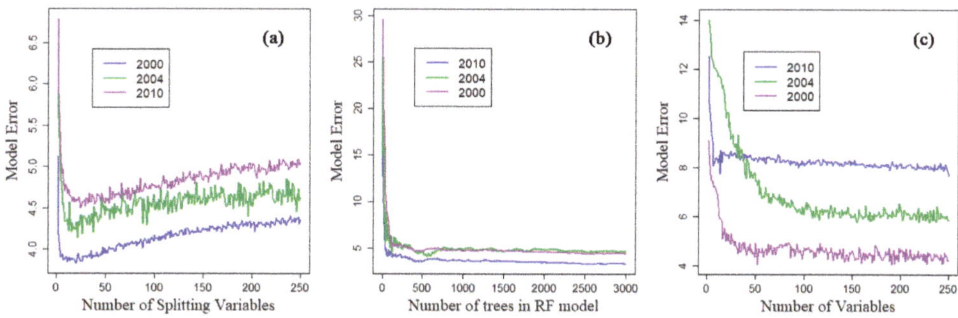

Figure 3. (**a**) Influence of *Mtry to* model error (Mg/ha), (**b**) Influence of *Ntree to* model error (Mg/ha), (**c**) Influence of variables number to model error (Mg/ha).

Table 5. Optimized parameters of random forest (RF) models in different years.

Year	Nodesize	Mtry	Ntree	Number of Variables
2000	5	20	2500	52
2004	5	21	2500	115
2010	5	24	2500	13

3.2.2. Variable Importance and Autocorrelation

The RF model in different years was ran 100 times each to observe changes in variables importance. We listed the first 20 variables with importance scores of different models in Figure 4. Among all of the variables used in the three models, most of the variables are texture features, while the original band and vegetation index account for a small proportion. Upon viewing of variable selection in different models, we found that 80% of the variables are the mean of the texture features in 2000 model. It reveals that the texture mean at different windows of the original band is important for estimating forest AGC. In the 2004 model, the sixth band of the original band occupies 45% of the overall variables, followed by the third band at 40%. Additionally, the texture window of 11*11 accounts for 65% of the total variables. It illustrated the importance of the texture features of the third and sixth band at 11 × 11 windows to the estimation. In the 2010 model, the mean of the texture features occupies 55% of the total and the sixth band has a selection probability of 35%, which exceeds other variables. As for the importance scores, in the model of 2000, the influence of Tb2w7ME on forest AGC estimation is the largest at an average of 14.83, followed by Tb2w5ME at 14.73. For 2004, Tb311HO has the largest influence of 8.66, followed by Tb3w9DI with score at 8.61. Tb6w7VA has the largest effect on forest carbon prediction at a score of 13.67 and is followed by Tb6w11ME at 12.01.

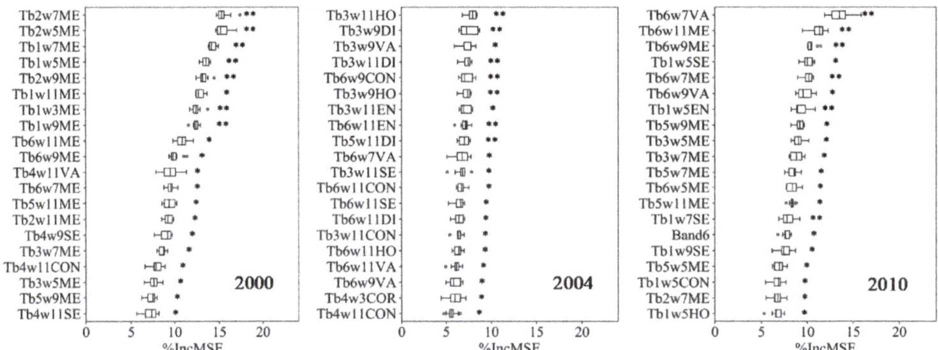

Figure 4. Importance of the first 20 variables measured in %InMSE (the percentage increase in the mean squared error) from 100 runs of the RF. Note: Tbiwjxx, a texture image developed using the texture measure xx (xx can be such texture measures as ME, VA, HO, CON, DI, EN, SE, COR) on spectral band i (bandi) with a window size of j × j pixel (wj).

We calculated the frequency of the top 100 variables in the different RF models to further understand the proportion of variables in the models. In the 2000 RF model, texture information accounted for 92%, of which the fourth band has the most texture information, while the highest window size is w9, and ME accounts for 28.3% of all texture information. Texture information in the 2004 model occupies 98% of the total variables. The number of textures in the third band is up to 24.5%. W7 and ME have a major advantage in terms of window size and texture value. In 2010, 93% of the texture information in the first band accounted for 27.6%, and w7 is also the window with the most selection. The ME value has the same status as the previous two models in selection to establish forest AGC models.

RF, which is widely used to estimate forest parameters while using remote sensing data, can effectively solve the multicollinearity problems of complex variables in traditional statistical regression models [20,76]. Figure 5 shows the collinear relationship between the first 20 variables. A weak correlation indicates that the RF model could solve the collinearity between variables and the overfitting problem to ensure the accuracy of the forest AGC estimation. We selected the optimal number of variables and the optimal variables to build models for different years based on the optimized parameters.

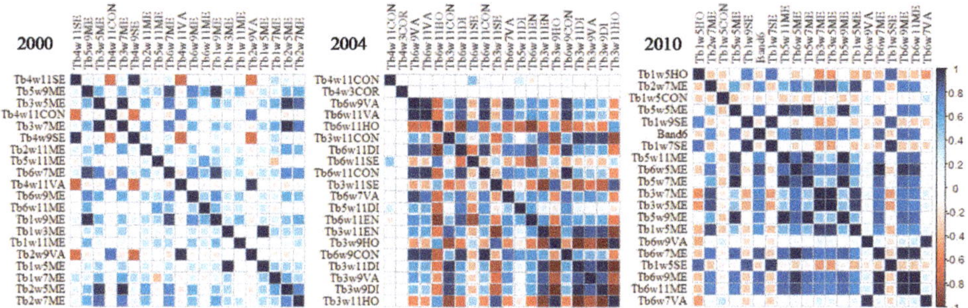

Figure 5. Autocorrelation of the first 20 variables in different models.

3.3. Estimation and Evaluation of Forest AGC

We extracted the AGC estimation for the forest covers based on the classification results of the maximum likelihood method and simulation results of the RF model established above (Figure 6).

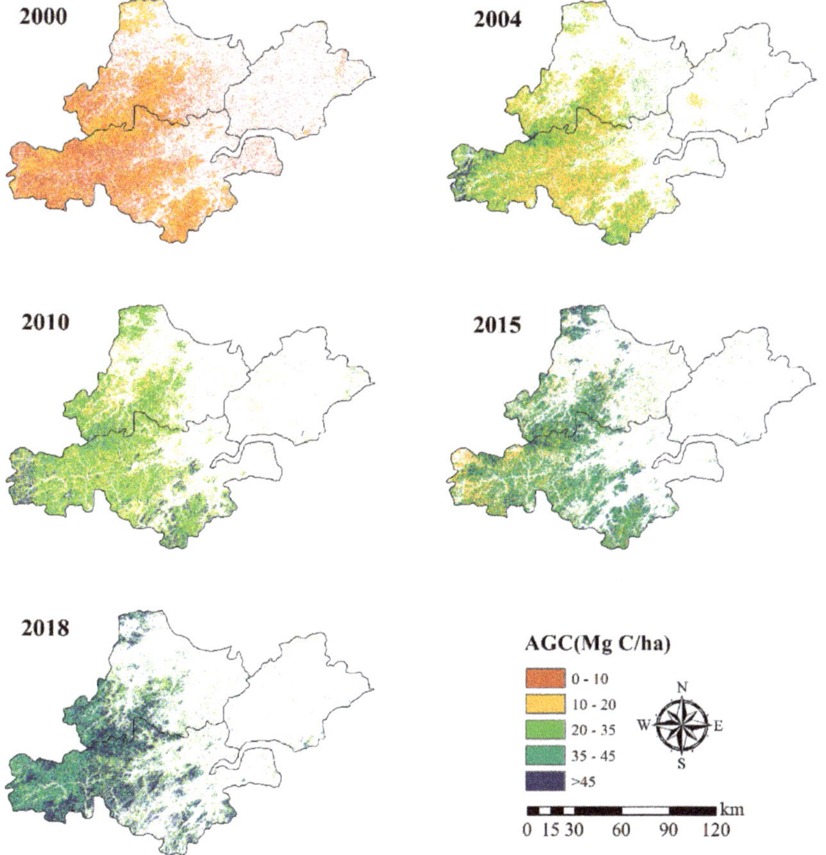

Figure 6. Spatial distribution of forest AGC from 2000 to 2018.

The model performance is evaluated with scatterplots of predicted AGC against the observed data (Figure 7). In the training phase (A) using data from 2000, 2004, and 2010, the models yielded R^2

that ranged from 0.69 to 0.73 ($p < 0.01$) and RMSE from 3.18 Mg/ha to 4.84Mg/ha. In the testing phase (B) of the model in 2000, 2004, and 2010, the R^2 value was in the range of 0.67–0.73. RMSE and RMSEr were in the range of 5.58–6.66 Mg/ha and 20.41–23.65. Forest AGC simulation in 2015 and 2018 was done by using the models in 2004 and 2010, respectively. The models yielded an R^2 of 0.57 and 0.32. RMSE were 5.59 and 5.76 Mg/ha and RMSEr were 23.65 and 20.77 for 2015 and 2018, respectively.

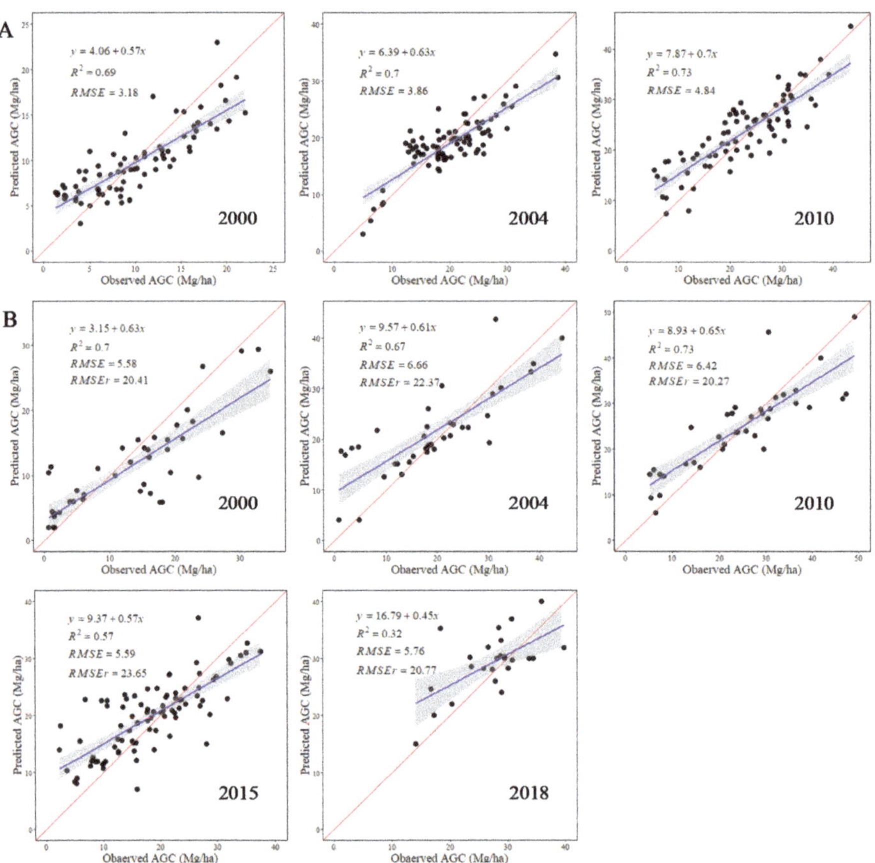

Figure 7. Accuracy evaluation of RF model at training phase (**A**) in 2000, 2004, 2010 and testing phase (**B**) in 2000, 2004, 2010, 2015, and 2018.

3.4. Spatiotemporal Evolution of Forest AGC

The total forest area in the study region decreased in the past two decades according to the land use classifications in different years. However, forest AGC has gradually increased from 2000 to 2018 in many areas, especially in the west of Hangzhou and southwest of Huzhou. We calculated statistics of forest area, forest AGC, and total AGC storage to understand the change and relationship between forest area, forest AGC, and the total forest AGC storage in Hang-Jia-Hu during the past two dacades (Figure 8). The total forest area only increased from 2000 to 2004 and then gradually decreased afterwards. The total amount of forest area was reduced by 77,713.92 ha with the largest declines occurring following 2004 and 2010. The forest AGC density increased 3.4 times, from 10.14 Mg/ha in 2000 to 44.59 Mg/ha in 2018, and the growth rate was the highest between 2015 and 2018. The total forest AGC storage did not decline due to the reduction of forest area, but it followed the same trend as forest AGC density. It increased from 641.38 Mg C in 2000 to 2472.51 Mg C in 2018, with a growth rate of 285%. This reflects that forest AGC density is more influential on to the total forest AGC storage when compared to the forest area.

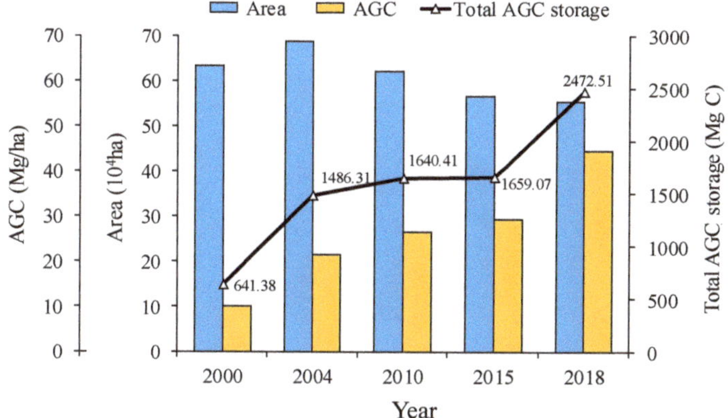

Figure 8. Forest area, forest AGC and total forest AGC storage from 2000 to 2018.

Figure 9 shows the interannual variability of forest AGC and total forest AGC storage of 15 sub-regions in Linan (LN), Yuhang (YH), Jiaxing (JX), Jiashan (Js), Anji (AJ), Fuyang (FY), Pinghu (PH), Deqing (DQ), Hangzhou (HangZ), Tongxiang (TX), Hanning (HN), Haiyan (HY), Huzhou (HuZ), Xioashan (XS), and Changxing (CX). Forest AGC did not show obvious variations across different regions (Figure 9a). Forest AGC was the lowest for all subregions in 2000 and the highest in 2018, with the remaining years fluctuating between 22.57 Mg/ha and 28.43 Mg/ha. The total forest AGC varied greatly across subregions mainly due to the variation in forest area. LN, AJ, and FY stood out with over 2×10^6 ha of AGC, while other subregions had less than 1×10^6 ha.

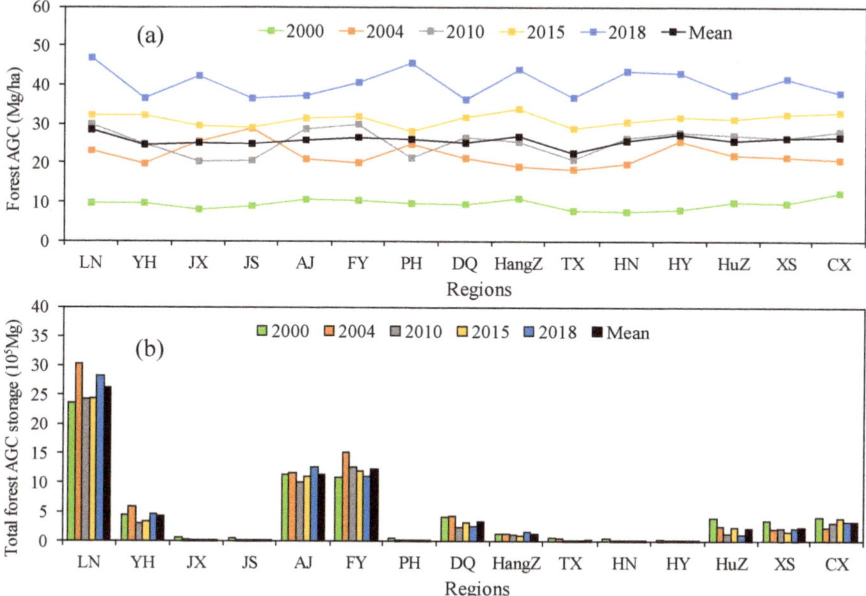

Figure 9. (**a**) Forest AGC of different regions in different years, (**b**) Total forest AGC storage of different regions in different years.

Figure 10 reveals the spatial distribution of forest AGC change from 2000 to 2018. Overall, forest AGC in most of the region is increasing, and the increase is mainly distributed in the western part of the study area. Furthermore, from the above data analysis, forest AGC is increasing during the past two decades. However, some areas showed a decrease in forest AGC, the reduction phenomenon. Such decline in forest AGC mainly occurred near the center of Huzhou and towards the south of Hangzhou in Fuyang District. The magnitude of decline was mainly between −10 Mg/ha and −20 Mg/ha. The reduction of forest AGC in these areas was mounted to 240,660.36 ha.

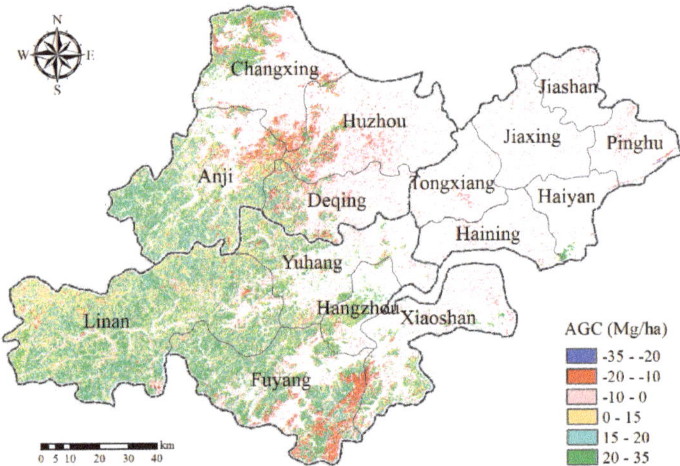

Figure 10. Spatial distribution change map of forest AGC.

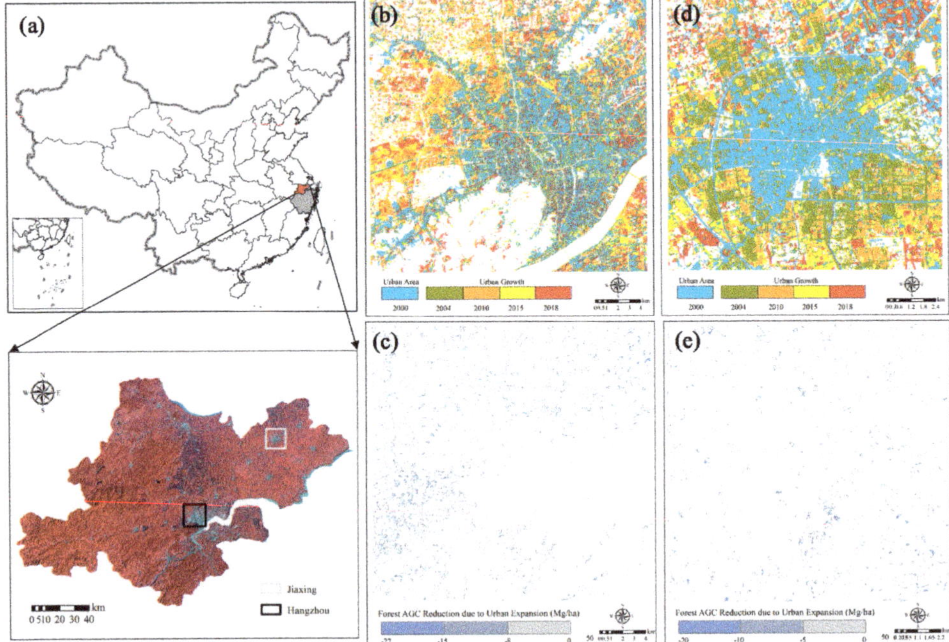

Figure 11. (**a**) Study area; (**b**) and (**d**) urban expansion in the central city of Hangzhou and Jiaxing; and, (**c**) and (**e**) forest AGC reduction due to urban expansion in central city of Hangzhou and Jiaxing.

Moreover, urban forests have received increasing attention in recent years due to their large C sequestration capability and the positive impact on the environment, especially in areas with rapid urban expansion. According to our land use classification, the urban area in the Hang-Jia-Hu region has expanded into suburbs of the cities over the past 20 years, especially around the metropolitan areas, such as Hangzhou and Jiaxing. Extensive decreases of forest AGC were particularly evident in Huzhou City, but not clearly distinguishable in other cities. Therefore, to reveal more details regarding the urban forest AGC change, we used 15-m pan-sharpened Landsat images to analyze the changes of land use in Hangzhou City and Jiaxing City (Figure 11). We found that the urban growth took place in the periphery of Jiaxing and mainly towards the northwest of Hangzhou. In Hangzhou, forest AGC decreased along with intense urban growth in the northwest plains of the City. The decline of forest area due to urbanization reached 3157 ha, which corresponded to an AGC storage loss of 25,309 Mg C. In Jiaxing, forest AGC reduced in all surrounding areas, where the pattern is consistent with the direction of the urban growth. The forest coverage in Jiaxing was reduced by 3078 ha and the forest AGC by 20,701 Mg C as a result of urban growth. It is clearly seen that, although the urban expansion mainly replaced the cultivated lands, it also significantly affected forests.

4. Discussion

In our modeling exercise using RF, we found that land surface texture information is the most important variables of forest AGC. We calculated the correlation between the variables of the corresponding optimal number of variables and forest AGC in different models, and presented a scatter plot of the first eight variables and forest AGC in three models of the training phase to further understand the specific correlation between variables and forest AGC (Figure 12). The results revealed that R^2 between the input variables in RF and forest AGC was in the range of 0.13 and 0.32 in the model of 2000, 2004, and 2010. It further reflected that the variables that are involved in RF are closely related with forest AGC, which means that texture information is one of the most essential factors in the process of simulating forest AGC. Furthermore, it also reflected that remote sensing technology, in combination with machine learning algorithms, are an effective means to monitor forest resources in a precise and timely manner at large spatial scales.

However, there are still defects and deficiencies. Firstly, the temporal mismatch of remote sensing and field survey data due to the lack of cloud free Landsat scenes corresponding to the exact times of ground observation can lead to estimation error in the results. Furthermore, different types of forest surfaces have similar spectral signatures, which could have resulted in misclassification [77]. The verification accuracy of the classification results indicates that the BMF is the main reason for the precision declining. Secondly, in the verification phase of simulation results, accuracy in 2015 and 2018 were all below the accuracy in the other three models by a degree. Parameter settings for different modeling due to different raw data vary from year to year. Additionally, different parameter settings will result in the model being only applicable to the simulation of forest AGC in the input data year, but not in other years. Finally, RF also has certain flaws in simulation due to the great requirements for input data. Spatiotemporal distribution of the simulation results show that forest AGC in different periods is concentrated in a certain range (fluctuating around the average of training data), which means that RF has a poor simulation effect on the data at both ends of a dataset, which is consistent with the research result of Gao et al. [65]. High simulation accuracy seems to show that RF is best suited to simulating AGC at certain range, and making the simulation results have the same data distribution as the input data is a critical step. Studies have shown that a combination of multi-source remote sensed data can improve the accuracy of estimation in forest biomass or forest AGC [78]. Furthermore, forest AGC can be simulated with other types of data, such as meteorological data, topography, and soil data [79]. The accuracy of forest AGC simulation can be further improved when considering multivariate remote sensing data and other various impact factors.

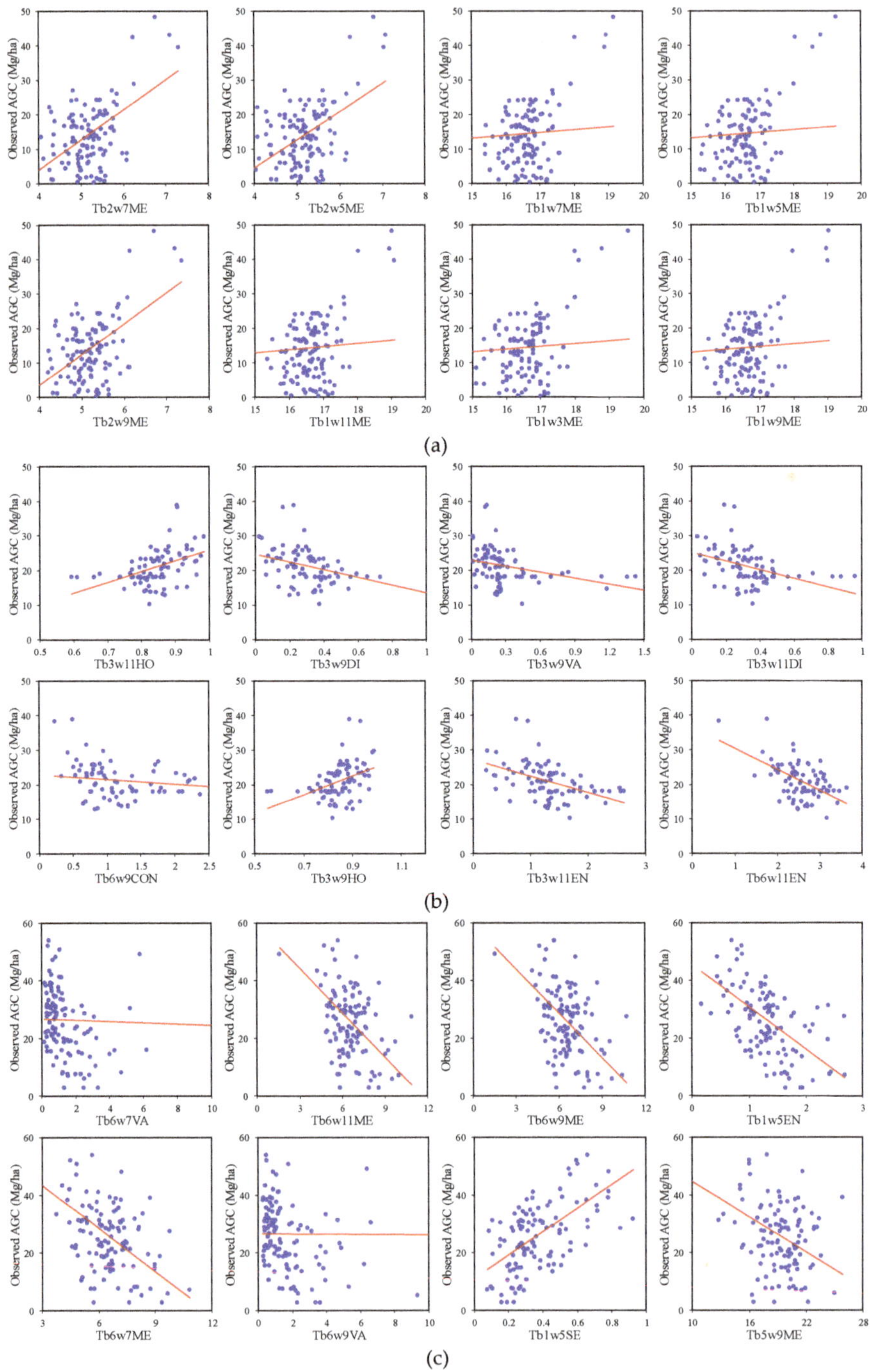

Figure 12. (**a**) Correlation between the first 8 variables and forest AGC in 2000 model. (**b**) Correlation between the first 8 variables and forest AGC in 2004 model. (**c**) Correlation between the first 8 variables and forest AGC in 2010 model.

5. Conclusions

Based on the remotely sensed and ground survey data, we used the optimized RF models to simulate the distribution patterns and temporal changes of forest AGC in the Hang-Jia-Hu region. Classification by using a maximum likelihood method can present well the land use cover. Additionally, our models estimated forest AGC for the study area in different years with relatively high accuracies. The standardized residuals of the testing samples in different models (Figure 13) are all in the range of −2 to 2, which further illustrates that the optimized RF model has good stability and reliability in predicting forest AGC. With the estimation results, we found that, during the study period from 2000 to 2018, although the total forest area has decreased, the total AGC storage has significantly increased as a result of the increased AGC density. However, forest AGC has decreased near the cities due to the urban expansion in spite of the increase of forest AGC over the entire region.

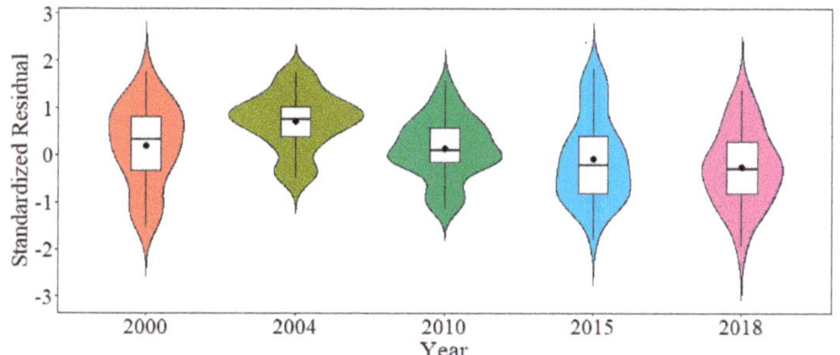

Figure 13. Standardized residual of the testing phase in different models.

Author Contributions: Conceptualization, H.D.; Data curation, X.L., L.D., J.Z., H.L., Z.H. and S.H.; Methodology, M.Z.; Supervision, G.Z.; Validation, F.M.; Writing—original draft, M.Z.; Writing—review & editing, H.D.

Funding: The authors gratefully acknowledge the support of the National Natural Science Foundation (No. U1809208, 31670644), the State Key Laboratory of Subtropical Silviculture (No. ZY20180201), the Joint Research fund of Department of Forestry of Zhejiang Province, the Chinese Academy of Forestry (2017SY04), the Zhejiang Provincial Collaborative Innovation Center for Bamboo Resources and High-efficiency Utilization (No. S2017011), and the National Natural Science Foundation (No.31901310).

Acknowledgments: The authors gratefully acknowledge the supports of various foundations. The authors are grateful to the Editor and anonymous reviewers whose comments have contributed to improving the quality.

Conflicts of Interest: The authors declare that they have no competing interests.

References

1. Federici, S.; Tubiello, F.N.; Salvatore, M.; Jacobs, H.; Schmidhuber, J. New estimates of CO_2 forest emissions and removals: 1990–2015. *For. Ecol. Manag.* **2015**, *352*, 89–98. [CrossRef]

2. Yan, M.; Tian, X.; Li, Z.; Chen, E.; Wang, X.; Han, Z.; Sun, H. Simulation of Forest Carbon Fluxes Using Model Incorporation and Data Assimilation. *Remote. Sens.* **2016**, *8*, 567. [CrossRef]

3. Piao, S.; Fang, J.; Zhu, B.; Tan, K. Forest biomass carbon stocks in China over the past 2 decades: Estimation based on integrated inventory and satellite data. *J.Geophys. Res.* **2015**, *110*, 195–221. [CrossRef]

4. Callewaert, P.; Lenders, S.; Gryze, S.D.; Morris, S.J.; Paustian, K. Measuring and Understanding Carbon Storage in Afforested Soils by Physical Fractionation. *Soil Sci. Soc. Am. J.* **2007**, *66*, 1981–1987.

5. Santini, N.S.; Adame, M.F.; Nolan, R.H.; Miquelajauregui, Y.; Piñero, D.; Mastretta-Yanes, A.; Cuervo-Robayo, Á.P.; Eamus, D. Storage of organic carbon in the soils of Mexican temperate forests. *For. Ecol. Manag.* **2019**, *446*, 115–125. [CrossRef]

6. Lin, B.; Ge, J. Valued forest carbon sinks: How much emissions abatement costs could be reduced in China. *J. Clean. Prod.* **2019**, *224*, 455–464. [CrossRef]

7. Pan, Y.; Birdsey, R.A.; Fang, J.; Houghton, R.; Kauppi, P.E.; Kurz, W.A.; Phillips, O.L.; Shvidenko, A.; Lewis, S.L.; Canadell, J.G.; et al. A Large and Persistent Carbon Sink in the World's Forests. *Science* **2011**, *333*, 988–993. [CrossRef]

8. Liu, G.; Bo, F.U.; Fang, J. Carbon dynamics of Chinese forests and its contribution to global carbon balance. *Acta Ecol. Sin.* **2000**, *20*, 733–740.

9. Goldewijk, K.K.; Van Minnen, J.G.; Kreileman, G.J.J.; Vloedbeld, M.; Leemans, R.; Minnen, J.G. Simulating the carbon flux between the terrestrial environment and the atmosphere. *Water Air Soil Pollut.* **1994**, *76*, 199–230. [CrossRef]

10. Labrecque, S.; Fournier, R.; Luther, J.; Piercey, D. A comparison of four methods to map biomass from Landsat-TM and inventory data in western Newfoundland. *For. Ecol. Manag.* **2006**, *226*, 129–144. [CrossRef]

11. Watson, R.T.; Noble, I.R.; Bolin, B.; Ravindranath, N.H.; Verardo, D.J.; Dokken, D.J.; Watson, R.T.; Noble, I.R.; Bolin, B.; Ravindranath, N.H. *Land Use, Land-Use Change and Forestry: A Special Report of the Intergovernmental Panel on Climate Change*; Cambridge University Press: Cambridge, UK, 2017.

12. Jingyun, F.; Zhaodi, G.; Huifeng, H.; Tomomichi, K.; Hiroyuki, M.; Yowhan, S. Forest biomass carbon sinks in East Asia, with special reference to the relative contributions of forest expansion and forest growth. *Glob. Chang. Biol.* **2014**, *20*, 2019–2030.

13. Mitchard, E.T.A. The tropical forest carbon cycle and climate change. *Nature* **2018**, *559*, 527–534. [CrossRef] [PubMed]

14. Liu, T.Y.; Mao, F.J.; Li, X.J.; Xing, L.Q.; Dong, L.F.; Zheng, J.L.; Zhang, M.; DU, H. Spatiotemporal dynamic simulation on aboveground carbon storage of bamboo forest and its influence factors in Zhejiang Province, China. *J. Appl. Ecol.* **2019**, *30*, 1743–1753.

15. Hai, R.; Hua, C.; Li, L.; Li, P.; Hou, C.; Wan, H.; Zhang, Q.; Zhang, P. Spatial and temporal patterns of carbon storage from 1992 to 2002 in forest ecosystems in Guangdong, Southern China. *Plant Soil* **2013**, *363*, 123–138.

16. Heather, K.; Mackey, B.G.; Lindenmayer, D.B. Re-evaluation of forest biomass carbon stocks and lessons from the world's most carbon-dense forests. *Proc. Natl. Acad. Sci. USA* **2009**, *106*, 11635–11640.

17. Chen, L.-C.; Guan, X.; Lib, H.-M.; Wang, Q.-K.; Zhang, W.-D.; Yang, Q.-P.; Wang, S.-L. Spatiotemporal patterns of carbon storage in forest ecosystems in Hunan Province, China. *For. Ecol. Manag.* **2019**, *432*, 656–666. [CrossRef]

18. He, B.; Miao, L.; Cui, X.; Wu, Z. Carbon sequestration from China's afforestation projects. *Environ. Earth Sci.* **2015**, *74*, 5491–5499. [CrossRef]

19. Andrews, D. The Carbon Story at the Shale Hills Critical Zone Observatory. In Proceedings of the ASA, CSSA and SSSA International Annual Meeting, Long Beach, CA, USA, 31 October–4 November 2010.

20. Li, X.; Du, H.; Mao, F.; Zhou, G.; Liang, C.; Xing, L.; Fan, W.; Xu, X.; Liu, Y.; Lu, C. Estimating bamboo forest aboveground biomass using EnKF-assimilated MODIS LAI spatiotemporal data and machine learning algorithms. *Agric. For. Meteorol.* **2018**, *256*, 445–457. [CrossRef]

21. Mao, F. Construction and Application of Spatiotemporal Carbon Cycle Model of Moso Bamboo Forest Ecosystem. Ph.D. Thesis, Zhejiang A & F University, Hangzhou, China, 2016.

22. Zhao, M.; Yang, J.; Zhao, N.; Liu, Y.; Wang, Y.; Wilson, J.P.; Yue, T. Estimation of China's forest stand biomass carbon sequestration based on the continuous biomass expansion factor model and seven forest inventories from 1977 to 2013. *For. Ecol. Manag.* **2019**, *448*, 528–534. [CrossRef]

23. Li, X.; Mao, F.; Du, H.; Zhou, G.; Xing, L.; Liu, T.; Han, N.; Liu, Y.; Zheng, J.; Dong, L.; et al. Spatiotemporal evolution and impacts of climate change on bamboo distribution in China. *Sci. Total Environ.* **2019**, *248*, 109265. [CrossRef]

24. Li, X.; Du, H.; Mao, F.; Zhou, G.; Han, N.; Xu, X.; Liu, Y.; Zheng, J.; Dong, L.; Zhang, M. Assimilating spatiotemporal MODIS LAI data with a particle filter algorithm for improving carbon cycle simulations for bamboo forest ecosystems. *Sci. Total Environ.* **2019**, *694*, 133803. [CrossRef]

25. He, L.H.; Wang, H.Y.; Lei, X.D. Parameter sensitivity of simulating net primary productivity of Larix olgensis forest based on BIOME-BGC model. *J. Appl. Ecol.* **2016**, *27*, 412.

26. Li, Y.; Du, H.; Mao, F.; Li, X.; Cui, L.; Han, N.; Xu, X. Information extracting and spatiotemporal evolution of bamboo forest based on Landsat time series data in Zhejiang Province. *Sci. Silvae Sin.* **2019**, *55*, 88–96.

27. Li, Y.; Ning, H.; Li, X.; Du, H.; Xing, L. Spatiotemporal Estimation of Bamboo Forest Aboveground Carbon Storage Based on Landsat Data in Zhejiang, China. *Remote Sens.* **2018**, *10*, 898. [CrossRef]

28. Zhang, J.S.; Pan, Y.Z.; Han, L.J.; Wei, S.U.; He, C.Y. Land Use/cover Change Detection with Multi-source Data. *J. Remote Sens.* **2007**, *11*, 500.

29. Hansen, M.C.; Defries, R.S.; Townshend, J.R.G.; Sohlberg, R. Global land cover classification at 1 km spatial resolution using a classification tree approach. *Int. J. Remote Sens.* **2000**, *21*, 1331–1364. [CrossRef]

30. Funkenberg, T.; Binh, T.T.; Moder, F.; Dech, S. The Ha Tien Plain—Wetland monitoring using remote-sensing techniques. *Int. J. Remote Sens.* **2014**, *35*, 2893–2909. [CrossRef]

31. Ning, H.; Du, H.; Zhou, G.; Xu, X.; Ge, H.; Liu, L.; Gao, G.; Sun, S. Exploring the synergistic use of multi-scale image object metrics for land-use/land-cover mapping using an object-based approach. *Int. J. Remote Sens.* **2015**, *36*, 3544–3562.

32. Olofsson, P.; Foody, G.M.; Herold, M.; Stehman, S.V.; Woodcock, C.E.; Wulder, M.A. Good practices for estimating area and assessing accuracy of land change. *Remote Sens. Environ.* **2014**, *148*, 42–57. [CrossRef]

33. Horton, A.A.; Walton, A.; Spurgeon, D.J.; Lahive, E.; Svendsen, C. Microplastics in freshwater and terrestrial environments: Evaluating the current understanding to identify the knowledge gaps and future research priorities. *Sci. Total Environ.* **2017**, *586*, 127–141. [CrossRef]

34. Chen, Y.; Guerschman, J.P.; Cheng, Z.; Guo, L. Remote sensing for vegetation monitoring in carbon capture storage regions: A review. *Appl. Energy* **2019**, *240*, 312–326. [CrossRef]

35. Jia, S.; Shi, S.; Jian, Y.; Lin, D.; Wei, G.; Chen, B.; Song, S. Analyzing the performance of PROSPECT model inversion based on different spectral information for leaf biochemical properties retrieval. *ISPRS J. Photogramm. Remote Sens.* **2018**, *135*, 74–83.

36. Graves, S.J.; Caughlin, T.T.; Asner, G.P.; Bohlman, S.A. A tree-based approach to biomass estimation from remote sensing data in a tropical agricultural landscape. *Remote Sens. Environ.* **2018**, *218*, 32–43. [CrossRef]

37. Kong, B.; Yu, H.; Du, R.; Wang, Q. Quantitative Estimation of Biomass of Alpine Grasslands Using Hyperspectral Remote Sensing. *Rangel. Ecol. Manag.* **2019**, *72*, 336–346. [CrossRef]

38. Wang, S.; Azzari, G.; Lobell, D.B. Crop type mapping without field-level labels: Random forest transfer and unsupervised clustering techniques. *Remote Sens. Environ.* **2019**, *15*, 303–317. [CrossRef]

39. Madry, A.; Makelov, A.; Schmidt, L.; Tsipras, D.; Vladu, A. Towards Deep Learning Models Resistant to Adversarial Attacks. *arXiv* **2017**, arXiv:1706.06083.

40. Yang, Q.; Liu, Y.; Chen, T.; Tong, Y. Federated Machine Learning: Concept and Applications. *ACM Trans. Intell. Syst. Technol.* **2019**, *10*, 1–19. [CrossRef]

41. Pietquin, O.; Tango, F. A Reinforcement Learning Approach to Optimize the longitudinal Behavior of a Partial Autonomous Driving Assistance System. In Proceedings of the European Conference on Artificial Intelligence, Montpellier, France, 27–31 August 2012.

42. Chen, X.; Ye, C.; Li, J.; Chapman, M.A. Quantifying the Carbon Storage in Urban Trees Using Multispectral ALS Data. *IEEE J. Sel. Top. Appl. Earth Obs. Remote Sens.* **2018**, *11*, 1–8. [CrossRef]

43. Duan, M.; Gao, Q.; Wan, Y.; Yue, L.; Guo, Y.; Zhao, G.; Liu, Y.; Qin, X. Biomass estimation of alpine grasslands under different grazing intensities using spectral vegetation indices. *Can. J. Remote Sens.* **2012**, *37*, 413–421. [CrossRef]

44. Brilli, L.; Chiesi, M.; Brogi, C.; Magno, R.; Arcidiaco, L.; Bottai, L.; Tagliaferri, G.; Bindi, M.; Maselli, F. Combination of ground and remote sensing data to assess carbon stock changes in the main urban park of Florence. *Urban For. Urban Green.* **2019**, *43*, 126377. [CrossRef]

45. Huang, B.F.F.; Boutros, P.C. The parameter sensitivity of random forests. *BMC Bioinform.* **2016**, *17*, 331. [CrossRef] [PubMed]

46. Verikas, A.; Gelzinis, A.; Bacauskiene, M. Mining data with random forests: A survey and results of new tests. *Pattern Recognit.* **2011**, *44*, 330–349. [CrossRef]

47. Khatami, R.; Mountrakis, G.; Stehman, S.V. A meta-analysis of remote sensing research on supervised pixel-based land-cover image classification processes: General guidelines for practitioners and future research. *Remote. Sens. Environ.* **2016**, *177*, 89–100. [CrossRef]

48. Bargiel, D. A new method for crop classification combining time series of radar images and crop phenology information. *Remote Sens. Environ.* **2017**, *198*, 369–383. [CrossRef]

49. Chen, L.; Zhou, G.; Du, H.; Liu, Y.; Mao, F.; Xu, X.; Li, X.; Cui, L.; Li, Y.; Zhu, D. Simulaiton of CO_2 Flux and Controlling Factors in Moso Bamboo Forest Using Random Forest Algorithm. *Sci. Silvae Sin.* **2018**, *54*, 1–12.

50. Gao, B.C.; Montes, M.J.; Davis, C.O.; Goetz, A.F.H. Atmospheric correction algorithms for hyperspectral remote sensing data of land and ocean. *Remote Sens. Environ.* **2009**, *113*, S17–S24. [CrossRef]

51. Wei, J.; Lee, Z.; Garcia, R.; Zoffoli, L.; Armstrong, R.A.; Shang, Z.; Sheldon, P.; Chen, R.F. An assessment of Landsat-8 atmospheric correction schemes and remote sensing reflectance products in coral reefs and coastal turbid waters. *Remote. Sens. Environ.* **2018**, *215*, 18–32. [CrossRef]

52. Yu, K.; Liu, S.; Zhao, Y. CPBAC: A quick atmospheric correction method using the topographic information. *Remote. Sens. Environ.* **2016**, *186*, 262–274. [CrossRef]

53. Guzzi, D.; Barducci, A.; Marcoionni, P.; Pippi, I. An atmospheric correction iterative method for high spectral resolution aerospace imaging spectrometers. In Proceedings of the IEEE International Geoscience and Remote Sensing Symposium, Cape Town, South Africa, 12–17 July 2009.

54. Rani, N.; Mandla, V.R.; Singh, T. Evaluation of atmospheric corrections on hyperspectral data with special reference to mineral mapping. *Geosci. Front.* **2016**, *8*, 797–808. [CrossRef]

55. Deng, F.; Rodgers, M.; Xie, S.; Dixon, T.H.; Charbonnier, S.; Gallant, E.A.; Vélez, C.M.L.; Ordoñez, M.; Malservisi, R.; Voss, N.K.; et al. High-resolution DEM generation from spaceborne and terrestrial remote sensing data for improved volcano hazard assessment—A case study at Nevado del Ruiz, Colombia. *Remote Sens. Environ.* **2019**, *233*, 111348. [CrossRef]

56. Mapelli, D.; Guarnieri, A.M.M.; Giudici, D. Generation and Calibration of High-Resolution DEM From Single-Baseline Spaceborne Interferometry: The "Split-Swath" Approach. *IEEE Trans. Geosci. Remote Sens.* **2014**, *52*, 4858–4867. [CrossRef]

57. Ren, H.; Zhou, G.; Feng, Z. Using negative soil adjustment factor in soil-adjusted vegetation index (SAVI) for aboveground living biomass estimation in arid grasslands. *Remote Sens. Environ.* **2018**, *209*, 439–445. [CrossRef]

58. Fatiha, B.; Abdelkader, A.; Latifa, H.; Mohamed, E. Spatio Temporal Analysis of Vegetation by Vegetation Indices from Multi-dates Satellite Images: Application to a Semi Arid Area in ALGERIA. *Energy Procedia* **2013**, *36*, 667–675. [CrossRef]

59. Taddeo, S.; Dronova, I.; Depsky, N. Spectral vegetation indices of wetland greenness: Responses to vegetation structure, composition, and spatial distribution. *Remote Sens. Environ.* **2019**, *234*, 111467. [CrossRef]

60. Bharati, M.H.; Liu, J.J.; MacGregor, J.F. Image texture analysis: Methods and comparisons. *Chemom. Intell. Lab. Syst.* **2004**, *72*, 57–71. [CrossRef]

61. Haralick, R.M. Statistical and structural approaches to texture. *Proc. IEEE* **1979**, *467*, 786–804. [CrossRef]

62. Gholizadeh, A.; Zizala, D.; Saberioon, M.; Boruvka, L. Soil Organic Carbon and Texture Retrieving and Mapping using Proximal, Airborne and Sentinel-2 Spectral Imaging. *Remote Sens. Environ.* **2018**, *218*, 89–103. [CrossRef]

63. Liu, J.; Pattey, E.; Jégo, G. Assessment of vegetation indices for regional crop green LAI estimation from Landsat images over multiple growing seasons. *Remote Sens. Environ.* **2012**, *123*, 347–358. [CrossRef]

64. Calvão, T.; Palmeirim, J.M. Mapping Mediterranean scrub with satellite imagery: Biomass estimation and spectral behaviour. *Int. J. Remote Sens.* **2004**, *25*, 3113–3126. [CrossRef]

65. Gao, Y.; Lu, D.; Li, G.; Wang, G.; Qi, C.; Liu, L.; Li, D. Comparative Analysis of Modeling Algorithms for Forest Aboveground Biomass Estimation in a Subtropical Region. *Remote Sens.* **2018**, *10*, 627. [CrossRef]

66. Lu, D.; Batistella, M. Exploring TM image texture and its relationships with biomass estimation in Rondônia, Brazilian Amazon. *Acta Amaz.* **2005**, *35*, 249–257. [CrossRef]

67. Qi, J.; Chehbouni, A.; Huete, A.R.; Kerr, Y.H.; Sorooshian, S.S. A modified soil adjusted vegetation index. *Remote Sens. Envrion.* **2015**, *48*, 119–126. [CrossRef]

68. Hagner, O.; Reese, H. A method for calibrated maximum likelihood classification of forest types. *Remote. Sens. Environ.* **2007**, *110*, 438–444. [CrossRef]

69. Cabral, A.I.R.; Silva, S.; Silva, P.C.; Vanneschi, L.; Vasconcelos, M.J. Burned area estimations derived from Landsat ETM+ and OLI data: Comparing Genetic Programming with Maximum Likelihood and Classification and Regression Trees. *ISPRS J. Photogramm. Remote Sens.* **2018**, *142*, 94–105. [CrossRef]

70. Breiman, L. Random Forests. *Mach. Learn.* **2001**, *45*, 5–32. [CrossRef]

71. Breiman, L. Bagging Predictors. *Mach. Learn.* **1996**, *24*, 123–140. [CrossRef]

72. Ao, Y.; Li, H.; Zhu, L.; Ali, S.; Yang, Z. The linear random forest algorithm and its advantages in machine learning assisted logging regression modeling. *J. Pet. Sci. Eng.* **2018**, *174*, 776–789. [CrossRef]

73. Brieuc, M.S.O.; Waters, C.D.; Drinan, D.P.; Naish, K.A. A Practical Introduction to Random Forest for Genetic Association Studies in Ecology and Evolution. *Mol. Ecol. Resour.* **2018**, *18*, 755–766. [CrossRef]

74. Gounaridis, D.; Koukoulas, S. Urban land cover thematic disaggregation, employing datasets from multiple sources and RandomForests modeling. *Int. J. Appl. Earth Obs. Geoinf.* **2016**, *51*, 1–10. [CrossRef]
75. Kai, L.; Yan, E. Co-mention network of R packages: Scientific impact and clustering structure. *J. Informetr.* **2018**, *12*, 87–100.
76. Lopatin, J.; Dolos, K.; Hernández, H.J.; Galleguillos, M.; Fassnacht, F.E. Comparing Generalized Linear Models and random forest to model vascular plant species richness using LiDAR data in a natural forest in central Chile. *Remote Sens. Environ.* **2016**, *173*, 200–210. [CrossRef]
77. Jiang, X.; Lu, D.; Moran, E.; Calvi, M.F.; Dutra, L.V.; Li, G. Examining impacts of the Belo Monte hydroelectric dam construction on land-cover changes using multitemporal Landsat imagery. *Appl. Geogr.* **2018**, *97*, 35–47. [CrossRef]
78. Zhao, P.; Lu, D.; Wang, G.; Liu, L.; Li, D.; Zhu, J.; Yu, S. Forest aboveground biomass estimation in Zhejiang Province using the integration of Landsat TM and ALOS PALSAR data. *Int. J. Appl. Earth Obs. Geoinf.* **2016**, *53*, 1–15. [CrossRef]
79. Du, H.; Mao, F.; Zhou, G.; Li, X.; Li, Y. Estimating and Analyzing the Spatiotemporal Pattern of Aboveground Carbon in Bamboo Forest by Combining Remote Sensing Data and Improved BIOME-BGC Model. *IEEE J. Sel. Top. Appl. Earth Obs. Remote Sens.* **2018**, *11*, 2282–2295. [CrossRef]

 *forests*

Article

Influence of Variable Selection and Forest Type on Forest Aboveground Biomass Estimation Using Machine Learning Algorithms

Yingchang Li [1], Chao Li [1], Mingyang Li [1,*] and Zhenzhen Liu [2]

[1] Co-Innovation Center for Sustainable Forestry in Southern China, College of Forestry, Nanjing Forestry University, Nanjing 210037, China; lychang@njfu.edu.cn (Y.L.); gislichao@njfu.edu.cn (C.L.)

[2] College of Forestry, Shanxi Agricultural University, Jinzhong 030801, China; lzz88312@sxau.edu.cn

* Correspondence: lmy196727@njfu.edu.cn; Tel.: +86-025-8542-7327

Received: 20 October 2019; Accepted: 21 November 2019; Published: 25 November 2019

Abstract: Forest biomass is a major store of carbon and plays a crucial role in the regional and global carbon cycle. Accurate forest biomass assessment is important for monitoring and mapping the status of and changes in forests. However, while remote sensing-based forest biomass estimation in general is well developed and extensively used, improving the accuracy of biomass estimation remains challenging. In this paper, we used China's National Forest Continuous Inventory data and Landsat 8 Operational Land Imager data in combination with three algorithms, either the linear regression (LR), random forest (RF), or extreme gradient boosting (XGBoost), to establish biomass estimation models based on forest type. In the modeling process, two methods of variable selection, e.g., stepwise regression and variable importance-base method, were used to select optimal variable subsets for LR and machine learning algorithms (e.g., RF and XGBoost), respectively. Comfortingly, the accuracy of models was significantly improved, and thus the following conclusions were drawn: (1) Variable selection is very important for improving the performance of models, especially for machine learning algorithms, and the influence of variable selection on XGBoost is significantly greater than that of RF. (2) Machine learning algorithms have advantages in aboveground biomass (AGB) estimation, and the XGBoost and RF models significantly improved the estimation accuracy compared with the LR models. Despite that the problems of overestimation and underestimation were not fully eliminated, the XGBoost algorithm worked well and reduced these problems to a certain extent. (3) The approach of AGB modeling based on forest type is a very advantageous method for improving the performance at the lower and higher values of AGB. Some conclusions in this paper were probably different as the study area changed. The methods used in this paper provide an optional and useful approach for improving the accuracy of AGB estimation based on remote sensing data, and the estimation of AGB was a reference basis for monitoring the forest ecosystem of the study area.

Keywords: aboveground biomass; variable selection; forest type; machine learning; subtropical forests

---

1. Introduction

The forest ecosystem plays a critical role in the global terrestrial carbon cycle, and it is the research topic of major scientific projects, such as the International Geosphere-Biosphere Program, the World Climate Research Programme, and an International Programme of Biodiversity Science [1,2]. Forest biomass can directly reflect the status and changes of forest ecosystem, and it is the basis for the rational utilization of forest resources and for improving the ecological environment [3,4]. Accurate and rapid estimation of forest biomass is particularly important for improving the efficiency of time, capital, and labor of forest resource investigation and studying the carbon cycle of the terrestrial ecosystem in large areas [5,6].

The traditional field measurement for forest aboveground biomass (AGB), which is more accurate for a small forest stand, cannot be used at the regional scale because it is too costly, labor intensive, and time consuming [7,8]. Remote sensing data, which have fast, real-time, dynamic, and regional-scale characteristics, are a frequently used data source for monitoring the dynamics of forests with the development of remote sensing technology [9,10]. Previous studies have shown that remote sensing data had a high correlation with AGB and can effectively predict and monitor forest biomass at the regional scale; thus, various types of remote sensing systems have been used for AGB estimation [11,12].

Among all available satellites, Landsat is currently the only satellite program to provide consistent, cross-calibrated data spanning more than 40 years for global surface observation [13,14]. The advantages of the global coverage reflective with increasing spectral and spatial fidelity, the unique record of the land surface and its change over time, the 40+ year coherent and temporally overlapping observatories and cross-sensor calibration, and free and open data access policy greatly stimulate new science and applications of Landsat [15,16]. Many countries have used the Landsat archive to carry out institutional systematic mapping and monitoring of forests in large areas, e.g., Canada used Landsat TM and ETM+ data in 2002 to produce the Earth Observation for Sustainable Development map of forests [17]; Australia used Landsat 5 and 7 data for national-scale carbon inventories [18]; and Brazil's National Institute for Space Research used Landsat data to monitor the annual deforestation rates of the Amazon since 1988 [19]. Landsat 8 was successfully launched on 11 February 2013, to ensure the continuity of the Landsat record. In addition to being consistent with the Landsat legacy, the significantly improved signal-to-noise ratio of Landsat 8 promises to enable better sensitivity of vegetation targets [16]. Therefore, Landsat 8 was used frequently to monitor the status, disturbance, and recovery of forests [20,21].

For remote sensing-based biomass estimation, multiple types of variables such as spectral bands, vegetation indices, and texture measures can be used as predictor variables for modeling [22,23]. The previous studies have testified the importance of selecting appropriate variables in improving AGB modeling [24,25]. Variable selection (also known as feature selection) can select a most effective variable subset from the full variable set to reduce variable space dimension, and improve the generalization and intelligibility of the model [26]. Variable selection is one of the most important steps in AGB modeling. Stepwise regression, which is the most commonly used method of variable selection of linear regression model, is simple and easy to perform [27]. Many variable selection algorithms (such as the random forest algorithm) include variable ranking based on some evaluation strategies as a principal or auxiliary selection mechanism because of their simplicity, scalability, and good empirical success [28,29].

In addition to variable selection, it is crucial to select a suitable algorithm to establish AGB estimation models. The traditional statistical regression algorithm, which can build a linear relationship between forest AGB and remote sensing data, is simple and easy to calculate. One of the traditional regression algorithms, the linear regression (LR) method was the most widely used method for AGB estimation in the previous studies [9,30]. However, the traditional statistical regression method cannot effectively express the complex relationship between forest AGB and remote sensing data under an indeterminate distribution of data. Therefore, the machine learning algorithms, such as K-nearest neighbor (KNN), support vector machine, artificial neural network, and decision tree, are applied to the remote sensing-based AGB estimation for improving the nonlinear estimation ability of the biomass model [31–34]. Previous studies have indicated that algorithms based on the decision tree, such as random forest (RF) and gradient boosting (GB), have an excellent performance in biomass estimation [35,36]. The RF is not only a variable selection algorithm but is also used as a nonlinear regression algorithm for AGB estimation because of its advantages of fewer adjustable parameters, high speed and efficiency, and the ability of variable importance calculation and permutation [37,38]. The extreme gradient boosting (XGBoost), as an advanced GB system, is widely used by data scientists and has provided state-of-the-art results for many fields, especially the financial field, such as credit risk assessment [39], but its potential has not been fully utilized in forestry.

The importance of field investigation for remote sensing-based AGB modeling is self-evident. Since 1973, China has conducted a continuous forest inventory, and in this process has established a comprehensive database covering many aspects of forest resources, involving forest health, timber production, and forest ecosystem services. The National Forest Continuous Inventory (NFCI), which is the first level of the forest inventory system of China, was designed to provide reliable data of the current status of and changes in the forests in the form of an integrated spatial database [40]. The NFCI survey is carried out every five years at the provincial scale. The sample plots have been systematically located at the graticule intersection of the national topographic map (scale of 1:100,000 or 1:50,000) [41]. Each tree with a diameter at breast height greater than or equal to 5 cm in the sample plot was tagged and permanently numbered for remeasurement in subsequent inventory periods. The NFCI is important for the formulation and refinement of state forest planning, management, and policy [42]. Therefore, the NFCI was widely used in many studies, including assessment and monitoring of forest status, conditions and changes, carbon sink and source identification, biomass estimation, and biodiversity [30,43].

In this paper, we used the NFCI data and Landsat 8 Operational Land Imager (OLI) data in combination with the LR and two machine learning algorithms, e.g., RF and XGBoost, to establish models for AGB estimation under the condition of known forest types and then created the AGB map for the study area using the optimal models. The specific objectives of this study were as follows: (1) to explore the influence of variable selection for the LR, RF, and XGBoost; (2) to validate the ability of the RF and XGBoost for estimating AGB; (3) to compare the accuracy of the LR, RF, and XGBoost models of different forest types; and (4) to draw the AGB map for the study area.

2. Study Area

Hunan Province (21.18×10^4 km^2, 24°38′ N–30°08′ N, 108°47′ E–114°15′ E) is situated in the south-central region of China (Figure 1). Most of the study areas are located in a subtropical monsoon humid climate zone, and the annual average temperature, rainfall, and sunshine duration are 14.80–18.50 °C, 1200–1800 mm, and 1238.7–1868.7 h, respectively. Therefore, the abundant resources of sunlight, water, and heat, with rain and heat over the same period, can promote the rapid growth of trees and enhance the ability of natural regeneration. The forestland area is 13.00×10^4 km^2, accounting for 61.37% of the study area; its forest coverage is 59.68%, and the total standing forest stock is 5.48×10^8 m^3 [44]; it is one of the key forest areas and major timber production bases in Southern China.

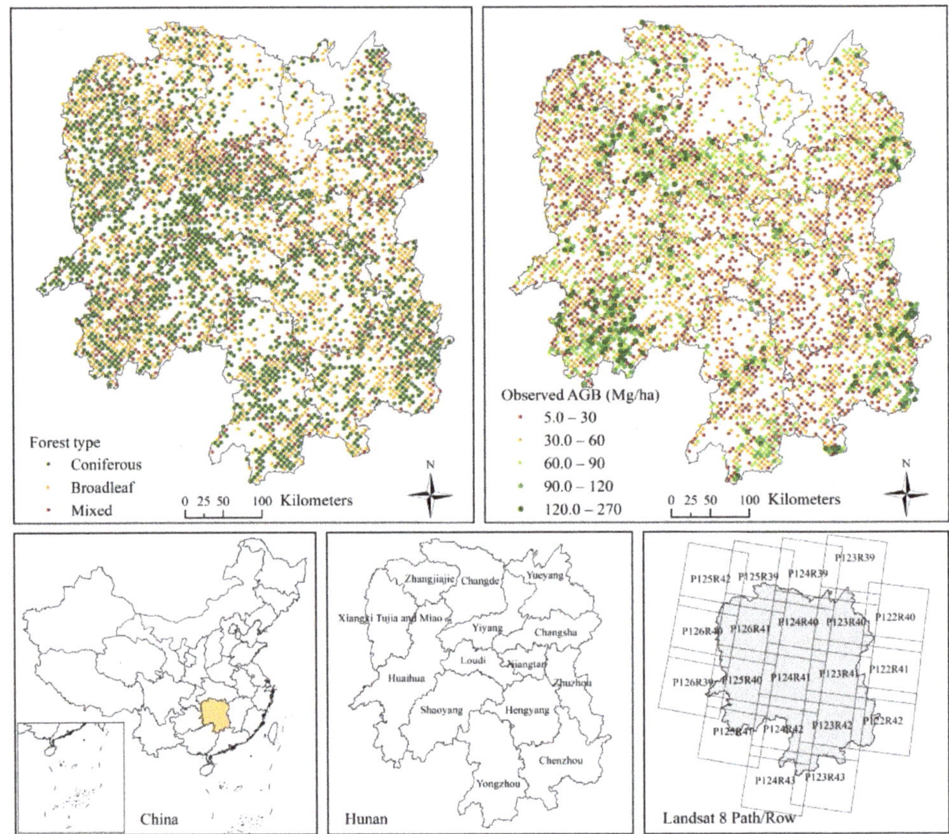

Figure 1. The location of the study area, including the forest types and observed AGB of the field plots, and the Landsat 8 scene numbers (P: path, R: row).

3. Data

3.1. Inventory Data

The eighth NFCI data of Hunan Province, which were surveyed in 2014, were used in this study. The size of each square plot is 25.82 × 25.82 m (approximately 0.0667 ha), and the plots were systematically allocated based on a grid of 4 km × 8 km. Note that the plots, which were situated on non-forestry land (such as cropland, water area, urban land, and bare land), or were covered by cloud in the remote sensing images, were eliminated. Finally, 3886 plots, which recorded around 149,000 trees, were used for modeling in this study (Figure 1).

The AGB of a tree was calculated by using the general one-variable aboveground biomass model, which can be expressed as [45]:

$$M_a = a \times D^{7/3} \tag{1}$$

$$a = 0.3 \times p \tag{2}$$

where M_a (kg) is the AGB of a tree, D (cm) is the diameter at breast height, a is the parameter of a tree species, and p (g/cm^3) is the basic wood density (Table A1). The plot AGB was converted to per hectare biomass (Mg/ha).

The plots were classified into three types, namely coniferous, broadleaf, and mixed forest, based on the species standing volume according to the technical regulation for forest continuous inventory of China (Table 1). In general, the average AGB of all plots with non-classification of forest types (abbreviated as "All" in all tables and figures) was 50.06 Mg/ha, within the range of 5.48–268.60 Mg/ha, with a standard deviation of 35.34 Mg/ha; the average AGB values of coniferous, broadleaf, and mixed forest were 48.71, 46.63, and 59.43 Mg/ha, respectively (Table 2).

Table 1. Classification standard of forest types.

Forest Type	Standard of Division
Coniferous	Pure coniferous forest (single coniferous species stand volume ≥ 65%) Coniferous mixed forest (coniferous species total stand volume ≥ 65%)
Broadleaf	Pure broadleaf forest (single broad-leaved species stand volume ≥ 65%) broadleaf mixed forest (broad-leaved species total stand volume ≥ 65%)
Mixed	Broadleaf-coniferous mixed forest (total stand volume of coniferous or broad-leaved species accounting for 35%–65%)

Compared with the digital elevation model, the high value of AGB is mainly distributed in the southeastern and western regions with a high altitude and steep slope and has a high vegetation coverage, low population density, less human interference, and poor economic condition. By contrast, the low value of AGB is mainly distributed in the low hills and valleys, with a low altitude and gentle slope; the conditions are opposite, especially in the middle region, which is the valley of Xiangjiang River with many towns and villages, and cropland. The spatial distribution trend of AGB is consistent with the topographic features and socio-economic conditions of the study area.

3.2. Landsat 8 Data

The Landsat Surface Reflectance products, which were derived from Landsat 8 OLI satellite images, were used in this study. The images, which were acquired in October 2015, were downloaded from the United States Geological Survey (USGS) website (https://earthexplorer.usgs.gov/, accessed on 20 October 2019). There were 30 screen images (Figure 1).

Radiometric and atmospheric correction of the Landsat Surface Reflectance images was performed by USGS [46]. For the areas of complex topography and with a great difference in elevation, terrain correction can effectively eliminate the shadow of the terrain as well as the difference in spectral features between a sunny slope and a shaded slope due to the topographic relief, preferably reflecting the true spectral feature of the object [47]. The terrain correction used the C-correction algorithm [48]. Then, the images were resampled to a pixel size of 25.82 m, the same as the inventory plot. The texture images were calculated using a grey-level co-occurrence matrix algorithm with 3 × 3, 5 × 5, and 7 × 7-pixel windows [49]. In addition, 20 vegetation indices were generated for this study (Appendix A). Landsat 8 OLI data were processed by the Environment for Visualizing Images software (Version 5.3.1, Boulder, CO, USA).

Finally, the remote sensing predictor variables, which were extracted for each plot center, included the primal images of 6 Landsat Surface Reflectance band images as well as the generated images of 20 vegetation index images and 144 texture images (Table 3).

Table 2. Distribution of the plot AGB values (Mg/ha) of the different forest types.

Forest Type	Count	Minimum	Maximum	Mean	Standard Deviation	Percentage of Different AGB Range (%)				
						5–30	30–60	60–90	90–120	120–270
Coniferous	1839	7.68	223.12	48.71	26.57	27.90	45.62	19.03	5.22	2.23
Broadleaf	1535	5.48	268.60	46.63	43.81	44.36	26.78	14.07	7.62	7.17
Mixed	512	18.60	219.95	59.43	34.21	20.31	38.87	24.61	9.38	6.84
All	3886	5.48	268.60	50.06	35.34	33.40	37.29	17.81	6.72	4.79

Table 3. Summary of predictor variables including Landsat Surface Reflectance band images, vegetation indices, and texture images for AGB estimation.

Variable Type	Variables Number	Variable Name	Description
Band Image	6	Band2, Band3, Band4, Band5, Band6, Band7	Landsat 8 Bands 2–7: Blue, Green, Red, NIR, SWIR1, SWIR2
Vegetation Index	20	ARVI	Atmospherically Resistant Vegetation Index
		DVI	Difference Vegetation Index
		EVI	Enhanced Vegetation Index
		GARI	Green Atmospherically Resistant Index
		GDVI	Green Difference Vegetation Index
		GNDVI	Green Normalized Difference Vegetation Index
		GRVI	Green Ratio Vegetation Index
		GVI	Green Vegetation Index
		IPVI	Infrared Percentage Vegetation Index
		LAI	Leaf Area Index
		MNLVI	Modified Non-Linear Vegetation Index
		MSRVI	Modified Simple Ratio Vegetation Index
		NDVI	Normalized Difference Vegetation Index
		NLVI	Non-Linear Vegetation Index
		OSAVI	Optimized Soil Adjusted Vegetation Index
		RDVI	Renormalized Difference Vegetation Index
		RVI	Ratio Vegetation Index
		SAVI	Soil Adjusted Vegetation Index
		TDVI	Transformed Difference Vegetation Index
		VARI	Visible Atmospherically Resistant Index
Texture Image	144	BiTjCon, BiTjDis, BiTjMea, BiTjHom, BiTjSeM, BiTjEnt, BiTjVar, BiTjCor	Landsat bands 2–7 texture measurement using gray-level co-occurrence matrix

Note: $BiTjXXX$ represents a texture image developed in the Landsat Surface Reflectance band i (2–7) using the texture measure XXX with a $j \times j$ (3, 5, 7) pixel window, where XXX is Con (contrast), Dis (dissimilarity), Mea (mean), Hom (homogeneity), SeM (angular second moment), Ent (entropy), Var (variance), or Cor (correlation).

3.3. Land Cover Image

The European Space Agency (ESA) Climate Change Initiative (CCI) project, of which the objective is to realize the full potential of the long-term global earth observation archives as a significant and timely contribution to the Essential Climate Variables databases required by United Nations Framework Convention on Climate Change, delivered consistent global Land Cover (LC) maps at a 300 m spatial resolution on an annual basis from 1992 to 2015 [50]. There is a highly positive result of the accuracy of the different classes: the highest user accuracy values are found for the classes of cropland (0.89–0.92), broadleaf forest (0.94–0.96), urban areas (0.86–0.88), bare (0.86–0.88), water bodies (0.92–0.96), and permanent snow and ice (0.96–0.97); the mixed and coniferous forest has a relatively low user accuracy value with 0.79–0.81 and 0.82–0.83, respectively [50]. The CCI-LC map for 2014 was downloaded from the ESA website (http://maps.elie.ucl.ac.be/CCI/viewer/index.php, accessed on 20 October 2019) for this study.

The typology of CCI-LC was defined using the Land Cover Classification System developed by the Food and Agriculture Organization of the United Nations, Rome, Italy. The map was consolidated into seven types based on the typology of CCI-LC: coniferous, broadleaf, and mixed forests, cropland, urban, water, and other types (non-forestry land, included bare land, grassland, etc.) (Figure 2). Then, the CCI-LC map was resampled to 25.82 m and snapped to the grid of Landsat 8 images.

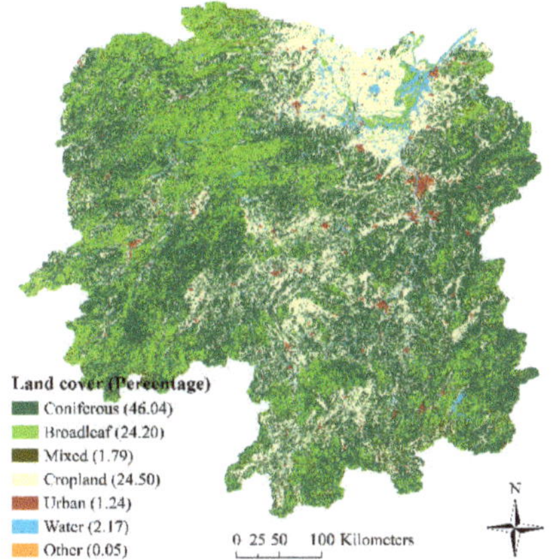

Figure 2. Classification of CCI-LC for the study area.

For validation of the accuracy and consistency of classification between NFCI and CCI-LC, the attribute of the CCI-LC map was extracted by the NFCI plot center. The result indicated that the producer accuracies of the CCI-LC map of coniferous, broadleaf, and mixed forests were 0.91, 0.88, and 0.82, respectively, and the user accuracies were 0.93, 0.91, and 0.92, respectively; the overall accuracy and kappa coefficient of coniferous, broadleaf and mixed forests were 0.92 and 0.88, respectively (Table 4). Therefore, the classified accuracy of the CCI-LC map can satisfy the research needs of this paper.

Table 4. Confusion matrix of classification between CCI-LC and NFCI.

Forest Type		Classification of CCI-LC							Producer Accuracy
		Coniferous	Broadleaf	Mixed	Cropland	Urban	Water	Other	
Classification	Coniferous	1649	76	33	29	7	3	11	0.91
based on	Broadleaf	62	1150	20	53	4	6	14	0.88
NFCI data	Mixed	54	43	627	31	2	5	7	0.82
User Accuracy		0.93	0.91	0.92	–	–	–	–	–

4. Methods

4.1. Algorithms of AGB Estimation

4.1.1. Linear Regression

The LR can quantitatively describe the correlation and significance between variables. The LR, which assumes a linear relationship between a response and a set of explanatory variables, can be expressed by the following model [30]:

$$Y = \alpha_0 + \alpha_1 X_1 + \alpha_2 X_2 + \cdots + \alpha_n X_n + \varepsilon \tag{3}$$

where Y is the value of AGB, X_1, X_2, ... , X_n are the predictor variables, α_0 is a constant, α_1, α_2, ... , α_n are the regression coefficients associated with the corresponding variables, n is the number of the predictor variables, and ε is the error term.

4.1.2. Random Forest

Decision trees are popular because they represent information in a way that is intuitive and easy to visualize and also have several other advantageous properties. The RF and XGBoost models, two ensemble techniques that combine the separate decision tree models to improve the ability of models, were considered in this paper.

RF is a classification and regression algorithm based on decision tree proposed by Breiman [51] in 2001. RF is one of the most common approaches to capture the complex relationship between a response and a set of explanatory variables with the following advantages: robustness to reduce over-fitting, ability to determine variable importance, higher accuracy, fewer parameters that need to be tuned, lower sensitivity to tuning of the parameters, fast training speed, and anti-noise property [25].

RF randomly collects a new dataset from the original sample dataset by bootstrapping. Generally, about 2/3 of the original sample data are selected in one bootstrap sample, and the remaining 1/3 of the data are used as out-of-bag data. Then, each bootstrap sample is used to establish a corresponding decision tree and combines multiple trees to improve the prediction performance [51]. When RF was used for regression, the mean of all decision tree prediction results was taken as the final prediction result. RF has been applied extensively as a classification algorithm [52] and has been used for time series forecasting in large-scale regression-based spatial applications [25,53].

4.1.3. Extreme Gradient Boosting

XGBoost, which was proposed by Chen et al. [54] and is very popular in data mining and machine learning competitions all over the world, is an improved gradient boosting decision tree (GBDT). Compared with the GBDT, XGBoost performs a second-order Taylor expansion for the objective function and uses the second derivative to accelerate the convergence speed of the model while training [55]. Unlike the independent decision trees of RF, XGBoost can correct the residual error to generate a new tree based on the previous tree [56].

The advantages of XGBoost include [54]:

(1) Using the second-order Taylor expression for the objective function, making the definition of the objective function more precise, and the optimal solution can be easily found;

(2) The addition of a regularization term into the objective function to control the complexity of the tree to obtain a simple model and to avoid overfitting;

(3) The use of sampling of the column feature to reduce the calculation amount and prevent overfitting; and

(4) The use of an effective cache-aware block structure for out-of-core tree learning to parallel and distributed computing makes learning faster for hundreds of millions of examples.

Generally, XGBoost is a highly scalable tree structure enhanced model, which can handle sparse and missing data well and can greatly improve the speed of the algorithm and compress computational memory in large-scale data training.

4.2. Methods of Variable Selection

Variable selection is the process of selecting the minimal and most effective variable subset from the original variable set to reduce the dimension of variable space and maximize the evaluation criteria [29]. Generally, the variable selection algorithm should determine four elements as follows: search starting point and direction, search strategy, evaluation function, and stopping criterion [57], but the algorithms mainly focus on the search strategy and evaluation function. In this paper, the stepwise regression approach was used to select the variable for LR, and the variable importance-based method was used for RF and XGBoost.

4.2.1. Stepwise Regression Approach

Stepwise regression is an important analysis method in LR analysis, which is mainly used to solve the problem of how to select explanatory variables when the number of explanatory variables is too many in the LR model so that all explanatory variables significantly impact the response variable in the regression equation [27]. Stepwise regression is used to introduce the explanatory variables one by one into the regression equation according to the contribution for the response variable. An introduced explanatory variable will be removed from the regression equation if it becomes non-significant due to the introduction of the subsequent new explanatory variable. After each explanatory variable is introduced or excluded, the F-test based on the sum of squares of partial regression is performed to ensure that only significant explanatory variables are included in the regression equation. This process is repeated until no non-significant explanatory variables are selected in the regression equation and no significant explanatory variables are removed from the regression equation to ensure that the final set of explanatory variables is optimal.

In this paper, stepwise regression was performed in SPSS software (Version 25, Armonk, NY, USA), and the probability of the F-test was set to 0.05 and 0.10 for entry and removal, respectively.

4.2.2. Variable Importance-Based Method

Each RF and XGBoost algorithm define two measures for variable importance, which can be used to rank variables. For RF, the first measure, which is computed from permuting out-of-bag data, is the percent increase in the mean square error (*%IncMSE*) of the prediction for each tree, and the second measure is the total decrease in node impurities (*IncNodePurity*) from splitting on the variable averaged over all trees, which is measured by the residual sum of squares [58]. Higher *%IncMSE* and *IncNodePurity* values indicate a more important predictor variable. For XGBoost, the first measure is calculated by the fractional contribution (*Gain*) of each feature to the model based on the total gain of this variable's splits, and the second measure is calculated by the relative number (*Frequency*) of times a feature be used in trees [59]. A higher percentage of *Gain* and *Frequency* means a more important predictor variable.

The acquisition of the optimal variable subset is a continuous search process, which would generally include four steps [60]:

(1) Subset generation: generate a candidate variable subset according to a certain search strategy. In this paper, the generalized sequential backward selection approach was used. The start point of the search is the original full variable set. The dataset was input into the RF and XGBoost models to obtain the variable importance and descending order, respectively, according to the measures. Then, a certain number (10%) of variables, which were the most unimportant, were removed to generate a variable subset.

(2) Subset evaluation: evaluate the prediction performance of the variable subset through an evaluation function. The generated subset was input into RF and XGBoost models, and the prediction

accuracy was evaluated using the coefficient of determination (R^2). In this paper, there were two evaluation results in each round, so the corresponding variable subset with high R^2 was compared and then selected as the selected variable subset in this round.

(3) Stopping criterion: determine when the variable search algorithm should stop. After the subset evaluation, the stopping criterion should be determined. If there is no stopping criterion, the search process cannot be stopped. In this paper, two stopping criteria were set: first if the number of variables of the subset was not larger than the set number, which was equal to the number of selected variables by stepwise regression for different forest types; and, second, if the R^2 of the prediction of the subset did not improve for three consecutive rounds.

(4) Subset validation: used to verify the validity of the selected variable subset. In this paper, a 10-fold cross-validation approach was performed to evaluate the performance of the variable subset in each round; therefore, the subset validation was not an independent step in the process.

In this paper, the modeling and variable selection of RF and XGBoost were implemented by the R packages *randomForest* [58] and *xgboost* [59], respectively. The workflow of variable selection is shown in Figure 3.

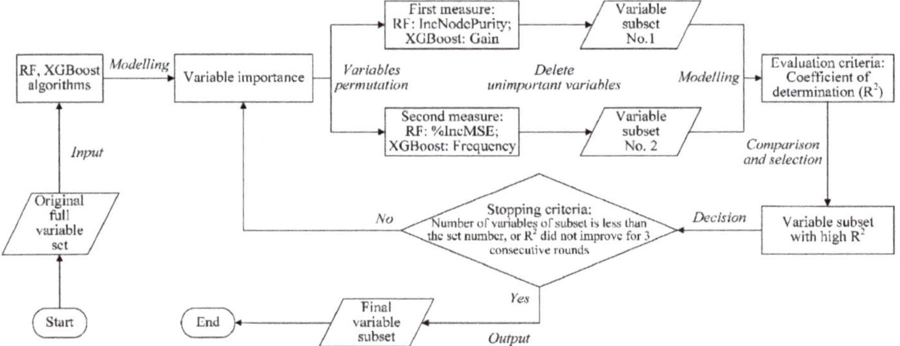

Figure 3. Workflow of the variable selection based on variable importance for RF and XGBoost models.

4.3. Variable Interactions

The two-way interactions between predictor variables graphically using the three-dimensional partial dependence plot, which was presented by Elith et al. [61], were used in this paper. In these plots, two of the predictor variables from the model, which are plotted on the x and y axes, are used to produce a grid of possible combinations of predictor variable values over the range of both variables, and the remaining predictor variables from the model are fixed at either their means (for continuous predictors) or their most common value (for categorical predictors). Model predictions are generated over this grid and plotted as the z-axis. The "*model.interaction.plot*" function of the R package *ModelMap* develops these plots, which can work with both continuous and categorical predictor variables [62].

4.4. Evaluation of AGB Models

The correlation test between the predictor variables and AGB was performed using the Pearson correlation coefficient in SPSS Statistics software.

In addition to the coefficient of determination (R^2), the root-mean-square error (RMSE) and the percentage root-mean-square error (RMSE%) were also used to evaluate the performance of the final models:

$$R^2 = 1 - \sum_{i=1}^{n}(y_i - \hat{y}_i)^2 / \sum_{i=1}^{n}(y_i - \bar{y}_i)^2 \tag{4}$$

$$RMSE = \sqrt{\sum_{i=1}^{n}(y_i - \hat{y}_i)^2/n} \tag{5}$$

$$RMSE\% = \frac{RMSE}{\overline{y}} \times 100 \tag{6}$$

where y_i is the observed AGB value, $\hat{y}_i$ is the predicted AGB value based on models, $\overline{y}$ is the arithmetic mean of all the observed AGB values, and n is the sample number. In general, a higher R^2 value and lower RMSE and RMSE% values indicate a better estimation performance of the model.

In addition, the difference of prediction between LR, RF, and XGBoost for different forest types was evaluated using the F-test.

5. Results

5.1. Role of Predictor Variables

5.1.1. Variable Importance

The result of the Pearson correlation coefficients between the predictor variables and AGB indicated that 144 variables had a significance level of 0.01 with the AGB, and the texture image variables had a significant correlation with the AGB. The variable with the highest correlation coefficient was *B4T7Mea*, with a value of −0.42.

Twenty-nine LR models were established using the selected predictor variables by stepwise regression for three forest types (i.e., coniferous, broadleaf, and mixed forest) and all plots with non-classification of forest types (Table 5). The results indicated that the performance of the models was improved when the count of predictor variables increased, and the models of different forest types worked better than the models of all forest plots (R^2 values of models for the coniferous, broadleaf, mixed, and all forest plots were 0.32, 0.37, 0.34, and 0.30, respectively).

The four best models (i.e., model numbers 7, 15, 21, and 29) were selected as the base to compare the performance of other types of models for the coniferous, broadleaf, and mixed forest (Table 6). The predictor variables of the LR models were different, and the collinearity statistics of the predictor variables were less than 5.50, which showed that the selected variables were effective. The predictor variables of these models were dominated by the image texture information. The standardized coefficients and the significance levels of the models showed that the texture-type variables contributed more than other variable types, which indicated that the texture variables were very important for the AGB estimation using the LR model in this study.

Figure 4 shows the selected predictor variables based on variable importance for the different forest types of RF and XGBoost models. The predictor variables of the RF and XGB models were not similar, and the main variables were the texture variables; the correlation and mean were included in all models, which indicated that the texture images had sufficient information to enhance the performance of models for AGB estimation. The texture of bands 5, 7, or both with a 7×7-pixel window were frequently involved in the models, indicating the significant roles of these two band texture variables in AGB estimation.

Table 5. Performance of LR models based on stepwise regression of different forest types.

Forest Type	Model No.	Count of Variables	R^2	RMSE	RMSE%	Forest Type	Model No.	Count of Variables	R^2	RMSE	RMSE%
Coniferous	1	1	0.28	34.24	70.29	Mixed	16	1	0.23	35.42	59.6
	2	2	0.29	32.64	67.01		17	2	0.28	33.09	55.68
	3	3	0.3	31.1	63.85		18	3	0.3	30.94	52.06
	4	4	0.31	30.79	63.21		19	4	0.32	30.25	50.9
	5	5	0.32	30.59	62.8		20	5	0.33	30.01	50.5
	6	6	0.32	30.26	62.12		21	6	0.34	29.65	49.89
	7	7	0.32	30.16	61.92						
Broadleaf	8	1	0.32	30.47	65.34	All	22	1	0.26	34.57	69.06
	9	2	0.34	29.73	63.76		23	2	0.27	34.33	68.58
	10	3	0.34	29.87	64.06		24	3	0.28	34.2	68.32
	11	4	0.35	28.92	62.02		25	4	0.28	33.44	66.8
	12	5	0.35	28.31	60.71		26	5	0.29	32.61	65.14
	13	6	0.36	27.97	59.98		27	6	0.29	31.9	63.72
	14	7	0.36	27.56	59.1		28	7	0.29	31.48	62.88
	15	8	0.37	27.32	58.59		29	8	0.3	31.12	62.17

Table 6. The predictor variable selection and estimation of the highest accuracy of LR models (Model Nos. 7, 15, 21, and 29 in Table 5) of different forest types.

Forest Type	Predictor Variable	Standardized Coefficients	Estimate (*t*-Test)	Significance (*p*-Value)	Collinearity Statistics	Forest Type	Predictor Variable	Standardized Coefficients	Estimate (*t*-Test)	Significance (*p*-Value)	Collinearity Statistics
Coniferous	B4T7Mea	−0.20	−6.74	0	1.66	Mixed	SAVI	0.22	4.24	0	1.41
	B5T7Cor	−0.10	−4.06	0	1.11		B3T7Cor	0.14	3.13	0	1.08
	B7T5Cor	0.07	2.8	0.01	1.15		B7T3Dis	0.11	2.24	0.03	1.38
	B4T3Ent	−0.19	−3.82	0	4.69		B6T3Cor	0.1	2.31	0.02	1.08
	B4T5Hom	−0.13	−2.59	0.01	4.96		B5T3Con	0.16	2.54	0.01	2.18
	B3T7Cor	0.05	2.1	0.04	1.06		B5T7Hom	0.13	1.89	0.05	2.44
	GVI	−0.05	−2.01	0.05	1.31						
Broadleaf	B5T7Mea	−0.13	−3.06	0	3.11	All	B4T7Mea	−0.14	−4.25	0	4.62
	LAI	0.29	7.26	0	2.63		B3T7Cor	0.08	4.8	0	1.04
	B3T7Cor	0.07	2.97	0	1.02		B4T7SeM	0.04	2.23	0.03	1.49
	B4T7SeM	0.3	2.87	0	5.14		B5T7Mea	−0.07	−2.42	0.02	3.4
	B4T5SeM	−0.25	−2.42	0.02	3.82		LAI	0.18	4.82	0	5.37
	B7T3Mea	0.18	2.81	0	4.99		B7T3Mea	0.1	3.66	0	3.4
	B6T7Mea	−0.14	−2.02	0.04	2.52		GDVI	−0.08	−2.17	0.03	5.15
	B6T3Var	−0.05	−1.83	0.05	1.09		B5T7Cor	−0.03	−1.97	0.05	1.03

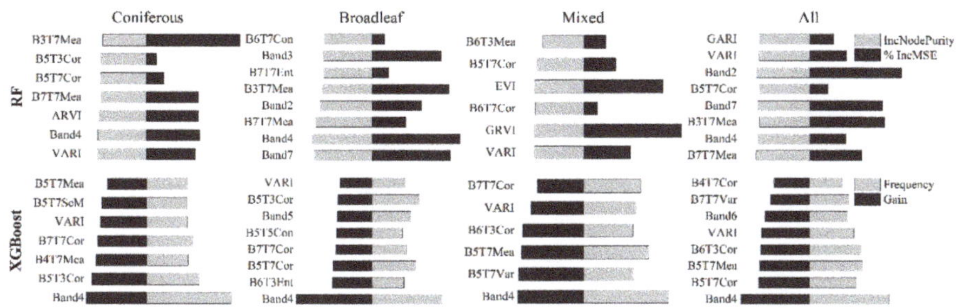

Figure 4. Variable importance of RF and XGBoost based on different forest types. The variable importance of each model was scaled to sum to 1.

However, the selected variables were different for the different forest types. The spectral bands, vegetation indices, and texture variables played a significant role in the broadleaf, mixed, and coniferous forest RF models, respectively. The species and canopy layers of broadleaf and mixed forests were multiple, which could be expressed by abundant spectral information; thus, the spectral and vegetation index variables could account sufficiently for the AGB estimation; the species composition of the coniferous forest was relatively single, which mainly consisted of fir and pine, and there was no obvious difference in the spectrum, whereas the texture information could well explain the AGB estimation [24,63]. This phenomenon of variable selection is more obvious in RF models than in XGBoost models. Unlike RF models, besides the texture variables, the spectral variables were also important for XGBoost models; especially *Band4*, which was the most important variable in all XGBoost models. Previous studies have shown that *Band4*, where the chlorophylls have peak absorption, had a strong relationship with biomass [64].

In addition, the relationship between the selected predictor variables was calculated using the Pearson correlation coefficient. We found that RF models mostly split the importance among the correlated multiple variables, whereas XGBoost models are inclined to centralize the importance at a single variable. For example, *Band4* was significantly correlated with *VARI* at a significance level of 0.01 with a value of −0.77; they had a similar importance in the RF model, but the importance was concentrated on *Band2* in the XGBoost model for the coniferous forest. This conclusion is the same as that reported by Freeman et al. [65].

5.1.2. Variable Interactions

The result indicated, surprisingly, that *Band4*, *VARI*, or both were involved in almost models, especially in XGBoost models (Figure 4). Figure 5 shows how *Band4* and *VARI* interact for the AGB estimation of the XGBoost models. We did not find significant interaction effects in these models, but we did find subtle interactions. These figures show that *Band4* mainly affected the interval with a low value (<300), but the effect of *VARI* was different. For the model of the coniferous forest, the high values of predicted AGB were mainly concentrated in the interval with a low value of *VARI* (<0.0), and there were some significant differences with the adjacent interval. In contrast, the high values of predicted AGB were dispersed in all intervals of *VARI* with a low interval of *Band4*, but there were hardly any high values in other intervals for the model of the broadleaf forest. However, the effects of *VARI* and *Band4* for the mixed forest model were significantly different from the models of the coniferous and broadleaf forests. Although the high values of predicted AGB were also concentrated in the interval with high values of *VARI* (>0.4), there were many higher values of predicted AGB that were distributed in other intervals of *Band4*. For the model of all forest plots, the effect of *Band4* and *VARI* was more similar to the combination of the models of coniferous, broadleaf, and mixed forests.

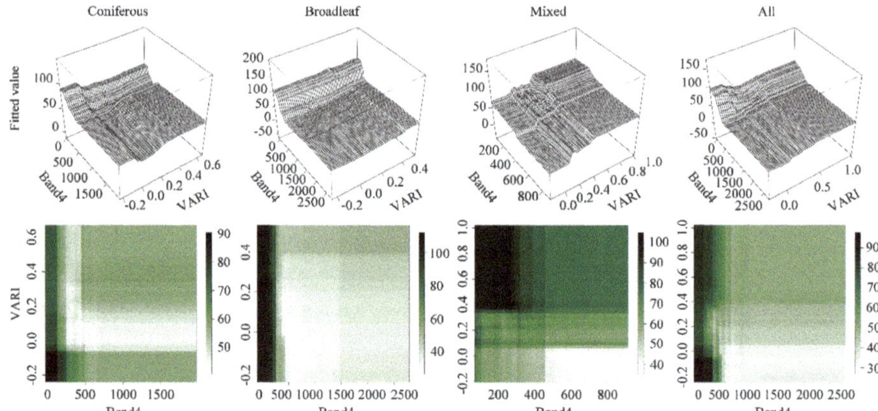

Figure 5. Interaction plots for *VARI* and *Band4* for the XGBoost models based on different forest types.

The interaction plots examine the effects of the two predictor variables with the remaining variables fixed at their mean value for continuous predictors (or the most common value for categorical predictors). The plots illustrated that they were seemingly more dependent on *VARI* than *Band4* in these two-way interactions, although *Band4* was the most important predictor in the estimation models. Compared with the model of all forest plots, each model of the coniferous, broadleaf, and mixed forests has distinct characteristics, which is beneficial for establishing AGB models with a high accuracy.

5.1.3. Performance of Variable Selection

The forward selection approach, which increases variables step-by-step, was used in stepwise regression; whereas the backward selection approach, which deletes variables step-by-step, was used in the variable selection of RF and XGBoost models. For the LR models, the performance of the models was improved when the number of predictor variables increased (Table 5).

Figure 6 illustrates how the R^2 values change with the number of selected variables for RF and XGBoost models. Each line represents an independent model, and the different colors indicate the different forest types. The result indicated that the R^2 values of models increased when the number of predictor variables decreased. Generally, the most dramatic change was the line of the mixed forest, followed by the lines of coniferous and broadleaf forests, whereas the line of all forest plots exhibited the smallest change in both RF and XGBoost models, although the variation degree of each line was different between RF and XGBoost models.

Contrasting the lines of the two algorithms, besides that the performance of XGBoost was better than that of RF, the influence of the number of predictor variables on the performance of models was also different. For RF, the influence of the number of predictor variables was relatively low, which manifested in a smooth change of the lines, whereas the influence was dramatic for XGBoost, with jagged peaks and valleys of the lines. This also indicated the variable selection was more important for XGBoost than for RF.

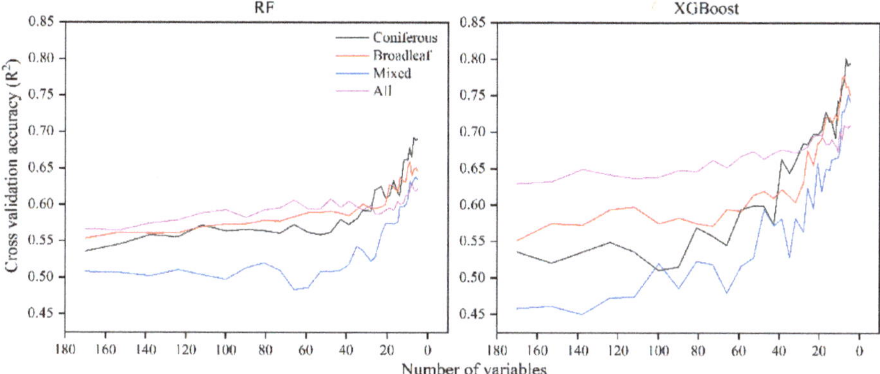

Figure 6. The accuracy of RF and XGBoost models with the selected variable number changing based on different forest types.

5.2. Evaluation of AGB Models

After the variable selection, we obtained the 12 best models of LR, RF, and XGBoost for different forest types. The performance of models was expressed by scatterplots, which showed the relationship between the predicted AGB values and observed AGB values (Figure 7). The plots showed that the RF model worked better than the LR model, and the XGBoost model worked better than the RF model for the same forest type. The results also indicated that the model of the broadleaf forest had the highest accuracy, followed by the models of mixed and coniferous forests, whereas the performance of the model for all forest plots was the worst for the same algorithm.

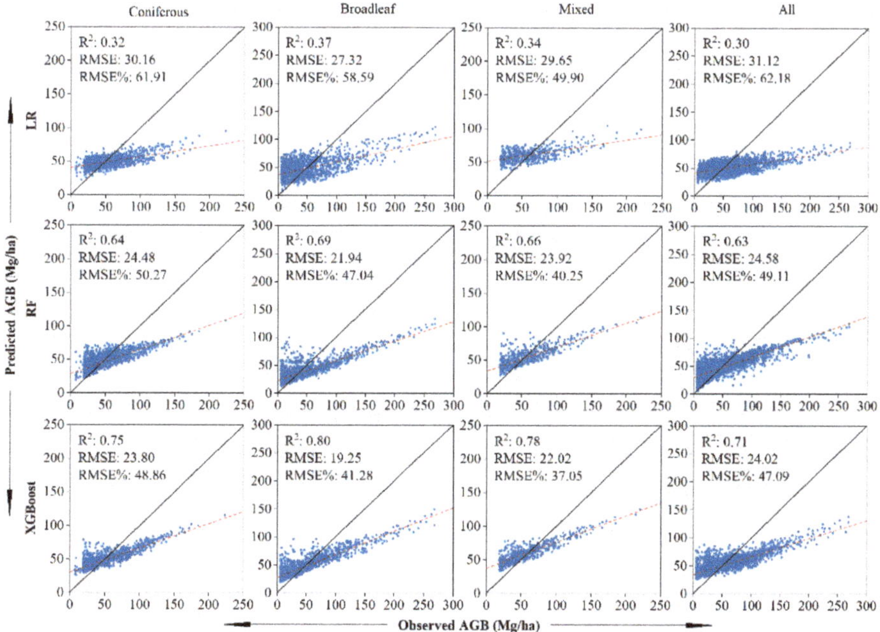

Figure 7. Scatter plot of the predicted and observed AGB of the LR, RF, and XGBoost models based on different forest types.

We found that problems of underestimation and overestimation, which also existed in the previous studies, were experienced by all models [30,66,67]. Intuitively, the predicted value was higher than the centerline when the biomass was low but lower when it was high in the figures. This means that the problem of overestimation and underestimation of remote sensing AGB estimation had no a fundamental solution, although the performance of models had a significant improvement based on forest type.

To further verify whether the models differed significantly, the F-test was used (Figure 8). The confidence level was set at 95%. In Figure 8, the numbers are the p-values, which are from the F-test, and the color of the checkerboard shows the levels of significance. The *F*-test results showed that there were significant differences of the predicted AGB between the LR, RF, and XGBoost models at a confidence level of 95%, although the p-values were different for these models (especially the p-value of all forest plots, which was higher than that of the other models).

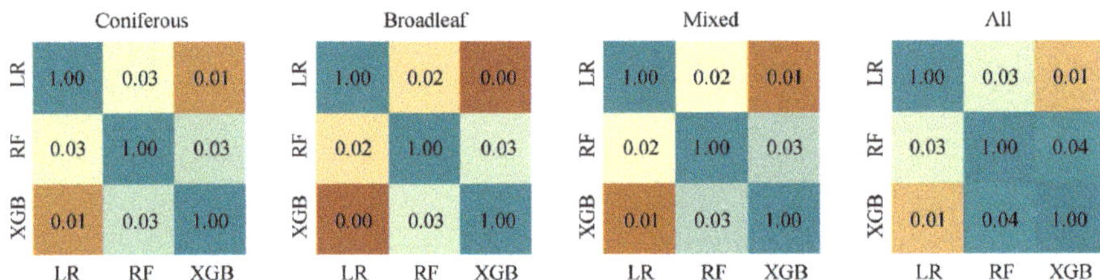

Figure 8. The comparisons (*F*-test) of the LR, RF, and XGBoost models based on forest type. The labels of the vertical and horizontal axes represent the models using a different algorithm. XGB represents the XGBoost model.

5.3. Mapping AGB

Finally, we drew the two AGB maps for the study area using the XGBoost models: first by estimating the AGB of the coniferous, broadleaf, and mixed forests according to the forest types in Figure 2, then combining these into one map (Figure 9a); second by estimating the AGB using all plots with non-classification of forest types (Figure 9b).

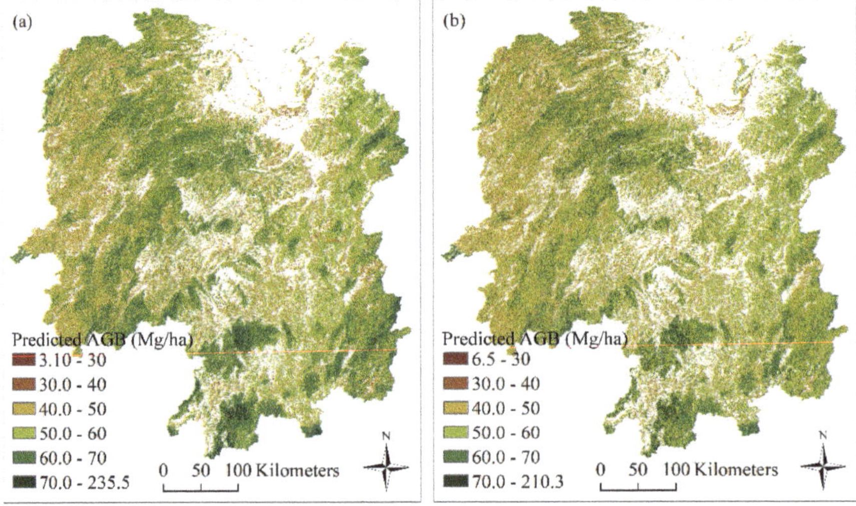

Figure 9. The predicted AGB using XGBoost models based on the different forest types, including (**a**) AGB map based on forest type and (**b**) AGB map based on all plots with non-classification of forest type.

In Figure 9, two maps of AGB had a similar trend in spatial distribution, which is consistent with the AGB distribution trend of the inventory plots in Figure 1. The results indicated that the ranges of predicted AGB for the two maps were different. The values ranged from 3.10 to 235.50 Mg/ha with a mean of 53.84 Mg/ha for the AGB map based on the forest type (Figure 9a), and the distribution and range of values were in close proximity to the inventory values in Figure 1. However, the range of values was from 6.50 to 210.30 Mg/ha with a mean of 52.39 Mg/ha for the AGB map based on all plots with non-classification of forest type (Figure 9b). In addition, the values of AGB in Figure 9a were higher than those in Figure 9b in the same area with a high value and were lower in the same area with a low value. This indicated that the ability of AGB estimation based on forest type was clearly improved for the high and low values. This improvement is also what we expected.

The degrees of overestimation and underestimation of the two maps were different, although the problems of overestimation and underestimation still existed. To further verify this conclusion, we sorted the values of the predicted AGB into three ranges based on the distribution of values: low (3 < predicted AGB < 25), medium (25 ≤ predicted AGB < 65), and high (65 ≤ predicted AGB < 236) values (Figure 10). The corresponding values of predicted AGB for the two maps and the AGB difference (Figure 9a minus Figure 9b) were obtained by the overlaying operations. In the low range of predicted AGB, most of the values of AGB prediction based on forest types (abbreviated as "Classification" in Figure 10) were lower than the values of AGB prediction of all plots with non-classification of forest type (abbreviated as "Non-classification" in Figure 10), and the values of the difference were mainly distributed from −10 to 2 Mg/ha (Figure 10a); therefore, the "Classification" map had a better prediction performance. In the medium range, the distribution of the AGB difference approximated a normal distribution, indicating that two maps had a similar performance for medium values of AGB (Figure 10b). In the high range, the values of "Classification" were clearly higher than those for "Non-classification", indicating that the "Classification" map also had better performance with respect to the high AGB values (Figure 10c). In summary, the map, which was predicted based on forest type, better estimated the AGB value than the map with the non-classification of forest type irrespective of high or low AGB.

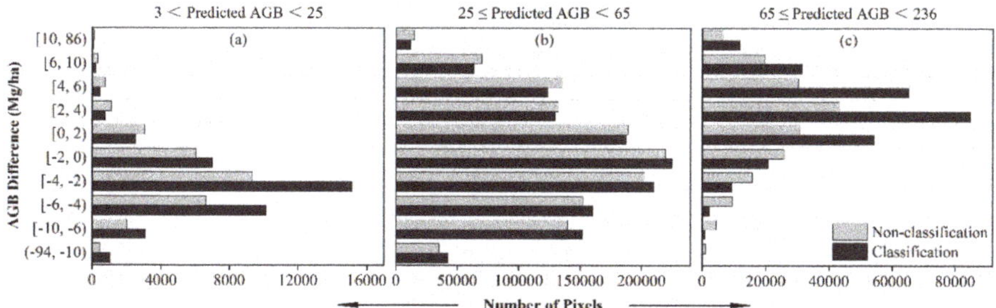

Figure 10. Histograms illustrating the difference in pixel number in three ranges. Note that the vertical axis labels represent the range of the prediction difference between two AGB maps (Figure 9a minus Figure 9b); Classification: values from Figure 10a, Non-classification: values from Figure 9b. (**a**) 3 < Predicted AGB ≤ 25, (**b**) 25 ≤ Predicted AGB < 65, (**c**) 65 ≤ Predicted AGB < 236.

6. Discussion

Through this experiment, we increased our understanding of the importance of variable selection, which can influence the performance of machine learning algorithms. Variable selection is one of the most important processes in modeling, which can reduce data dimension and the storage space of data, speed the estimation process, and improve the interpretability and performance of models [29].

Multiple predictor variables, such as spectral bands, vegetation indices, and textures, were extracted from remote sensing images and were used for modeling AGB in this paper. However, these predictors cannot all be used for modeling due to their high correlations and high numbers. The performance of models was significantly impacted by the number of selected predictor variables (Figure 6, Table 5). Through the variable selection, the number of predictor variables was reduced from hundreds to several, which makes it easier to interpret the model. In this study, the Red (*Band4*), NIR (*Band5*), SWIR (*Band7*) bands, and the derived variables played a more important role than other bands. In models of AGB estimation, the SWIR band is more sensitive to the shadow of vegetation and humidity of soil and is less influenced by the atmospheric conditions [16,22,68]; the NIR band of Landsat 8, of which the wavelength range was adjusted to 0.845–0.885 μm to exclude the effect of water-vapor absorption at 0.825 μm, is more sensitive to vegetation of different types [13,69]; and the red band is usually used to distinguish the vegetation type [21,64,70]. We cannot ignore the fact that the vegetation index variables were also selected frequently, especially *VARI*, which exists in all XGBoost models. Compared with other vegetation indices, *VARI* is minimally sensitive to the atmospheric effect, and the estimation error of vegetation affected by the atmosphere is less than 10% in a large area [71,72].

In addition, the textures, which are dominant in all models, were also critical for AGB estimation, although the importance of the texture predictor variables was different from that in previous studies [24,30,66]. However, for the different forest types, texture variables and spectral variables played different roles in AGB models (Figure 4). For example, the spectral variables were more important than texture variables in the RF model of the broadleaf forest, while the texture variables were more important in the RF model of the coniferous forest. This illustrated that the role of texture variables and spectral variables was dependent on forest structure: in the broadleaf and mixed forest with multiple species and complex structure, the models tended to select the spectral variables, while in the coniferous forest with relatively fewer species and simple structure, the models tended to choose texture variables [63,64,73].

Due to the different characters between spectral variables and texture variables, their combination is beneficial to improve the performance of AGB models, and this improvement was evident in all models. Moreover, the influence of variable selection was different for RF and XGBoost. We found that the accuracy of the XGBoost algorithm varied greatly with the number of selected variables compared with RF (Figure 6). The RF algorithm can be regarded as a parallel ensemble algorithm because the decision tree of RF is independent; thus, RF is not sensitive to inclusion of the noisy predictor variables [74,75], whereas the decision tree of XGBoost is generated based on the previous tree; thus, the noisy predictor variables will influence the accuracy of the subsequent new tree [76,77].

The LR algorithm, which assumes a linear relationship between predictor and predicted variables, was used frequently in most early biomass estimation studies due to the interpretability of LR [30,78]. However, the relationship between remote sensing data and AGB is complex; thus, the traditional statistical regression algorithm cannot efficiently describe the relationship between them. Therefore, machine learning algorithms such as random forest and gradient boosting, which can establish a complex non-linear relationship between vegetation information and remote sensing images with an indeterminate distribution of data, were introduced to improve the accuracy of AGB estimation [79].

In our study, we extracted 170 predictor variables from Landsat 8 images; then, a few variables were selected from these through the variable selection process to build RF and XGBoost models (Figure 4). We found that the machine learning algorithms prevented overfitting and significantly improved the estimation accuracy compared with the LR models, and the result also indicated that the XGBoost model worked better than the RF model (Figure 7). The XGBoost algorithm, which is a highly flexible algorithm with the ability to correct the residual error to generate a new tree based on the previous tree, provided an improvement in processing a regularized learning objective to avoid overfitting [54].

Before this study, few studies had used the XGBoost algorithm to estimate AGB. Li et al. [30] used a linear mixed-effects model and linear dummy variable model to estimate AGB in the western Hunan Province of China; the R^2 values of total vegetation were 0.41; Zhu et al. [6] used multiple algorithms (LR, KNN, logistic regression) to estimate AGB for the Xiangjiang River, and the results indicated the machine learning algorithm had a good performance for AGB estimation. In contrast, the results obtained by machine learning methods in this study were better, and the XGBoost algorithm had a good performance in AGB estimation and could reduce underestimation and overestimation to some extent.

In this paper, we established the models based on forest type to improve the accuracy of AGB estimation, and the results indicated this method was valuable. We found that the models based on forest type had a better performance at the lower and higher values compared to the models of all plots with non-classification of forest types, especially XGBoost (Figures 7 and 9). In addition, the problem of overestimation and underestimation, which are the main factors influencing AGB modeling performance, was not completely solved, although the performance of models had a significant improvement compared with the previous studies. As to this problem, it is decided by the algorithm itself on one hand. The decision trees, which are the key components of the RF and XGBoost methods, cannot extrapolate outside the training set. On the other hand, it is related to the remote sensing data. For plots with low AGB values, the shrubs, grass, and bare soil will influence the reflectance of bands; the pixel of Landsat 8 with relatively low spatial resolution (30 × 30 m) is a mixed pixel, which cannot accurately express the spectral information of land cover. For plots with high AGB values, the saturation in multispectral sensors such as Landsat 8 OLI is the main reason for underestimation of AGB [80,81]. Therefore, remote sensing data with higher spatial and radiometric resolution such as LIDAR data and hyperspectral data, or the approach of mixed pixel decomposition, may be solutions for AGB estimation. Meanwhile, a modeling approach based on the AGB range may be a useful method for improving the prediction of AGB, but it needs more sample plots.

The subtropical forests of China are distributed in 13 provinces, including Zhejiang, Jiangsu, Anhui, Fujian, Jiangxi, Hubei, Hunan, Guangdong, Guangxi, Hainan, Guizhou, Sichuan, and Yunnan. They are one of the dominant distribution areas of forest resources in China. It is necessary to monitor the subtropical forest change because the forests have been influenced by the improved silviculture, woody encroachment, climate change, and human activities. The forest biomass estimation based on traditional field measurements is a relatively accurate method, but it is impossible to implement for such large areas of subtropical forests. Therefore, remote sensing-based estimation of forest biomass change is a very important method. The NFCI data, which has been checked and revised many times by the state and provincial forest departments before it is released, is the only available data with highest quality in the provincial scale at present. However, the residual atmospheric effects and calibration errors in satellite data cannot be completely eliminated. Therefore, until the more effective satellite data are available, we can only hope to improve the accuracy of forest biomass estimation by using new modeling methods. Despite certain inaccuracies, the performance of the biomass estimation method used in this study exceeds our expectations, and the selected modeling method of XGBoost seems to be more effective. The results show that the NFCI data in combination with Landsat 8 can be successfully applied to biomass estimation. Although the XGBoost models had the relatively high RMSE and RMSE% values, the total accuracies of models were significantly increased with the variable selection, and it is still manifested that the methods in this paper were very important and useful for the provincial-scale accurate estimation of forest biomass, and these methods can also be used to other similar areas. In addition, we must admit that there are still many sections that could be improved in our research, such as methods of variable selection, variable data cleaning, and parameter optimization for machine learning. We will do further research in these aspects in the future.

7. Conclusions

In this study, we selected the subtropical region of Hunan Province, China, as a case study area to analyze the AGB estimation based on forest type using different modeling algorithms, namely, LR, RF, and XGBoost. The results indicated the following: (1) Variable selection is a very important part of machine learning algorithms. In this study, variable selection had a significant effect on the performance of XGBoost compared with that of RF. (2) Machine learning algorithms have advantages in AGB estimation. In this paper, the XGBoost and RF models significantly improved the estimation accuracy compared with the LR models, and the XGBoost algorithm reduced overestimation and underestimation to a certain extent, although the problem was not fully eliminated. (3) The method of AGB estimation based on forest type is a very useful approach to improve the accuracy of AGB estimation, and the models had a better performance at the lower and higher values compared with the models using all plots with non-classification of forest types. In this paper, we provided a new approach when establishing similar models. The result and conclusion may be different for other areas, but we hope to pay attention to variable selection when using machine learning algorithms in the future and will try to use various remote sensing data and algorithms to improve the accuracy of biomass estimation.

Author Contributions: Y.L. and M.L.; data curation: Y.L., C.L., and Z.L.; formal analysis: Y.L., C.L., and Z.L.; funding acquisition: M.L.; methodology: Y.L. and M.L.; project administration: M.L.; resources: M.L.; software: Y.L., and C.L.; supervision: M.L.; validation: Y.L., C.L., and Z.L.; visualization: Y.L. and Z.L.; writing—original draft: Y.L.; writing—review and editing: Y.L., M.L., C.L., and Z.L.

Acknowledgments: This study was financially supported by the National Natural Science Foundation of China (no. 31770679), and Top-notch Academic Programs Project of Jiangsu Higher Education Institutions, China (TAPP, PPZY2015A062). The authors would like to thank our editors, as well as the anonymous reviewers for their valuable comments, and also LetPub (www.letpub.com) for linguistic assistance during the preparation of this manuscript.

Conflicts of Interest: The authors declare no conflict of interest.

Appendix A

Table A1. The wood density of the tree species or groups.

Tree Species/Groups	Wood Density (p)	Tree Species/Groups	Wood Density (p)
Abies	0.3464	*Pinus massoniana*	0.4476
Betula	0.4848	*Pinus tabulaeformis*	0.4243
Cinnamomum	0.4600	*Pinus taiwanensis*	0.4510
Cryptomeria	0.3493	*Pinus yunnanensis*	0.3499
Cunninghamia lanceolata	0.3098	*Populus*	0.4177
Cupressus	0.5970	*Quercus*	0.5762
Eucalyptus	0.5820	*Robinia pseudoacacia*	0.6740
Fraxinus mandshurica	0.4640	*Salix*	0.4410
Larix	0.4059	*Schima superba*	0.5563
Liquidambar formosana	0.5035	*Tilia*	0.3200
Paulownia	0.2370	*Ulmus*	0.4580
Picea	0.3730	Other conifers	0.3940
Pinus armandii	0.3930	Other pines	0.4500
Pinus densata	0.4720	Other hardwood broadleaves	0.6250
Pinus elliottii	0.4118	Other softwood broadleaves	0.4430

Note: The total relative error of the tree species or groups was 2.10%, not exceeding the common allowance of 3%, and the average of the absolute relative error was 6.37%, less than the error allowance of 10% [45].

References

1. Brown, S. Measuring carbon in forests: Current status and future challenges. *Environ. Pollut.* **2002**, *116*, 363–372. [CrossRef]
2. Gower, S.T. Patterns and mechanisms of the forest carbon cycle. *Annu. Rev. Environ. Resour.* **2003**, *28*, 169–204. [CrossRef]
3. Houghton, R.A. Aboveground forest biomass and the global carbon balance. *Glob. Chang. Biol.* **2005**, *11*, 945–958. [CrossRef]
4. Houghton, R.A.; Hall, F.; Goetz, S.J. Importance of biomass in the global carbon cycle. *J. Geophys. Res. Biogeosci.* **2009**, *114*, 1–13. [CrossRef]

5. Lu, D.; Batistella, M.; Moran, E. Satellite estimation of aboveground biomass and impacts of forest stand structure. *Photogramm. Eng. Remote Sens.* **2005**, *71*, 967–974. [CrossRef]

6. Zhu, J.; Huang, Z.; Sun, H.; Wang, G. Mapping forest ecosystem biomass density for xiangjiang river basin by combining plot and remote sensing data and comparing spatial extrapolation methods. *Remote Sens.* **2017**, *9*, 241. [CrossRef]

7. West, P.W. *Tree and Forest Measurement*, 3rd ed.; Springer: Berlin/Heidelberg, Germany, 2015; ISBN 978-3-319-14707-9.

8. Crosby, M.K.; Matney, T.G.; Schultz, E.B.; Evans, D.L.; Grebner, D.L.; Londo, H.A.; Rodgers, J.C.; Collins, C.A. Consequences of landsat image strata classification errors on bias and variance of inventory estimates: A forest inventory case study. *IEEE J. Sel. Top. Appl. Earth Obs. Remote Sens.* **2017**, *10*, 243–251. [CrossRef]

9. Lu, D. The potential and challenge of remote sensing-based biomass estimation. *Int. J. Remote Sens.* **2006**, *27*, 1297–1328. [CrossRef]

10. Avitabile, V.; Herold, M.; Heuvelink, G.B.M.; Simon, L.; Phillips, O.L.; Asner, G.P.; Armston, J.; Peter, S.; Banin, L.; Bayol, N.; et al. An integrated pan-tropical biomass map using multiple reference datasets. *Glob. Chang. Biol.* **2016**, *22*, 1406–1420. [CrossRef]

11. Deng, S.; Katoh, M.; Guan, Q.; Yin, N.; Li, M. Estimating forest aboveground biomass by combining ALOS PALSAR and WorldView-2 data: A case study at Purple Mountain National Park, Nanjing, China. *Remote Sens.* **2014**, *6*, 7878–7910. [CrossRef]

12. Cao, L.; Coops, N.C.; Innes, J.L.; Sheppard, S.R.J.; Fu, L.; Ruan, H.; She, G. Estimation of forest biomass dynamics in subtropical forests using multi-temporal airborne LiDAR data. *Remote Sens. Environ.* **2016**, *178*, 158–171. [CrossRef]

13. Loveland, T.R.; Irons, J.R. Landsat 8: The plans, the reality, and the legacy. *Remote Sens. Environ.* **2016**, *185*, 1–6. [CrossRef]

14. Wulder, M.A.; Loveland, T.R.; Roy, D.P.; Crawford, C.J.; Masek, J.G.; Woodcock, C.E.; Allen, R.G.; Anderson, M.C.; Belward, A.S.; Cohen, W.B.; et al. Current status of Landsat program, science, and applications. *Remote Sens. Environ.* **2019**, *225*, 127–147. [CrossRef]

15. Loveland, T.R.; Dwyer, J.L. Landsat: Building a strong future. *Remote Sens. Environ.* **2012**, *122*, 22–29. [CrossRef]

16. Roy, D.P.; Wulder, M.A.; Loveland, T.R.; Woodcock, C.E.; Allen, R.G.; Anderso, M.C.; Helder, D.; Irons, J.R.; Johnson, D.M.; Kennedy, R.; et al. Landsat-8: Science and product vision for terrestrial global change research. *Remote Sens. Environ.* **2014**, *145*, 154–172. [CrossRef]

17. Wulder, M.A.; White, J.C.; Cranny, M.; Hall, R.J.; Luther, J.E.; Beaudoin, A.; Goodenough, D.G.; Dechka, J.A. Monitoring Canada's forests. Part 1: Completion of the EOSD land cover project. *Can. J. Remote Sens.* **2008**, *34*, 549–562. [CrossRef]

18. Lehmann, E.A.; Wallace, J.F.; Caccetta, P.A.; Furby, S.L.; Zdunic, K. Forest cover trends from time series Landsat data for the Australian continent. *Int. J. Appl. Earth Obs. Geoinf.* **2013**, *21*, 453–462. [CrossRef]

19. Shimabukuro, Y.E.; Batista, G.T.; Mello, E.M.K.; Moreira, J.C.; Duarte, V. Using shade fraction image segmentation to evaluate deforestation in landsat thematic mapper images of the Amazon Region. *Int. J. Remote Sens.* **1998**, *19*, 535–541. [CrossRef]

20. Banskota, A.; Kayastha, N.; Falkowski, M.J.; Wulder, M.A.; Froese, R.E.; White, J.C. Forest monitoring using landsat time series data: A review. *Can. J. Remote Sens.* **2014**, *40*, 362–384. [CrossRef]

21. Chrysafis, I.; Mallinis, G.; Gitas, I.; Tsakiri-Strati, M. Estimating Mediterranean forest parameters using multi seasonal Landsat 8 OLI imagery and an ensemble learning method. *Remote Sens. Environ.* **2017**, *199*, 154–166. [CrossRef]

22. Lu, D. Aboveground biomass estimation using Landsat TM data in the Brazilian Amazon. *Int. J. Remote Sens.* **2005**, *26*, 2509–2525. [CrossRef]

23. Ouma, Y.O. Optimization of second-order grey-level texture in high-resolution imagery for statistical estimation of above-ground biomass. *J. Environ. Inf.* **2006**, *8*, 70–85. [CrossRef]

24. Lu, D.; Batistella, M. Exploring TM image texture and its relationships with biomass estimation in Rondônia, Brazilian Amazon. *Acta Amaz.* **2005**, *35*, 249–257. [CrossRef]

25. Shen, W.; Li, M.; Huang, C.; Tao, X.; Wei, A. Annual forest aboveground biomass changes mapped using ICESat/GLAS measurements, historical inventory data, and time-series optical and radar imagery for Guangdong province, China. *Agric. For. Meteorol.* **2018**, *259*, 23–38. [CrossRef]

26. Yu, K.; Wu, X.; Ding, W.; Pei, J. Scalable and accurate online feature selection for big data. *ACM Trans. Knowl. Discov. Data* **2016**, *11*, 1–39. [CrossRef]

27. Wang, Y.; Wen, L.; Chen, M. *Dictionary of Mathematics*, 5th ed.; Science Press: Beijing, China, 2017; ISBN 9787030533364.

28. Genuer, R.; Poggi, J.M.; Tuleau-Malot, C. Variable selection using random forests. *Pattern Recognit. Lett.* **2010**, *31*, 2225–2236. [CrossRef]

29. Bolón-Canedo, V.; Sánchez-Maroño, N.; Alonso-Betanzos, A. Feature selection for high-dimensional data. *Prog. Artif. Intell.* **2016**, *5*, 65–75. [CrossRef]

30. Li, C.; Li, Y.; Li, M. Improving forest aboveground biomass (AGB) estimation by incorporating crown density and using landsat 8 OLI images of a subtropical forest in Western Hunan in Central China. *Forests* **2019**, *10*, 104. [CrossRef]

31. Reese, H.; Nilsson, M.; Sandstro, P. Applications using estimates of forest parameters derived from satellite and forest inventory data. *Comput. Electron. Agric.* **2002**, *37*, 37–55. [CrossRef]

32. Baccini, A.; Laporte, N.; Goetz, S.J.; Sun, M.; Dong, H. A first map of tropical Africa's above-ground biomass derived from satellite imagery. *Environ. Res. Lett.* **2008**, *3*, 1–9. [CrossRef]

33. Nelson, R.; Montesano, P.; Ranson, K.J.; Kharuk, V.; Sun, G.; Kimes, D.S. Estimating Siberian timber volume using MODIS and ICESat/GLAS. *Remote Sens. Environ.* **2009**, *113*, 691–701. [CrossRef]

34. Monnet, J.-M.; Chanussot, J.; Berger, F. Support vector regression for the estimation of forest stand parameters using airborne laser scanning. *IEEE Geosci. Remote Sens. Lett.* **2011**, *8*, 580–584. [CrossRef]

35. Blackard, J.A.; Finco, M.V.; Helmer, E.H.; Holden, G.R.; Hoppus, M.L.; Jacobs, D.M.; Lister, A.J.; Moisen, G.G.; Nelson, M.D.; Riemann, R.; et al. Mapping U.S. forest biomass using nationwide forest inventory data and moderate resolution information. *Remote Sens. Environ.* **2008**, *112*, 1658–1677. [CrossRef]

36. Carreiras, J.M.B.; Vasconcelos, M.J.; Lucas, R.M. Understanding the relationship between aboveground biomass and ALOS PALSAR data in the forests of Guinea-Bissau (West Africa). *Remote Sens. Environ.* **2012**, *121*, 426–442. [CrossRef]

37. Karlson, M.; Ostwald, M.; Reese, H.; Sanou, J.; Tankoano, B.; Mattsson, E. Mapping tree canopy cover and aboveground biomass in sudano-sahelian woodlands using landsat 8 and random forest. *Remote Sens.* **2015**, *7*, 10017–10041. [CrossRef]

38. Zald, H.S.J.; Wulder, M.A.; White, J.C.; Hilker, T.; Hermosilla, T.; Hobart, G.W.; Coops, N.C. Integrating landsat pixel composites and change metrics with lidar plots to predictively map forest structure and aboveground biomass in Saskatchewan, Canada. *Remote Sens. Environ.* **2016**, *176*, 188–201. [CrossRef]

39. Carmona, P.; Climent, F.; Momparler, A. Predicting failure in the U.S. banking sector: An extreme gradient boosting approach. *Int. Rev. Econ. Financ.* **2019**, *61*, 304–323. [CrossRef]

40. Lei, X.; Tang, M.; Lu, Y.; Hong, L.; Tian, D. Forest inventory in China: Status and challenges. *Int. For. Rev.* **2009**, *11*, 52–63. [CrossRef]

41. Zeng, W.; Tomppo, E.; Healey, S.P.; Gadow, K.V. The national forest inventory in China: History—Results—International context. *For. Ecosyst.* **2015**, *2*, 23. [CrossRef]

42. Xie, X.; Wang, Q.; Dai, L.; Su, D.; Wang, X.; Qi, G.; Ye, Y. Application of China's National Forest Continuous Inventory Database. *Environ. Manage.* **2011**, *48*, 1095–1106. [CrossRef]

43. Fang, J.; Chen, A.; Peng, C.; Zhao, S.; Ci, L. Changes in forest biomass carbon storage in China between 1949 and 1998. *Science* **2001**, *292*, 2320–2322. [CrossRef]

44. Hunan Provincial People's Government Natural Resources of Hunan Province. Available online: http://www.enghunan.gov.cn/AboutHunan/HunanFacts/NaturalResources/201507/t20150707_1792317.html (accessed on 1 November 2019).

45. Zeng, W. Developing one-variable individual tree biomass models based on wood density for 34 tree species in China. *For. Res. Open Access* **2018**, *7*, 1–5.

46. USGS Landsat Surface Reflectance Data. Available online: https://pubs.usgs.gov/fs/2015/3034/pdf/fs20153034.pdf (accessed on 27 March 2019).

47. Richter, R. Correction of Atmospheric and Topographic Effects for High Spatial Resolution Satellite Imagery. *Int. J. Remote Sens.* **1997**, *18*, 1099–1111. [CrossRef]

48. Teillet, P.M.; Guindon, B.; Goodenough, D.G. On the slope-aspect correction of multispectral scanner data. *Can. J. Remote Sens.* **1982**, *8*, 84–106. [CrossRef]

49. Haralick, R.M.; Shanmugam, K.; Dinstein, I. Textural features for image classification. *IEEE Trans. Syst. Man. Cybern.* **1973**, *SMC-3*, 610–621. [CrossRef]

50. ESA Land Cover CCI Product User Guide. Available online: http://maps.elie.ucl.ac.be/CCI/viewer/download/ESACCI-LC-Ph2-PUGv2_2.0.pdf (accessed on 10 April 2017).

51. Breiman, L. Random forest. *Mach. Learn.* **2001**, *45*, 5–32. [CrossRef]

52. Yu, Y.; Li, M.; Fu, Y. Forest type identification by random forest classification combined with SPOT and multitemporal SAR data. *J. For. Res.* **2018**, *29*, 1407–1414. [CrossRef]

53. Tyralis, H.; Papacharalampous, G.; Tantanee, S. How to explain and predict the shape parameter of the generalized extreme value distribution of streamflow extremes using a big dataset. *J. Hydrol.* **2019**, *574*, 628–645. [CrossRef]

54. Chen, T.; Guestrin, C. XGBoost: A Scalable Tree Boosting System. In Proceedings of the 22nd ACM SIGKDD International Conference on Knowledge Discovery and Data Mining (KDD '16), San Francisco, CA, USA, 13–17 August 2016; pp. 785–794.

55. He, H.; Zhang, W.; Zhang, S. A novel ensemble method for credit scoring: Adaption of different imbalance ratios. *Expert Syst. Appl.* **2018**, *98*, 105–117. [CrossRef]

56. Friedman, J.H. Stochastic gradient boosting. *Comput. Stat. Data Anal.* **2002**, *38*, 367–378. [CrossRef]

57. Guyon, I.; Andre, E. An introduction to variable and feature selection. *J. Mach. Learn. Res.* **2003**, *3*, 1157–1182.

58. Liaw, A.; Wiener, M. RandomForest: Breiman and Cutler's Random Forests for Classification and Regression. Available online: https://cran.r-project.org/package=randomForest (accessed on 25 March 2018).

59. Chen, T.; He, T.; Benesty, M.; Khotilovich, V.; Tang, Y. xgboost: Extreme Gradient Boosting. Available online: https://cran.r-project.org/package=xgboost (accessed on 1 August 2019).

60. Dash, M.; Liu, H. Feature selection for classification. *Intell. Data Anal.* **1997**, *1*, 131–156. [CrossRef]

61. Elith, J.; Leathwick, J.R.; Hastie, T. A working guide to boosted regression trees. *J. Anim. Ecol.* **2008**, *77*, 802–813. [CrossRef]

62. Freeman, E.; Frescino, T. ModelMap: Modeling and Map Production Using Random Forest and Related Stochastic Models. Available online: https://cran.r-project.org/web/packages/ModelMap/index.html (accessed on 11 September 2018).

63. Suganuma, H.; Abe, Y.; Taniguchi, M.; Tanouchi, H.; Utsugi, H.; Kojima, T.; Yamada, K. Stand biomass estimation method by canopy coverage for application to remote sensing in an arid area of Western Australia. *For. Ecol. Manag.* **2006**, *222*, 75–87. [CrossRef]

64. Lu, D.; Chen, Q.; Wang, G.; Liu, L.; Li, G.; Moran, E. A survey of remote sensing-based aboveground biomass estimation methods in forest ecosystems. *Int. J. Digit. Earth* **2016**, *9*, 63–105. [CrossRef]

65. Freeman, E.A.; Moisen, G.G.; Coulston, J.W.; Wilson, B.T. Random forests and stochastic gradient boosting for predicting tree canopy cover: Comparing tuning processes and model performance. *Can. J. For. Res.* **2016**, *46*, 323–339. [CrossRef]

66. Gao, Y.; Lu, D.; Li, G.; Wang, G.; Chen, Q.; Liu, L.; Li, D. Comparative analysis of modeling algorithms for forest aboveground biomass estimation in a subtropical region. *Remote Sens.* **2018**, *10*, 627. [CrossRef]

67. Kajisa, T.; Murakami, T.; Mizoue, N.; Kitahara, F.; Yoshida, S. Estimation of stand volumes using the k-nearest neighbors method in Kyushu, Japan. *J. For. Res.* **2008**, *13*, 249–254. [CrossRef]

68. Ustin, S.L.; Roberts, D.A.; Gamon, J.A.; Asner, G.P.; Green, R.O. Using imaging spectroscopy to study ecosystem processes and properties. *Bioscience* **2004**, *54*, 523. [CrossRef]

69. USGS Landsat 8 (L8) Data Users Handbook. Available online: https://prd-wret.s3-us-west-2.amazonaws.com/assets/palladium/production/atoms/files/LSDS-1574_L8_Data_Users_Handbook_v4.pdf (accessed on 20 September 2004).

70. Barsi, J.; Lee, K.; Kvaran, G.; Markham, B.; Pedelty, J. The spectral response of the Landsat-8 operational land imager. *Remote Sens.* **2014**, *6*, 10232–10251. [CrossRef]

71. Gitelson, A.A.; Stark, R.; Grits, U.; Rundquist, D.; Kaufman, Y.; Derry, D. Vegetation and soil lines in visible spectral space: A concept and technique for remote estimation of vegetation fraction. *Int. J. Remote Sens.* **2002**, *23*, 2537–2562. [CrossRef]

72. Gitelson, A.A.; Kaufman, Y.J.; Stark, R.; Rundquist, D. Novel algorithms for remote estimation of vegetation fraction. *Remote Sens. Environ.* **2002**, *80*, 76–87. [CrossRef]

73. Kelsey, K.; Neff, J. Estimates of aboveground biomass from texture analysis of landsat imagery. *Remote Sens.* **2014**, *6*, 6407–6422. [CrossRef]

74. Dietterich, T.G. An experimental comparison of three methods for constructing ensembles of decision trees. *Mach. Learn.* **2000**, *40*, 139–157. [CrossRef]

75. Díaz-Uriarte, R.; Alvarez de Andrés, S. Gene selection and classification of microarray data using random forest. *BMC Bioinf.* **2006**, *7*, 1–13. [CrossRef]

76. Sheridan, R.P.; Wang, W.M.; Liaw, A.; Ma, J.; Gifford, E.M. Extreme gradient boosting as a method for quantitative structure—Activity relationships. *J. Chem. Inf. Model.* **2016**, *56*, 2353–2360. [CrossRef]

77. Georganos, S.; Grippa, T.; Vanhuysse, S.; Lennert, M.; Shimoni, M.; Wolff, E. Very high resolution object-based land use–land cover urban classification using extreme gradient boosting. *IEEE Geosci. Remote Sens. Lett.* **2018**, *15*, 607–611. [CrossRef]

78. Dong, J.; Kaufmann, R.K.; Myneni, R.B.; Tucker, C.J.; Kauppi, P.E.; Liski, J.; Buermann, W.; Alexeyev, V.; Hughes, M.K. Remote sensing estimates of boreal and temperate forest woody biomass: Carbon pools, sources, and sinks. *Remote Sens. Environ.* **2003**, *84*, 393–410. [CrossRef]

79. Ali, I.; Greifeneder, F.; Stamenkovic, J.; Neumann, M.; Notarnicola, C. Review of machine learning approaches for biomass and soil moisture retrievals from remote sensing data. *Remote Sens.* **2015**, *7*, 16398–16421. [CrossRef]

80. Moghaddam, M.; Dungan, J.L.; Acker, S. Forest variable estimation from fusion of SAR and multispectral optical data. *IEEE Trans. Geosci. Remote Sens.* **2002**, *40*, 2176–2187. [CrossRef]

81. Mutanga, O.; Skidmore, A.K. Narrow band vegetation indices overcome the saturation problem in biomass estimation. *Int. J. Remote Sens.* **2004**, *25*, 3999–4014. [CrossRef]

Article

Comparative Analysis of Seasonal Landsat 8 Images for Forest Aboveground Biomass Estimation in a Subtropical Forest

Chao Li [1], Mingyang Li [1,*], Jie Liu [2], Yingchang Li [1] and Qianshi Dai [3]

[1] Co-Innovation Center for Sustainable Forestry in Southern China, College of Forestry, Nanjing Forestry University, Nanjing 210037, China; gislichao@njfu.edu.cn (C.L.); lychang@njfu.edu.cn (Y.L.)
[2] College of Landscape Architecture, Nanjing Forestry University, Nanjing 210037, China; liujienl@njfu.edu.cn
[3] Central South Inventory and Planning Institute of National Forestry and Grassland Administration, Changsha 410014, China; daiqianshi@126.com
* Correspondence: lmy196727@njfu.edu.cn; Tel.: +86-025-8542-7327

Received: 10 December 2019; Accepted: 25 December 2019; Published: 31 December 2019

Abstract: To effectively further research the regional carbon sink, it is important to estimate forest aboveground biomass (AGB). Based on optical images, the AGB can be estimated and mapped on a regional scale. The Landsat 8 Operational Land Imager (OLI) has, therefore, been widely used for regional scale AGB estimation; however, most studies have been based solely on peak season images without performance comparison of other seasons; this may ultimately affect the accuracy of AGB estimation. To explore the effects of utilizing various seasonal images for AGB estimation, we analyzed seasonal images collected using Landsat 8 OLI for a subtropical forest in northern Hunan, China. We then performed stepwise regression to estimate AGB of different forest types (coniferous forest, broadleaf forest, mixed forest and total vegetation). The model performances using seasonal images of different forest types were then compared. The results showed that textural information played an important role in AGB estimation of each forest type. Stratification based on forest types resulted in better AGB estimation model performances than those of total vegetation. The most accurate AGB estimations were achieved using the autumn (October) image, and the least accurate AGB estimations were achieved using the peak season (August) image. In addition, the uncertainties associated with the peak season image were largest in terms of AGB values <25 Mg/ha and >75 Mg/ha, and the quality of the AGB map depicting the peak season was poorer than the maps depicting other seasons. This study suggests that the acquisition time of forest images can affect AGB estimations in subtropical forest. Therefore, future research should consider and incorporate seasonal time-series images to improve AGB estimation.

Keywords: aboveground biomass; Landsat 8 OLI; seasonal images; stepwise regression; map quality; subtropical forest

---

1. Introduction

As an important characteristic of forest ecosystems, forest aboveground biomass (AGB) provides a basis for ecosystem and forestry research; AGB estimation further provides data critical to estimating the forest carbon sink [1,2]. In recent years, accurate and rapid AGB estimation has, therefore, become crucial for forest ecosystem and global climate change research.

Traditionally, high precision AGB field measurement methodologies have involved extensive field surveys [3]. However, these methods are time-consuming, laborious and destructive; in addition, they cannot be used to analyze the spatial distribution and dynamic change of AGB on a large scale [4]. Today, remote sensing-based methodologies are more commonly used to estimate AGB as they rapidly provide near real-time, dynamic and regional scale data, and the characteristics of the obtained images are strongly correlated with AGB [5]. Remote sensing data can be divided into two categories: Passive remote sensing (optical sensors, thermal and microwave) and active remote sensing (radar and light detection and ranging (LiDAR)) [5–7]. Optical sensors such as Landsat, Systeme Probatoire d'Observation de la Terre (SPOT), the moderate-resolution imaging spectroradiometer (MODIS), QuickBird and the Advanced Very High-Resolution Radiometer (AVHRR) have been widely used for AGB estimation because of their coverage, repetitive observation and cost-effectiveness [6,8]. Of these sensors, Landsat images are the most commonly used for remote sensing-based AGB estimations because the sensors provide a long-term data record and have medium spatial resolution, wide spatial coverage and high spectral sensitivity [9]. In many countries, the spatial resolution obtained using Landsat is similar to the size of sample plots in national forest inventories; therefore, using Landsat to estimate AGB can reduce spatial errors associated with matching pixels to sample plots [10].

The information derived from Landsat images significantly correlates with AGB because these images provide valuable information regarding the forest canopy [11]. In fact, previous studies have shown that individual spectral bands, vegetation indices, transformed images (using principal component analysis (PCA)) and textural images are strongly correlated with AGB and can, therefore, be used to effectively estimate AGB [12–15]. Furthermore, many statistical models can be used in developing remote sensing-based AGB models. These models can be divided into two categories: (i) Parametric models (linear, nonlinear, etc.) [16–18] and (ii) nonparametric models (random forest, RF; artificial neural networks, ANN; support vector machines, SVM; etc.) [14,19–21]. Multiple linear regression models, however, are most frequently used in AGB research.

Optical sensors mainly provide information about the forest canopy [11]. The canopy structure of subtropical forests significantly varies between seasons, and even between months [6,22,23]. These variations can cause differences in remote sensing data [24]. Therefore, AGB estimation can vary widely when time-series images are used to model AGB in the same study area [25]. Previous studies have used a single Landsat image (taken during the peak growing season or at a time close to when the ground survey of national forest inventory plots took place) to estimate AGB [21,26–28]. These images, however, do not always accurately reflect forest characteristics. For example, dense canopy cover during the peak growing season often results in extremely saturated images [25,29,30], which ultimately affects AGB estimation accuracy. Some studies have, therefore, utilized time-series of Landsat images to estimate AGB, e.g., Zhu and Liu [25], Safari et al. [31] and Powell et al. [32]. These studies, however, focused on particular forest type or a regional forest. Therefore, there exists a knowledge gap regarding whether time-series Landsat images affect the accuracy of AGB estimations in different forest types and whether the estimations differ among forest types.

Given this gap in knowledge, this study explores the use of seasonal Landsat 8 Operational Land Imager (OLI) images in estimating AGB in a subtropical forest in northern Hunan, China, using stepwise regression. The main objectives of this study were to: (1) Explore the potential variables of seasonal time-series data for different forest types when estimating AGB; (2) investigate the potential of seasonal time-series data in improving the accuracy of AGB estimations in different forest types; and (3) investigate the uncertainties associated with using seasonal time-series data to estimate AGB.

2. Materials and Methods

2.1. Study Area

The study area is located in Hunan Province, central China (path/row: 124/40), and comprises an inclined transition zone from the hills of central Hunan to Dongting Plain. The climate is a typical subtropical monsoon humid climate [33] with an average annual temperature and annual precipitation of 16.5 °C and 1200–1700 mm, respectively. The study area is further characterized by four distinct seasons: Spring (March to May), summer (June to August), autumn (September to November) and winter (December to February). Chinese fir (*Cunninghamia lanceolate* (Lamb.) Hook.) and Chinese red pine (*Pinus massoniana*) plantations, evergreen broadleaf, deciduous and mixed forests dominate this area with scattered bamboo and shrub lands [34]. A total of 303 forest plots were inventoried in 2014 by the National Forest Continuous Inventory (NFCI) in China (Figure 1).

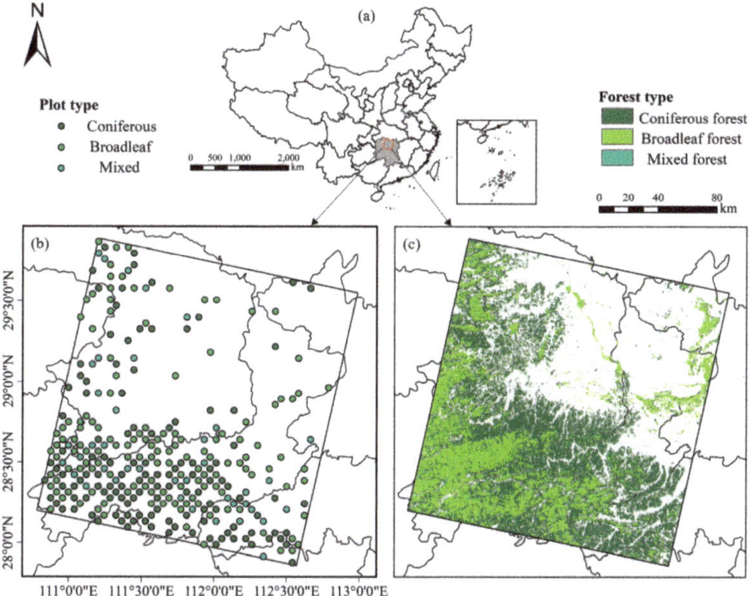

Figure 1. Study area (red box) in Hunan Province, China (**a**); the spatial distribution of sample plots (**b**); and the distribution of forest types (**c**).

2.2. Calculation of Plot-Level AGB

A total of 303 sample plots were used in this research including, 125 CFF (coniferous forest) plots, 138 BLF (broadleaf forest) plots, and 40 MXF (mixed forest) plots (Table 1). The area of the sample plots is 0.067 ha, and the plots were systematically allocated based on a 4 × 8 km grid (NFCI). The AGB values of the study plots were calculated according to tree species or species groups described in a previous study [35]. Statistical information regarding the sample plot data based on different forest types is summarized in Table 1. The AGB values of all sample plots ranged from 5.01 Mg/ha to 151.06 Mg/ha with an average AGB of 48.27 Mg/ha. The mixed forest had the highest mean (±standard deviation) AGB (52.69 ± 29.45 Mg/ha).

Table 1. Summary of the sample plots by forest type (CFF, coniferous forest; BLF, broadleaf forest; MXF, mixed forest; TV, total vegetation).

Forest Type	No. of Sample Plots	Minimum (Mg/ha)	Mean (Mg/ha)	Maximum (Mg/ha)	Standard Deviation
CFF	125	5.01	47.63	118.33	22.89
BLF	138	6.08	47.58	135.08	28.12
MXF	40	22.33	52.69	151.06	29.45
TV	303	5.01	48.27	151.06	26.24

2.3. Remote Sensing Data and Information Extraction

To explore the effectiveness of utilizing seasonal images to estimate AGB, we acquired four cloud-free Landsat 8 OLI images which covered different forest states within the study area from spring to winter during 2013 and 2014 (Table 2). These four Landsat 8 OLI images were Landsat surface reflectance data downloaded from the United States Geological Survey (USGS) website (https://earthexplorer.usgs.gov/). Landsat 8 OLI surface reflectance data are generated using the land surface reflectance code (LaSRC), which utilizes the coastal aerosol band to perform aerosol inversion tests, uses auxiliary climate data from MODIS, and a unique radiative transfer model [36].

Table 2. Landsat 8 Operational Land Imager (OLI) imagery acquired for this study.

Remote Sensing Data	Month	Acquisition Date	Cloud Cover (%)	Image Type
	January	14 January 2014	0.04	L1TP
Landsat 8 OLI (124/40)	April	4 April 2014	0.01	L1TP
	August	7 August 2013	0.61	L1TP
	October	10 October 2013	0.17	L1TP

To estimate forest AGB in the study area, we calculated and extracted 165 spectral variables: Six original bands, 12 vegetation indices, the first three bands from principal component analysis, and 144 texture variables using a gray-level co-occurrence matrix (Table 3).

Table 3. Summary of the predictor variables.

Predictor Variable	Formula	Reference
Landsat 8 OLI original bands 2–7		[31]
Normalized Difference Vegetation Index (NDVI)	$(NIR - R)/(NIR + R)$	[37]
Atmospherically Resistant Vegetation Index (ARVI)	$(NIR - 2R + B)/(NIR + 2R - B)$	[38]
Corrected Normalized Difference Vegetation Index (CNDVI)	$NDVI * (1 - (SWIR1 - SWIRmin)/(SWIRmax - SWIRmin))$	[39]
Difference Vegetation Index (DVI)	$NIR - R$	[40]
Enhanced Vegetation Index (EVI)	$(NIR - R)/(NIR + R + B)$	[30]
Generalized Difference Vegetation Index (GDVI)	$(NIR^2 - R^2)/(NIR^2 + R^2)$	[41]
Linearized NDVI (LNDVI)	$4/\pi * \arctan(NDVI)$	[42]
Normalized Difference Water Index (NDWI)	$(NIR - SWIR2)/(NIR + SWIR2)$	[43]
Normalized Green Difference Vegetation Index (NGDI)	$(NIR - G)/(NIR + G)$	[44]
Red-green Vegetation Index (RGVI)	$(R - G)/(R + G)$	[10]
Soil-adjusted Vegetation Index (SAVI)	$(1 + L) * (NIR - R)/(NIR + R + L)$	[45]
Simple Ratio (SR)	NIR/R	[46]
Principal Component Analysis (PCA) The first three PCs from principal component analysis		[6]
Texture (window sizes: $3 \times 3, 5 \times 5, 7 \times 7$ pixels) Contrast, Correlation, Dissimilarity, Entropy, Homogeneity, Angular second moment, Mean, and Variance		[47]

2.4. Vegetation Classification Data

The European Space Agency (ESA) Climate Change Initiative (CCI) was developed to address climate change at a global level [48]. As part of this initiative, the ESA has derived and consolidated global CCI land cover (CCI-LC) information including annual landcover maps from 1992 to 2015. For the present study, we obtained the CCI-LC data of the study area from MERIS and SPOT satellite images at 300 m spatial resolution [49]. Further, the 2014 CCI-LC map of the study area was downloaded from the ESA website (http://maps.elie.ucl.ac.be/CCI/viewer/index.php) to obtain forest stratifications (coniferous forest (CFF), broadleaf forest (BLF), and mixed forest (MXF)) for AGB estimation.

2.5. AGB Estimation Model

Pearson product-moment correlation coefficient was used to analyze the relationships between plot AGB and spectral variables, and the spectral variables which had significant correlations with AGB were used as independent variables. Stepwise regression analysis is a frequently used approach in AGB research to determine and select the spectral variables which best contribute to forest AGB. Stepwise regression ultimately results in a regression model containing the variables which best explain the dependent variable (AGB in this study). During the stepwise regression, multicollinearity, which creates highly sensitive parameter estimators with inflated variances and leads to improper model selection, was analyzed between each pair of selected spectral variables using the variance inflation factor (VIF). In this study, if the VIF of a spectral variable exceeded ten, this spectral variable was considered seriously collinear with other variables [50,51].

The stepwise regression model developed in this study assumed that a linear relationship exists between independent (spectral variables) and dependent variables (AGB of different forest types). The model is defined in Equation (1) and describes the relationship between AGB and spectral variables:

$$y = a + b_1 x_1 + b_2 x_2 + \cdots + b_n x_n + \varepsilon, \tag{1}$$

where y is AGB, a is the constant term, $x_1, \ldots, x_n$ represent the independent variables, $b_1, \ldots, b_n$ represent the parameters of the independent variables, and ε is the error.

To analyze the accuracy of the AGB models derived from the seasonal images for different forest types, the following workflow was used (Figure 2).

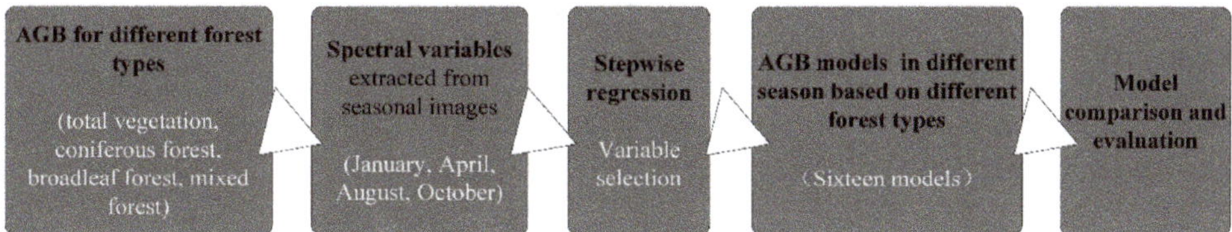

Figure 2. The workflow for aboveground biomass (AGB) models derived from different scenarios.

2.6. Model Comparison and Evaluation

Model performance was evaluated using '10-fold' cross validation [52], and predicted AGB values were compared to observed AGB values using three accuracy indicators: Coefficient of determination (R^2), root mean square error (RMSE and RMSE %) and bias. Accuracy indicator Equations (2)–(5) are as follows:

$$R^2 = 1 - \sum_{i=1}^{n} (y_i - \hat{y}_i)^2 / \sum_{i=1}^{n} \left(y_i - \overline{y}_i\right)^2, \tag{2}$$

$$RMSE = \sqrt{\sum_{i=1}^{n} (y_i - \hat{y}_i)^2 / n}, \tag{3}$$

$$RMSE\% = \frac{RMSE}{\overline{y}} \times 100, \tag{4}$$

$$Bias = (y_i - \hat{y}_i) / \overline{y}, \tag{5}$$

where y_i is the observed AGB value, $\hat{y}_i$ is the predicted AGB value based on models, $\overline{y}$ is the arithmetic mean of all observed AGB values, and n is the sample number. In general, a higher R^2 value and lower RMSE and RMSE% values indicate a greater accuracy of the model.

We generated ten predicted forest AGB maps using the results of 10-fold cross validation, and the average of these AGB maps was taken as the final spatial distribution of AGB. In addition, the standard deviation (Stdev) of spatial AGB predictions was calculated to analyze the uncertainty of each pixel [53,54]. Larger Stdev values indicate higher estimation uncertainty and smaller Stdev values indicate lower estimation uncertainty.

3. Results

3.1. Comparison of AGB Estimates Using Seasonal Images of Total Vegetation

The variables of AGB models for total vegetation using seasonal images were selected using stepwise regression according to the correlation between AGB, the dependent variable and spectral variables. We found that four variables were included in the AGB models for January, April and October, whereas six variables were included in the AGB model for August (Table 4). The selected variables of these models indicated that the textural images of Landsat 8 OLI played an important role in forest AGB estimation of total vegetation regardless of the season.

Table 4. The selected variables for AGB estimation models for different months based on the total vegetation.

Month	Selected Variables for Total Vegetation
January	b6_EN3Jan, b3_EN3Jan, b2_EN3Jan, b5_Jan
April	b7_SEM3Apr, b5_Apr, b2_VA7Apr, NDWI_Apr
August	b6_COR5Aug, b4_COR7Aug, b5_SEM7Aug, b5_HO7Aug, b2_CON5Aug, b4_SEM5Aug
October	b5_SEM5Oct, b4_CON7Oct, SR Oct, b6_COR3Oct

Note: bi_M, original band i; NDWI_Apr, normalized difference water index of April; SR_Oct, simple ratio of october; bi_XYjM, textural image developed from spectral band i with a window size of jxj pixels using texture entropy (EN), angular second moment (SEM), variance (VA), correlation (COR), contrast (CON) or homogeneity (HO).

Based on 10-fold cross validation, the results of model fitting are shown in Table 5. We found that the use of seasonal Landsat 8 imagery resulted in different AGB estimates. For total vegetation, the stepwise regression model of the October image showed the highest R^2 value (0.39) and the lowest RMSE (21.67 Mg/ha; 44.1% of the mean) and bias (−0.19 Mg/ha) values. The model based on the peak season (August) image showed the lowest R^2 value (0.27), followed by the January and April models. Overall, the results demonstrated that the acquisition time of Landsat 8 imagery significantly influenced AGB estimation, and that the peak season (August) image showed inferior performance compared to that of the other AGB estimation models.

Table 5. Summary of the accuracy assessment results for the seasonal models based on the total vegetation.

Month	R^2	RMSE (Mg/ha)	RMSE %	Bias (Mg/ha)
January	0.31	21.90	44.8	1.08
April	0.34	21.95	45.0	0.81
August	0.27	22.15	45.7	0.54
October	0.39	21.67	44.1	−0.19

Note: R^2, coefficient of determination; RMSE, root mean squared error; RMSE%, relative root mean squared error.

The relationship between the predicted AGB and observed total vegetation AGB for different seasons using stepwise regression model is shown as scatterplots in Figure 3a1–d1. Each month, we detected overestimations when the plot AGB value was lower than 30 Mg/ha, and underestimations when the plot AGB value was higher than approximately 90 Mg/ha. The August model showed the largest bias (Figure 3c2). The bias calculated for each prediction model showed a skewed distribution (Figure 3a2–d2), but when the bias was less than −25 Mg/ha or greater than 25 Mg/ha, bias frequencies of the October model were smaller than those of the other three months.

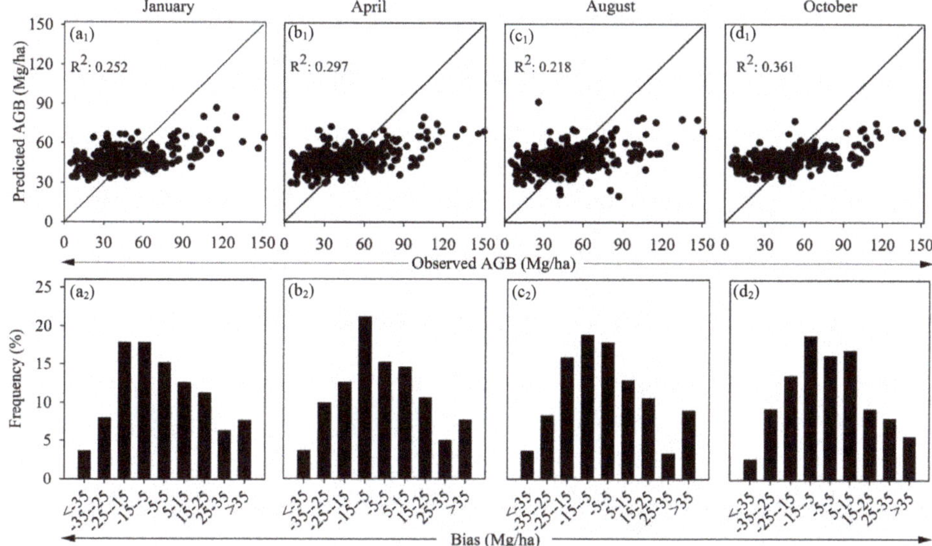

Figure 3. Model performances were evaluated using 10-fold cross validation and predicted AGB values were compared to observed AGB values using accuracy indicators. Scatterplots depict the relationship between predicted and observed AGB estimation values in each month (**a1–d1**). Histograms depict model biases (**a2–d2**).

The above analysis was based on the overall performance of different stepwise regression models generated for each month, but it cannot provide detailed information regarding the effect of different forest types on estimation of total vegetation AGB. Table 6 summarizes the RMSE and RMSE% results for different forest types. For CFF and BLF, the RMSE and RMSE% were lowest when the October image was used for AGB estimation. For MXF, the RMSE and RMSE% were lowest when the April image was used for AGB estimation. While the October model resulted in lower R^2 and RMSE values than the April model, the April model performed better in MXF AGB estimation than the October model.

Table 6. Summary of RMSE (Mg/ha) and RMSE% results from different seasonal images under non-stratified conditions.

Month	RMSE (Mg/ha)			RMSE%		
	CFF	**BLF**	**MXF**	**CFF**	**BLF**	**MXF**
January	20.72	23.14	26.96	42.70	47.42	53.67
April	19.65	22.60	21.06	41.04	48.07	38.10
August	18.50	24.26	27.68	39.32	50.25	54.12
October	18.02	22.69	23.47	37.58	47.31	41.99

Note: CFF, coniferous forest; BLF, broadleaf forest; MXF, mixed forest; RMSE, root mean squared error; RMSE%, relative root mean squared error.

3.2. Comparison of AGB Estimates Using Seasonal Images of Different Forest Types

The independent variables selected by the AGB models using seasonal images of the three forest types are summarized in Table 7. The selected variables varied among each forest type in different months. However, in general, texture measures were involved in all AGB models, indicating that when considering different forest types and months, textural information significantly contributed to improving the AGB predictions in this study.

Table 7. The selected variables for AGB estimation models in different months based on different forest types.

Month	Selected Variables for Different Forest Types		
	CFF	**BLF**	**MXF**
January	b3_VA7Jan, b4_HO5Jan, b2_COR7Jan, DVI_Jan, b4_CON3Jan	b5_SEM5Jan, b7_EN3Jan, b6_SEM3Jan, b3_EN3Jan b6_SEM5Jan, b7_SEM7Jan,	b3_COR3Jan, b6_EN3Jan, b6_SEM7Jan, b6_CON3Jan, b5_COR5Jan
April	b2_VA3Apr, b2_VA5Apr, b2_HO7Apr, b5_EN5Apr, b6_SEM7Apr	b7_SEM3Apr, b2_ME7Apr SR_Apr, b6_ME5Apr, b2_CON5Apr, b2_EN7Apr	b6_COR3Apr, b7_VA3Apr, b2_HO3Apr, b5_COR3Apr, b5_COR7Apr
August	b6_COR5Aug, b5_DI5Aug, b5_DI7Aug, b4_COR3Aug, b7_SEM5Aug, b6_HO5Aug	b6_COR5Aug, b5_HO5Aug, b5_SEM5Aug, b5_HO3Aug, b5_EN3Aug, b6_COR3Aug	b4_VA3Aug, b7_EN3Aug, b7_COR3Aug, b3_SEM7Aug, b6_HO3Aug
October	b6_COR3Oct, SR_Oct b7_COR3Oct, b3_ME7Oct b6_SEM3Oct, b6_EN3Oct	b5_SEM3Oct, PCA3_Oct, b4_CON7Oct, b7_ME7Oct, b2_SEM7Oct, b5_EN3Oct	b6_VA5Oct, b6_VA7Oct, b5_EN3Oct, b5_COR3Oct, b6_SEM3Oct

Note: CFF, coniferous forest; BLF, broadleaf forest; MXF, mixed forest; bi_M, original band I of month M; DVI_Jan, difference vegetation index of January; SR_M, simple ratio of month M; PCA3_Otc, band 3 of principal component analysis in October; bi_XYjM, textural image developed from spectral band i with a window size of jxj pixels of month M using texture entropy (EN), angular second moment (SEM), variance (VA), correlation (COR), contrast (CON), mean (ME), dissimilarity (DI) or homogeneity (HO).

We further compared the AGB models derived using seasonal images of three forest types (Table 8). For the different forest types, we found that regardless of month, MXF model performances were better than those of CFF and BLF. The performances of the CFF and BLF models did not significantly differ. Regarding all model performances, R^2 value differences ranged from 0.13 for the BLF models to 0.2 for the MXF models, RMSE (RMSE%) value differences ranged from 1.85 Mg/ha (3.03%) for the BLF models to 3.92 Mg/ha (7.44%) for the MXF models. Overall, the model obtained using data from the October image had the least bias, whereas the August model performed had the largest bias; the January and April models were intermediate. The R^2 values were all less than 0.55 and the RMSE% were all larger than 35%; these results indicated that though the performances of the models for different forest types were better than those of total vegetation, nearly half of the AGB variation cannot be explained. When compared to the previously constructed total vegetation models (Table 5 vs. Table 8), the models based on different forest types resulted in larger R^2 and lower RMSE (RMSE%) values and performed better overall, indicating that consideration of forest type can improve AGB estimation.

Table 8. Summary of the accuracy assessment results for the seasonal models based on different forest types.

Month	Forest Types	Accuracy Indicators			
		R^2	RMSE (Mg/ha)	RMSE%	Bias (Mg/ha)
January	CFF	0.35	18.53	39.07	0.12
	BLF	0.38	21.89	45.65	−0.26
	MXF	0.48	20.82	39.52	−0.003
April	CFF	0.43	17.40	35.97	0.22
	BLF	0.45	21.18	44.35	−0.09
	MXF	0.52	20.16	38.26	−0.003
August	CFF	0.31	19.33	40.10	−0.18
	BLF	0.33	23.03	47.38	−0.10
	MXF	0.35	23.37	44.36	−0.36
October	CFF	0.47	17.23	35.95	0.04
	BLF	0.46	21.34	44.38	−0.02
	MXF	0.55	19.45	36.92	−0.0001

Note: CFF, coniferous forest; BLF, broadleaf forest; MXF, mixed forest; R^2, coefficient of determination; RMSE, root mean squared error; RMSE%, relative root mean squared error.

We further compared the AGB models derived using seasonal images of three forest types (Table 8). For the different forest types, we found that regardless of month, MXF model performances were better than those of CFF and BLF. The performances of the CFF and BLF models did not significantly differ. Regarding all model performances, R^2 value differences ranged from 0.13 for the BLF models to 0.2 for the MXF models, RMSE (RMSE%) value differences ranged from 1.85 Mg/ha (3.03%) for the BLF models to 3.92 Mg/ha (7.44%) for the MXF models. Overall, the model obtained using data from the October image had the least bias, whereas the August model performed had the largest bias; the January and April models were intermediate. The R^2 values were all less than 0.55 and the RMSE% were all larger than 35%; these results indicated that though the performances of the models for different forest types were better than those of total vegetation, nearly half of the AGB variation cannot be explained. When compared to the previously constructed total vegetation models (Table 5 vs. Table 8), the models based on different forest types resulted in larger R^2 and lower RMSE (RMSE%) values and performed better overall, indicating that consideration of forest type can improve AGB estimation.

The relationship between the predicted and observed AGB values of the three forest types in different seasons using stepwise regression models is shown as scatterplots in Figure 4. Overestimations occurred in plots with AGB values lower than approximately 30 Mg/ha for CFF, BLF, and MXF, whereas underestimations occurred in plots with AGB values higher than approximately 100 Mg/ha for each forest type. The scatter plot constructed using the October data better fit the line y = x, whereas the scatter plot constructed using the August data was more discrete with serious over- and underestimation issues (Figure 4). January and April prediction model biases showed skewed distributions (Figure 4), the model bias of October represented a normal distribution, and the model bias of August was discrete. Further, the October bias values mostly ranged from −15 Mg/ha to 15 Mg/ha, and there were lower proportions of bias values <−25 Mg/ha or >25 Mg/ha.

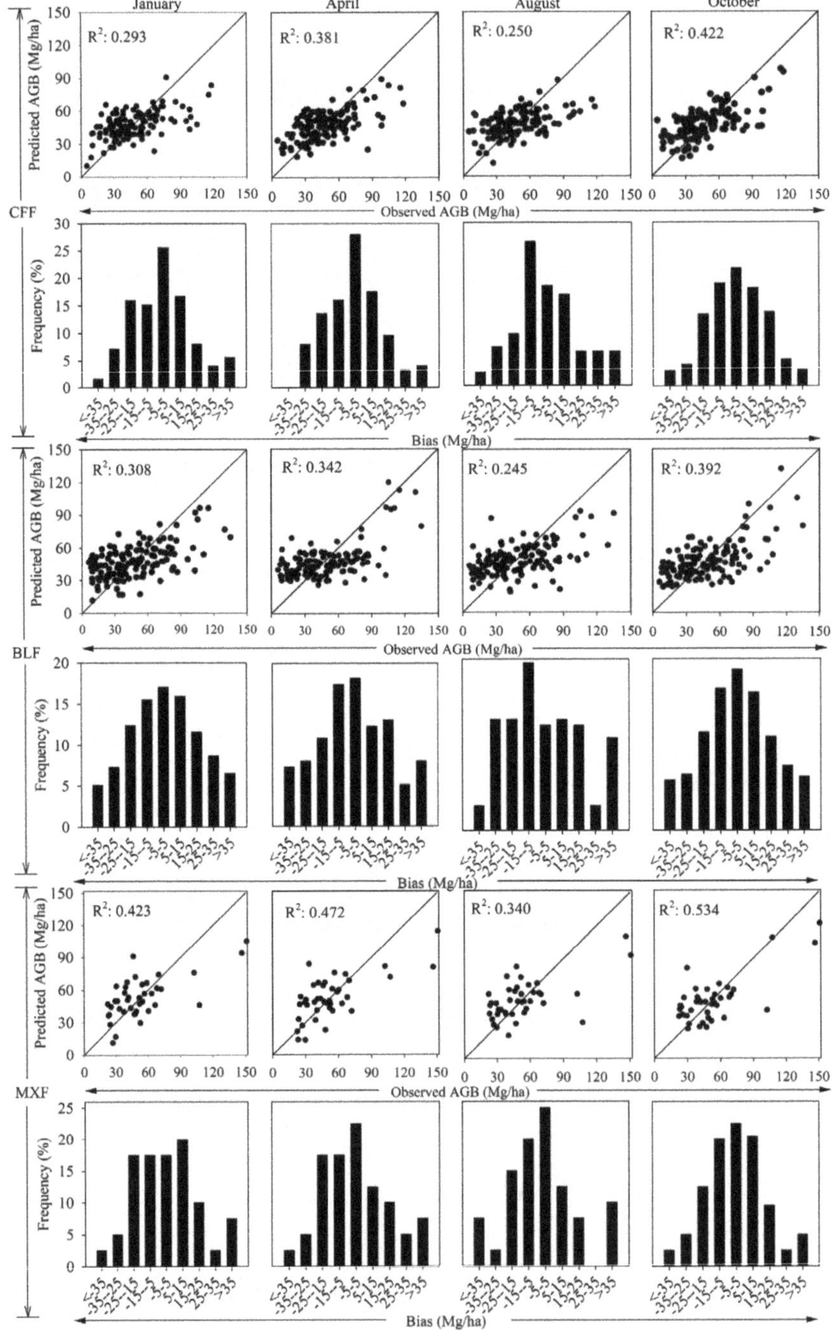

Figure 4. Model performances were evaluated using 10-fold cross validation and predicted AGB values were compared to observed AGB values using accuracy indicators. Scatterplots depict the relationship between predicted and observed AGB estimation values in each month for each forest type (**top**). Histograms depict model biases in each month for each forest type (**bottom**). CFF, coniferous forest; BLF, broadleaf forest; MXF, mixed forest.

3.3. AGB Distribution Maps and Map Quality

In addition to model diagnostics, we predicted AGB distribution maps and AGB standard deviation (Stdev) maps within the study area. Using seasonal images, we constructed AGB spatial distribution maps based on total vegetation and different forest types (Figure 5). AGB distribution patterns in different months varied, supporting our previous results (Sections 3.1 and 3.2) which suggested that seasonal model performances differed. Further, AGB distribution patterns for total vegetation were narrow (within the range of 25 Mg/ha to 75 Mg/ha; Figure 5a), whereas AGB distribution patterns for different forest types were discrete (within the range of 0 Mg/ha to 100 Mg/ha; Figure 5b). These results indicate that models constructed based on forest types can achieve relatively low (<25 Mg/ha) and high (>75 Mg/ha) AGB values and thus alleviate over- and underestimation. In addition, October distribution maps were more heterogeneous than those of the other months, further indicating that October model performances were superior.

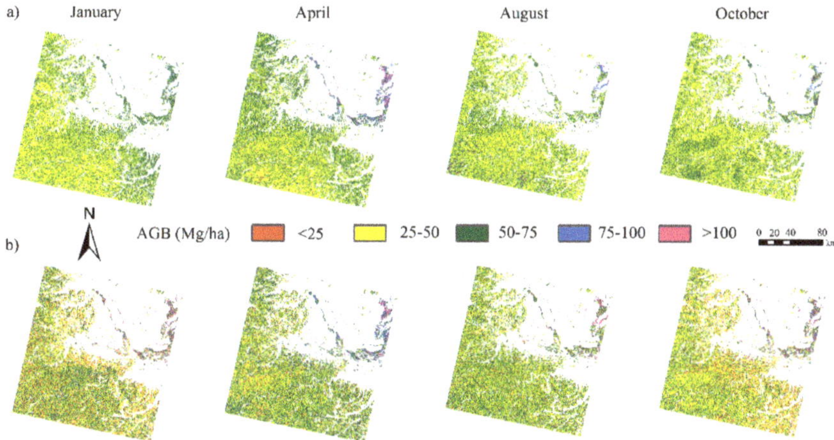

Figure 5. Spatial distribution of forest aboveground biomass (AGB) using seasonal images under different scenarios: (**a**) The total vegetation; (**b**) forest types including coniferous forest, broadleaf forest and mixed forest.

Stdev maps of each scenario are shown in Figure 6. For both total vegetation and different forest types, the model uncertainties for October were smaller compared with those for January, April and August, indicating that the October AGB maps were more accurate. Moreover, model uncertainties for August were larger compared with those for the other three months, indicating that the August AGB maps were the least accurate. The Stdev values of different AGB ranges for different forest types were further calculated and analyzed (Figure 7). When mapping the AGB of both the total vegetation and the different forest types, the Stdev values were greater when the AGB values were <25 Mg/ha or >75 Mg/ha. In this case, the Stdev of the August models were the largest, followed by the January, April and October models. This result indicated that AGB maps exhibiting these AGB values (<25 Mg/ha or >75 Mg/ha) showed the largest uncertainty when utilizing the August image. In addition, when attained AGB values were >75 Mg/ha, all Stdev values for each scenario were larger than three, further indicating a large amount of uncertainty associated with these particular AGB values.

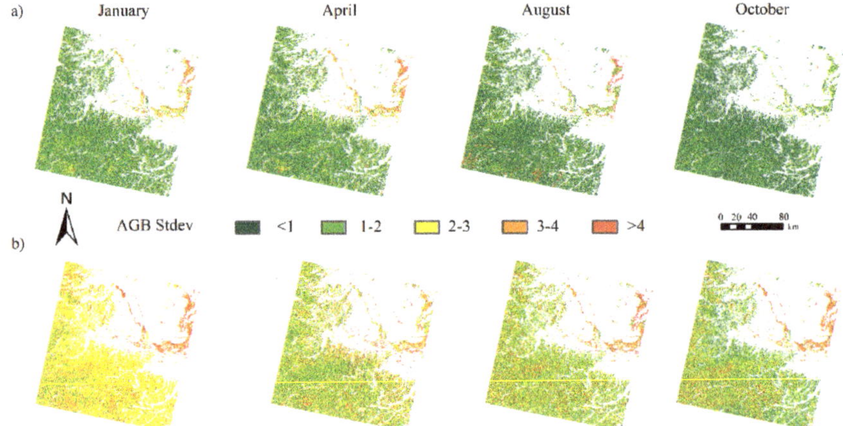

Figure 6. Standard deviation (Stdev) maps of AGB values using seasonal images under different scenarios: (**a**) The total vegetation; (**b**) forest types including coniferous forest, broadleaf forest and mixed forest.

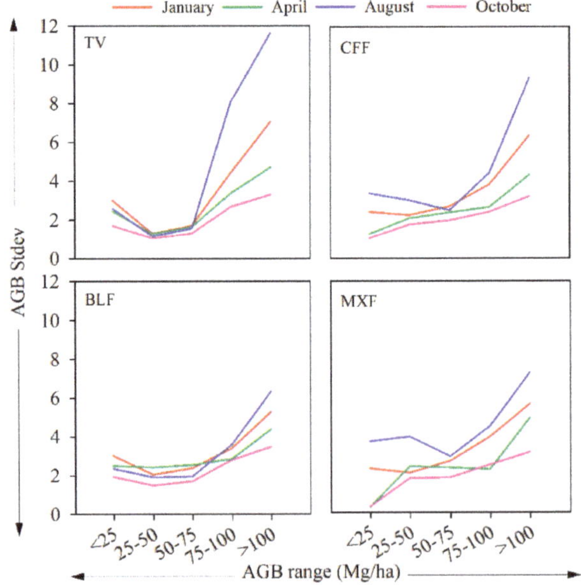

Figure 7. The standard deviation (Stdev) of AGB within different AGB value ranges (TV, total vegetation; CFF, coniferous forest; BLF, broadleaf forest; MXF, mixed forest).

4. Discussion

Forests are complex ecosystems containing variable species composition and structure; therefore, the image information (especially textural information) of these ecosystems also varies considerably [55,56]. Previous studies utilizing Landsat images to estimate AGB were unable to determine which spectral variables were best able to predict AGB [6,57]. In this study, the selected spectral variables used for AGB models of different months and different forest types varied. Nonetheless, we found that for all forest types, textural images played an important role in AGB estimation, in accordance with previous research [58]. The selected variables belonged to various original bands (bands 2 to 7), indicating that all original bands can be used to estimate AGB in this study. These results differed from earlier research in which the shortwave infrared (SWIR) bands (e.g., Landsat TM spectral bands 5 and 7) were more important in AGB estimation than other bands [59–61]. In addition, in previous research utilizing Landsat imagery, spectral information (e.g., vegetation index, original band) was often selected to estimate the AGB of coniferous forest given that the structure of coniferous forest was simple and the importance of spectral information over textural information. On the other hand, textural information has often been used in the study of broadleaf forest and mixed forest given that those forests often have multiple canopy layers and more complex structures. In our study area, because of the low level of forest management, the forest structure was complex; therefore, in this study, for each forest type, textural information was mostly used to estimate AGB, regardless of which seasonal image was utilized.

In this study, stepwise regression was used to estimate AGB of different forest types based on Landsat 8 OLI seasonal images. We found that in our study area, the best month for AGB estimation was October given that the R^2 values of different forest types were higher than 0.39. Overall, this result indicates that the October image can explain more than 39% of the information regarding the estimated AGB for each forest type. The less accuracy month for AGB estimation was August given that the R^2 value for total vegetation was only 0.27. Stepwise regression is a widely used methodology of fitting regression models based on the correlation between dependent and independent variables. During this procedure, the significance of an introduced variable is tested, and the variable that is of least significant is discarded [62]. While selection of variables depends upon the degree of linear correlation, selection of variables with low correlation is possible; this can ultimately affect the accuracy of the model. The forest characteristics of different forest types were heterogeneous. Different forest types were different in spectral characteristics caused by the heterogeneity of the stand structures and species compositions. The correlations among the spectral variables and AGB of different forest types were also different. In this case, the performances of models for different forest types were significantly different. In our study, among all forest types analyzed, we found that the MXF models achieved the best results for AGB estimation. This indicates that the image information was most strongly correlated with MXF compared with other forest types, and therefore, the image can better reflect the condition of the mixed forest. However, when the forest types were considered in AGB estimation, model accuracy was further affected by the number of plots [59]. In this study, there were 135 CFF plots and 128 BLF plots, whereas there only 40 MXF plots. Therefore, MXF models may have been more accurate given the far fewer number of plots analyzed compared with the models for the other forest types.

In this study, Landsat 8 OLI seasonal images were used to estimate AGB. The four seasonal images utilized were associated with four seasons of the study area (January (winter), April (spring), August (summer) and October (autumn)). The results showed that utilization of the peak season (August) image resulted in inadequate AGB estimation compared with the other seasons, in accordance with results reported by Zhu and Liu (2014) [25]. These researchers further found that the normalized difference vegetation index (NDVI)-based AGB estimates of the forest senescing period were better than those of the peak season in a temperate forest of southeastern Ohio, USA [25]. Furthermore, in accordance with our results, previous researchers detected over- and underestimations when utilizing Landsat 8 OLI imagery to estimate AGB in a subtropical forest in western Hunan, China [58]. In our study, these uncertainties were common among all seasonal images analyzed. The observed underestimations within the higher range of AGB values may have been a consequence of image saturation issues affecting model performance [56,63]. Regarding AGB values within the lower range, model performance was likely affected by mixed pixels, thus resulting in AGB overestimation [64]. While uncertainties were detected among all time-series images, underestimation associated with the peak season (August) within the high AGB range (>75 Mg/ha) was more serious than that associated with the other seasons. Taken together, these results suggest that image saturation more strongly influenced AGB estimation results for August than it did for the other seasons, further indicating that the uncertainties were less in the other seasons. In addition, the overestimation associated with the peak season was greater than that associated with the other seasons.

5. Conclusions

In this study, seasonal Landsat 8 OLI imagery was utilized to estimate forest AGB in a subtropical forest in northern Hunan Province, China. Study plots were classified according to forest types (CFF, BLF, MXF and total vegetation) and stepwise regression was used to select appropriate variables and thus effectively model AGB based on the seasonal images. Subsequently, models of the different scenarios (different forest types in different seasons) were compared. Given the variables selected during stepwise regression, we concluded that seasonal image textural information was most significantly correlated with AGB, and that the study area is made up of forests with complex structures. The method of AGB estimation based on forest type is very useful for improving the accuracy of AGB estimation because the model performances for the different forest types (CFF, BLF and MXF) are better than those for the total vegetation, regardless of season. The time-series images, which reflect various seasons, can affect AGB estimations, with the autumn image (October) potentially yielding the most accurate AGB estimations and the peak season (August) image being of poorer quality in a subtropical forest. We also explored the accuracies of seasonal images in remote sensing-based AGB estimation. We hope to provide new insight into using Landsat images to improve the accuracy of biomass estimation.

Future research will focus on the mechanism underlying the cause of these differences when utilizing seasonal Landsat 8 OLI images in AGB estimation of different forest types.

Author Contributions: Conceptualization, C.L. and M.L.; Data Curation, C.L., Y.L. and Q.D.; Formal Analysis, C.L., Y.L. and J.L.; Funding Acquisition, M.L.; Methodology, C.L. and M.L.; Project Administration, M.L.; Resources, M.L. and Q.D.; Software, C.L. and Y.L.; Supervision, M.L.; Validation, C.L., Y.L. and J.L.; Visualization, C.L. and J.L.; Writing—Original Draft, C.L.; Writing—Review & Editing, C.L., M.L., J.L., Y.L. and Q.D. All authors have read and agreed to the published version of the manuscript.

Funding: This research was funded by National Natural Science Foundation (NO. 31770679), and Top–notch Academic Programs Project of Jiangsu Higher Education Institutions, China (TAPP, PPZY2015A062).

Acknowledgments: The authors are grateful to the ESA (http://maps.elie.ucl.ac.be/CCI/viewer/index.php) and USGS (http://glovis.usgs.gov/) for providing the Land Cover map and Landsat 8 OLI images.

Conflicts of Interest: The authors declare no conflict of interest.

References

1. Houghton, R.A. Aboveground forest biomass and the global carbon balance. *Glob. Chang. Biol.* **2005**, *11*, 945–958. [CrossRef]
2. Houghton, R.A.; Hall, F.; Goetz, S.J. Importance of biomass in the global carbon cycle. *J. Geophys. Res.* **2009**, *114*. [CrossRef]
3. West, P.W. *Tree and Forest Measurement*, 2nd ed.; Springer: New York, NY, USA, 2009; ISBN 9783540959656.
4. Vashum, K.T.; Jayakumar, S. Methods to Estimate Above-Ground Biomass and Carbon Stock in Natural Forests—A Review. *J. Ecosyst. Ecogr.* **2012**. [CrossRef]
5. Galidaki, G.; Zianis, D.; Gitas, I.; Radoglou, K.; Karathanassi, V.; Tsakiri–Strati, M.; Woodhouse, I.; Mallinis, G. Vegetation biomass estimation with remote sensing: Focus on forest and other wooded land over the Mediterranean ecosystem. *Int. J. Remote Sens.* **2017**, *38*, 1940–1966. [CrossRef]
6. Lu, D.; Chen, Q.; Wang, G.; Liu, L.; Li, G.; Moran, E. A survey of remote sensing-based aboveground biomass estimation methods in forest ecosystems. *Int. J. Digit. Earth* **2014**, *9*, 63–105. [CrossRef]
7. Rodríguez-veiga, P.; Wheeler, J.; Louis, V.; Tansey, K. Quantifying Forest Biomass Carbon Stocks from Space. *Curr. For. Rep.* **2017**, *3*, 1–18. [CrossRef]
8. Kumar, L.; Sinha, P.; Taylor, S.; Alqurashi, A.F. Review of the use of remote sensing for biomass estimation to support renewable energy generation. *J. Appl. Remote Sens.* **2015**, *9*, 097696. [CrossRef]
9. Cohen, W.B.; Goward, S.N. Landsat's Role in Ecological Applications of Remote Sensing. *Bioscience* **2004**, *54*, 535–545. [CrossRef]

10. Sun, H.; Qie, G.; Wang, G.; Tan, Y.; Li, J.; Peng, Y.; Ma, Z.; Luo, C. Increasing the accuracy of mapping urban forest carbon density by combining spatial modeling and spectral unmixing analysis. *Remote Sens.* **2015**, *7*, 15114–15139. [CrossRef]

11. Roy, P.S.; Ravan, S.A. Biomass estimation using satellite remote sensing data—An investigation on possible approaches for natural forest. *J. Biosci.* **1996**, *21*, 535–561. [CrossRef]

12. Safari, A.; Sohrabi, H. Ability of landsat-8 OLI derived texture metrics in estimating aboveground carbon stocks of coppice Oak Forests. *Int. Arch. Photogramm. Remote Sens. Spat. Inf. Sci.* **2016**, *XLI-B8*, 751–754. [CrossRef]

13. Zhao, Q.; Yu, S.; Zhao, F.; Tian, L.; Zhao, Z. Comparison of machine learning algorithms for forest parameter estimations and application for forest quality assessments. *For. Ecol. Manag.* **2019**, *434*, 224–234. [CrossRef]

14. Ou, G.; Li, C.; Lv, Y.; Wei, A.; Xiong, H.; Xu, H.; Wang, G. Improving aboveground biomass estimation of Pinus densata forests in Yunnan using Landsat 8 imagery by incorporating age dummy variable and method comparison. *Remote Sens.* **2019**, *11*, 738. [CrossRef]

15. Zhang, Y.; Li, F.; Liu, F. Forest biomass estimation based on remote sensing method for north Daxingan mountains. *Adv. Mater. Res.* **2011**, *339*, 336–341. [CrossRef]

16. Yan, E.; Lin, H.; Wang, G.; Sun, H. Improvement of Forest Carbon Estimation by Integration of Regression Modeling and Spectral Unmixing of Landsat Data. *IEEE Geosci. Remote Sens. Lett.* **2015**, *12*, 2003–2007. [CrossRef]

17. Fernández-Manso, O.; Fernández-Manso, A.; Quintano, C. Estimation of aboveground biomass in Mediterranean forestsby statistical modelling of ASTER fraction images. *Int. J. Appl. Earth Obs. Geoinf.* **2014**, *31*, 45–56. [CrossRef]

18. Zhao, P.; Lu, D.; Wang, G.; Wu, C.; Huang, Y.; Yu, S. Examining spectral reflectance saturation in landsat imagery and corresponding solutions to improve forest aboveground biomass estimation. *Remote Sens.* **2016**, *8*, 469. [CrossRef]

19. Shen, W.; Li, M.; Huang, C.; Wei, A. Quantifying Live Aboveground Biomass and Forest Disturbance of Mountainous Natural and Plantation Forests in Northern Guangdong, China, Based on Multi-Temporal Landsat, PALSAR and Field Plot Data. *Remote Sens.* **2016**, *8*, 595. [CrossRef]

20. Tian, X.; Su, Z.; Chen, E.; Li, Z.; van der Tol, C.; Guo, J.; He, Q. Estimation of forest above-ground biomass using multi-parameter remote sensing data over a cold and arid area. *Int. J. Appl. Earth Obs. Geoinf.* **2012**, *14*, 160–168. [CrossRef]

21. Wu, C.; Shen, H.; Shen, A.; Deng, J.; Gan, M.; Zhu, J.; Xu, H.; Wang, K. Comparison of machine-learning methods for above-ground biomass estimation based on Landsat imagery. *J. Appl. Remote Sens.* **2016**, *10*, 035010. [CrossRef]

22. Blodgett, C.; Jakubauskas, M. *Remote Sensing of Coniferous Forest Structure in Grand Teton National Park*; University of Wyoming National Park Service Research Center: Laramie, WY, USA, 1995; Volume 19.

23. Delissio, L.J.; Primack, R.B. The impact of drought on the population dynamics of canopy-tree seedlings in an aseasonal Malaysian rain forest. *J. Trop. Ecol.* **2003**, *19*, 489–500. [CrossRef]

24. Rautiainen, M.; Heiskanen, J. Seasonal Dynamics of Boreal Forest Structure and Reflectance. In Proceedings of the AGU Fall Meeting, San Francisco, CA, USA, 13–17 December 2010.

25. Zhu, X.; Liu, D. Improving forest aboveground biomass estimation using seasonal Landsat NDVI time-series. *ISPRS J. Photogramm. Remote Sens.* **2015**, *102*, 222–231. [CrossRef]

26. Tian, X.; Li, Z.; Su, Z.; Chen, E.; van der Tol, C.; Li, X.; Guo, Y.; Li, L.; Ling, F. Estimating montane forest above-ground biomass in the upper reaches of the Heihe River Basin using Landsat-TM data. *Int. J. Remote Sens.* **2014**, *35*, 7339–7362. [CrossRef]

27. Zhu, J.; Huang, Z.; Sun, H.; Wang, G. Mapping forest ecosystem biomass density for Xiangjiang River Basin by combining plot and remote sensing data and comparing spatial extrapolation methods. *Remote Sens.* **2017**, *9*, 241. [CrossRef]

28. Shao, Z.; Zhang, L. Estimating forest aboveground biomass by combining optical and SAR data: A case study in genhe, inner Mongolia, China. *Sensors* **2016**, *16*, 834. [CrossRef]

29. Mutanga, O.; Adam, E.; Cho, M.A. High density biomass estimation for wetland vegetation using worldview-2 imagery and random forest regression algorithm. *Int. J. Appl. Earth Obs. Geoinf.* **2012**, *18*, 399–406. [CrossRef]

30. Huete, A.; Didan, K.; Miura, T.; Rodriguez, E.P.; Gao, X.; Ferreira, L.G. Overview of the radiometric and biophysical performance of the MODIS vegetation indices. *Remote Sens. Environ.* **2002**, *83*, 195–213. [CrossRef]

31. Safari, A.; Sohrabi, H.; Powell, S.; Shataee, S. A comparative assessment of multi-temporal Landsat 8 and machine learning algorithms for estimating aboveground carbon stock in coppice oak forests. *Int. J. Remote Sens.* **2017**, *38*, 6407–6432. [CrossRef]

32. Scott, L.; Powell; Cohen, W.B.; Healey, S.P.; Kennedy, R.E.; Moisen, G.G.; Pierce, K.B.; Ohmann, J.L. Quantification of live aboveground forest biomass dynamics with Landsat time-series and field inventory data: A comparison of empirical modeling approaches. *Remote Sens. Environ.* **2010**, *114*, 1053–1068. [CrossRef]

33. Chen, A.; He, X.; Guan, H.; Cai, Y. Trends and periodicity of daily temperature and precipitation extremes during 1960–2013 in Hunan Province, central south China. *Theor. Appl. Climatol.* **2018**, *132*, 71–88. [CrossRef]

34. Li, W.; Li, F. *Research of Forest Resources in China*, 1st ed.; China Forestry Publishing House: Beijing, China, 1996; ISBN 7503817224.

35. Li, H.; Lei, Y.; Zeng, W.; Chen, Y.; Huang, G. *Estimation and Evaluation of Forestry Biomass Carbon Storage in China*; China Forestry Press: Beijing, China, 2010.

36. Vermote, E.; Justice, C.; Claverie, M.; Franch, B. Preliminary analysis of the performance of the Landsat 8/OLI land surface reflectance product. *Remote Sens. Environ.* **2016**, *185*, 46–56. [CrossRef]

37. Rouse, W.; Haas, R.H.; Deering, D.W. Monitoring vegetation systems in the Great Plains with ERTS, NASA SP-351. In *Third ERTS-1 Symposium*; NASA: Washington, DC, USA, 1974; Volume 1, pp. 309–317.

38. Lu, D.; Mausel, P.; Brondízio, E.; Moran, E. Relationships between forest stand parameters and Landsat TM spectral responses in the Brazilian Amazon Basin. *For. Ecol. Manag.* **2004**, *198*, 149–167. [CrossRef]

39. Nemani, R.; Pierce, L.; Running, S.; Band, L. Forest ecosystem processes at the watershed scale: Sensitivity to remotely-sensed leaf area index estimates. *Int. J. Remote Sens.* **1993**, *14*, 2519–2534. [CrossRef]

40. Clevers, J.G.P.W. The derivation of a simplified reflectance model for the estimation of leaf area index. *Remote Sens. Environ.* **1988**, *25*, 53–69. [CrossRef]

41. Wu, W. The Generalized Difference Vegetation Index (GDVI) for dryland characterization. *Remote Sens.* **2014**, *6*, 1211–1233. [CrossRef]

42. Ünsalan, C.; Boyer, K.L. Linearized vegetation indices based on a formal statistical framework. *IEEE Trans. Geosci. Remote Sens.* **2004**, *42*, 1575–1585. [CrossRef]

43. Gao, B.-C. NDWI-A Normalized Difference Water Index for Remote Sensing of Vegetation Liquid Water from Space. *Remote Sens. Environ.* **1996**, *58*, 257–266. [CrossRef]

44. Gitelson, A.A.; Merzlyak, M.N. Remote sensing of chlorophyll concentration in higher plant leaves. *Adv. Space Res.* **1998**, *22*, 689–692. [CrossRef]

45. HUETE, A.R. A Soil-Adjusted Vegetation Index (SAVI). *Remote Sens. Environ.* **1988**, *25*, 295–309. [CrossRef]

46. Birth, G.S.; McVey, G.R. Measuring the Color of Growing Turf with a Reflectance Spectrophotometer 1. *Agron. J.* **1968**, *60*, 640–643. [CrossRef]

47. Lu, D.; Batistella, M. Exploring TM image texture and its relationships with biomass estimation in Rondônia, Brazilian Amazon. *Acta Amaz.* **2005**, *35*, 249–257. [CrossRef]

48. Bontemps, S.; Defourny, P.; Radoux, J.; Van Bogaert, E.; Lamarche, C.; Achard, F.; Mayaux, P.; Boettcher, M.; Brockmann, C.; Kirches, G.; et al. Consistent Global Land Cover Maps for Climate Modeling Communities: Current Achievements of the ESA's Land Cover CCI. In Proceedings of the ESA Living Planet Symposium, Edinburgh, UK, 9–13 September 2013; pp. 9–13.

49. Liu, X.; Yu, L.; Li, W.; Peng, D.; Zhong, L.; Li, L.; Xin, Q.; Lu, H.; Yu, C.; Gong, P. Comparison of country-level cropland areas between ESA-CCI land cover maps and FAOSTAT data. *Int. J. Remote Sens.* **2018**, *39*, 6631–6645. [CrossRef]

50. Gómez, R.S.; Pérez, J.G.; Del, M.; López, M. Collinearity diagnostic applied in ridge estimation through the variance inflation factor. *J. Appl. Stat.* **2016**, *43*, 1831–1849. [CrossRef]

51. Brien, R.M.O. A Caution Regarding Rules of Thumb for Variance Inflation Factors. *Qual. Quant.* **2007**, *41*, 673–690. [CrossRef]

52. Burman, B.Y.P. A comparative study of ordinary cross-validation, r-fold cross-validation and the repeated learning-testing methods. *Biometrika* **1989**, *76*, 503–514. [CrossRef]

53. Freeman, E.A.; Moisen, G.G.; Coulston, J.W.; Wilson, B.T. Random forests and stochastic gradient boosting for predicting tree canopy cover: Comparing tuning processes and model performance. *Can. J. For. Res.* **2015**, *46*, 323–339. [CrossRef]

54. Fassnacht, F.E.; Hartig, F.; Latifi, H.; Berger, C.; Hernández, J.; Corvalán, P.; Koch, B. Importance of sample size, data type and prediction method for remote sensing-based estimations of aboveground forest biomass. *Remote Sens. Environ.* **2014**, *154*, 102–114. [CrossRef]

55. White, J.C.; Gómez, C.; Wulder, M.A.; Coops, N.C. Characterizing temperate forest structural and spectral diversity with Hyperion EO-1 data. *Remote Sens. Environ.* **2010**, *114*, 1576–1589. [CrossRef]

56. Singh, M.; Malhi, Y.; Bhagwat, S. Biomass estimation of mixed forest landscape using a Fourier transform texture-based approach on very-high-resolution optical satellite imagery. *Int. J. Remote Sens.* **2014**, *35*, 3331–3349. [CrossRef]

57. Wang, X.; Shao, G.; Chen, H.; Lewis, B.J.; Qi, G.; Yu, D.; Zhou, L.; Dai, L. An application of remote sensing data in mapping landscape-level forest biomass for monitoring the effectiveness of forest policies in northeastern china. *Environ. Manag.* **2013**, *52*, 612–620. [CrossRef]

58. Li, C.; Li, Y.; Li, M. Improving forest aboveground biomass (AGB) estimation by incorporating crown density and using Landsat 8 OLI images of a subtropical forest in western Hunan in central China. *Forests* **2019**, *10*, 104. [CrossRef]

59. Gao, Y.; Lu, D.; Li, G.; Wang, G.; Chen, Q.; Liu, L.; Li, D. Comparative analysis of modeling algorithms for forest aboveground biomass estimation in a subtropical region. *Remote Sens.* **2018**, *10*, 627. [CrossRef]

60. Chrysafis, I.; Mallinis, G.; Gitas, I.; Tsakiri-Strati, M. Estimating Mediterranean forest parameters using multi seasonal Landsat 8 OLI imagery and an ensemble learning method. *Remote Sens. Environ.* **2017**, *199*, 154–166. [CrossRef]

61. Lu, D. Aboveground biomass estimation using Landsat TM data in the Brazilian Amazon. *Int. J. Remote Sens.* **2005**, *26*, 2509–2525. [CrossRef]

62. Bonate, P.L. *Pharmacokinetic-Pharmacodynamic Modeling and Simulation*; Springer: New York, NY, USA, 2011; ISBN 9781441994844.

63. Lu, D. The potential and challenge of remote sensing-based biomass estimation. *Int. J. Remote Sens.* **2006**, *27*, 1297–1328. [CrossRef]

64. Main-Knorn, M.; Moisen, G.G.; Healey, S.P.; Keeton, W.S.; Freeman, E.A.; Hostert, P. Evaluating the remote sensing and inventory-based estimation of biomass in the western carpathians. *Remote Sens.* **2011**, *3*, 1427–1446. [CrossRef]

 *forests*

Article

Estimating Urban Vegetation Biomass from Sentinel-2A Image Data

Long Li [1,2], Xisheng Zhou [1,3,*], Longqian Chen [1], Longgao Chen [4], Yu Zhang [4] and Yunqiang Liu [1]

[1] School of Environmental Science and Spatial Informatics, China University of Mining and Technology, Daxue Road 1, Xuzhou 221116, China; long.li@cumt.edu.cn or long.li@vub.be (L.L.); chenlq@cumt.edu.cn (L.C.); yunqiang.liu@cumt.edu.cn (Y.L.)

[2] Department of Geography, Earth System Sciences, Vrije Universiteit Brussel, Pleinlaan 2, 1050 Brussels, Belgium

[3] Jiangsu Institute of Urban Planning and Design, Caochangmen Avenue 88, Nanjing 210036, China

[4] School of Geography, Geomatics, and Planning, Jiangsu Normal University, Shanghai Road 101, Xuzhou 221116, China; longgao.chen@jsnu.edu.cn (L.C.); yuzhang@jsnu.edu.cn (Y.Z.)

* Correspondence: xisheng.zhou@cumt.edu.cn; Tel.: +86-516-8359-1327

Received: 18 December 2019; Accepted: 19 January 2020; Published: 21 January 2020

Abstract: Urban vegetation biomass is a key indicator of the carbon storage and sequestration capacity and ecological effect of an urban ecosystem. Rapid and effective monitoring and measurement of urban vegetation biomass provide not only an understanding of urban carbon circulation and energy flow but also a basis for assessing the ecological function of urban forest and ecology. In this study, field observations and Sentinel-2A image data were used to construct models for estimating urban vegetation biomass in the case study of the east Chinese city of Xuzhou. Results show that (1) Sentinel-2A data can be used for urban vegetation biomass estimation; (2) compared with the Boruta based multiple linear regression models, the stepwise regression models—also multiple linear regression models—achieve better estimations (RMSE = 7.99 t/hm^2 for low vegetation, 45.66 t/hm^2 for broadleaved forest, and 6.89 t/hm^2 for coniferous forest); (3) the models for specific vegetation types are superior to the models for all-type vegetation; and (4) vegetation biomass is generally lowest in September and highest in January and December. Our study demonstrates the potential of the free Sentinel-2A images for urban ecosystem studies and provides useful insights on urban vegetation biomass estimation with such satellite remote sensing data.

Keywords: urban vegetation; biomass estimation; Sentinel-2A; stepwise regression; Xuzhou

---

1. Introduction

According to the World Urbanized Prospects, urban residents are expected to compose 68% of the global population by 2050 [1], and this would bring increasingly intensive urban heat island (UHI) effects, environmental degradation, and ecological damage. As an important carrier of urban ecosystems, urban vegetation—which refers to all naturally growing and human-planted vegetation within an urban area [2,3]—brings considerable ecological, economic, and social benefits [4]. These include improving urban microclimates, mitigating UHI effects, increasing surface runoffs, maintaining the urban carbon–oxygen balance, and equally importantly, enhancing the quality of urban life by providing spaces for relaxation and recreation [5–8]. As such, the focus of urban eco-environmental studies has been long on urban vegetation, particularly the biomass of urban vegetation [9]. Urban vegetation biomass is an effective indicator of the capacity of carbon storage and sequestration, and ecological effect of an urban ecosystem [10,11]; it is, therefore, important to estimate urban vegetation biomass in urban eco-environmental management.

Traditional biomass measurement is simply to remove and weigh all the biomass occurring in quadrats, which is a labor-intensive and time-consuming practice [12,13]. This method does not allow quick monitoring and, more importantly, to some extent, might be destructive to the phenomenon being investigated. Remote sensing, however, provides an alternative to biomass measurement largely because it makes objective and mostly non-destructive observations of vegetated areas at various spatial and temporal resolutions. While vegetation biomass cannot be directly derived from remote sensing image data, remote sensing based estimation requires the use of sample plots to acquire field measurements for allometric growth equations based modeling and image interpretation for estimation (e.g., [14]). Vegetation biomass estimation with remote sensing has been summarized and reviewed in previous studies [15–17]. While optical sensor, radar, and lidar data can be used for biomass estimation separately or jointly [18–22], multispectral data is the most frequently used data type [15]. Although it has been widely recognized for its advantages, remote sensing has been mostly used to measure the biomass of individual vegetation types in natural forest [23,24], grassland [25–27], wetlands [28,29], and deserts [30] but rarely the biomass of urban vegetation [14,31].

Sentinel satellites are an Earth observation satellite constellation developed by the European Space Agency (ESA) as part of the Copernicus Program. Sentinel-2 is a wide-swath, high-resolution, multispectral imaging mission with two twin satellites (Sentinel-2A and Sentinel-2B), supporting land and climate-change monitoring [32]. Sentinel-2A was launched in June 2015 and has offered free image data at the ESA's website as of December 2015. The Sentinel-2 MSI (multispectral imager) samples 13 different spectral bands ranging from the visible to shortwave infrared of electromagnetic spectrum, four bands at 10 m, six bands at 20 m, and three bands at 60 m spatial resolution [32]. It has now been used for a variety of forestry applications such as fire damage monitoring [33,34], forest storage estimation [35,36], and canopy cover calculation [37]. While some researchers have combined Sentinel-2A with radar data for biomass estimation [24], using such free optical sensor data alone has not been assessed. Testing the capability of Sentinel-2A data to estimate urban vegetation biomass would be interesting as Sentinel-2A data is being increasingly important for land monitoring, particularly for forestry.

In this study, we therefore focus on the modeling of urban vegetation biomass estimation from Sentinel-2A image data. Quadrat biomass was calculated using the allometric biomass equations with field measurements, and then vegetation biomass models were constructed with remote sensing derived variables. Specific objectives are testing the capability of Sentinel-2A data to estimate urban vegetation biomass and examining whether vegetation type-specific modeling can improve estimation accuracy.

2. Study Area

Bordering the provinces of Shandong, Henan, and Anhui, Xuzhou (33°43′~34°58′ N, 116°22′~118°40′ E) (Figure 1) is a national key railway hub located in the northwestern part of Jiangsu province, east China [38]. It has a monsoon-influenced humid subtropical climate with an annual mean daily temperature of 14.5 °C and an annual total precipitation of 832 mm [39]. As a typical forested city, Xuzhou has received multiple titles and awards such as the National Forest City in 2012, the National Ecological Gardening City in 2015, and particularly the UN-Habitat Scroll of Honor Award in 2018 [40], which is attributed largely to the implementation of several greening and ecological restoration programs in recent decades. Although the importance of urban vegetation to cities is generally acknowledged here, no research has been conducted to estimate and assess the urban vegetation biomass for Xuzhou.

The area within the third ring road of Xuzhou (indicated by the red line in Figure 1a) was selected for this research, covering a geographical area of ~108.51 km^2. The area within the third ring road is traditionally considered as the urbanized part of Xuzhou and home to the majority of Xuzhou's urban residents. Its urban green areas have expanded remarkably in recent years and would be an ideal area for this research. The study area is flat in the central area with thick soil and hilly in the north, east, and south parts with thin humus-poor soil. The soil type is leached cinnamon soil, weak alkaline with pH ranging from 7.63 to 8.07 [41].

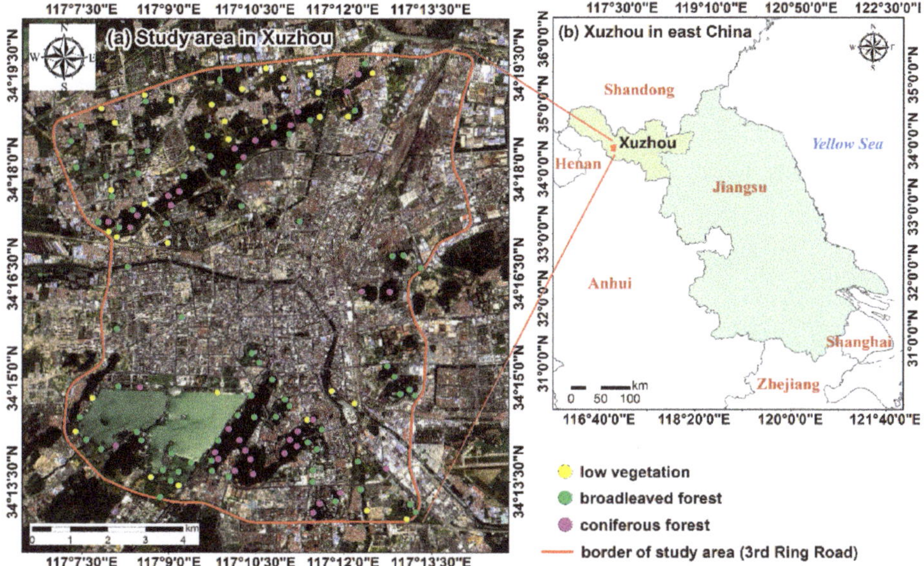

Figure 1. The location of the study area: (**a**) the border of the study area (i.e., the third ring road of Xuzhou) and the sites for field investigations (yellow for low vegetation, green for broadleaved forest, and purple for coniferous forest); (**b**) Xuzhou in east China.

According to our fieldwork, most of the trees in the study area are coniferous, consisting largely of arborvitae trees (*Platycladus orientalis*). These evergreen trees were mainly planted during the 1950s and 1960s with 700–3000 trees per hectare [41]. They are usually 5–12 m high (avg. 8.36 m) with diameters at breast height (DBH) ranging from 5 to 15 cm (avg. 12.47 cm) [41]. Broadleaved forest is dominated by poplar (*Populus euramevicana*), black locust (*Robinia pseudoacacia*), and paper mulberry (*Broussonetia papyrifera*) trees. While the poplar trees are usually large (avg. DBH = 21.40 cm) and high (avg. height = 20 m) and concentrated along rivers and roads, the black locust and paper mulberry trees are scattered in parks and small hills. Shrubs are mostly found in parks, including colorful and decorative species such as *Buxus megistophylla*, and *Berberis thunbergii*. Grassland is relatively small in urban Xuzhou, usually in parks and residential/institutional properties. Typical grass includes *Setaria viridis*, *Ophiopogon bodinieri*, *Iris tectorum*, and *Allium macrostemon*.

3. Materials and Methods

3.1. Remote Sensing Data

In this study, we used Sentinel-2A image data—freely obtained from ESA's website—for urban vegetation biomass estimation. These L1C-level data, which have already been radiometrically calibrated, were acquired in six different months of 2017 (Table 1). The image quality is generally good with a mean cloudiness of less than 10%. Although the January and May images were more cloud-contaminated, the study area remains cloud-free in the images—the images are therefore still usable. For data preprocessing, they were first atmospherically corrected and then re-sampled to 10-m, both using SNAP (SentiNel Application Platform), an image processing package developed by ESA for processing Sentinel data [42]. Lastly, the study area was extracted from the image data in ENVI 5.1 software for further processing.

Table 1. Remote sensing image data used for urban vegetation estimation.

Image ID	Acquisition Time	Cloudiness
S2A_MSIL1C_20170115T030041_N0204_R032_T50SNC_20170115T030235	15-Jan-2017	40.88%
S2A_MSIL1C_20170326T025541_N0204_R032_T50SNC_20170326T030153	26-Mar-2017	0.10%
S2A_MSIL1C_20170525T025551_N0205_R032_T50SNC_20170525T030448	25-May-2017	13.40%
S2A_MSIL1C_20170724T025551_N0205_R032_T50SNC_20170724T030446	24-July-2017	1.74%
S2A_MSIL1C_20170922T025541_N0205_R032_T50SNC_20170922T030440	22-Sept-2017	0.82%
S2B_MSIL1C_20171206T030059_N0206_R032_T50SNC_20171206T063334	6-Dec-2017	0.02%

3.2. Urban Vegetation Classification

Based on our preliminary field investigations, we decided to classify the vegetation of the study area into three coarse categories, namely low vegetation (mostly shrubs and grass), broadleaved forest (mostly poplar, black locust, and paper mulberry), and coniferous forest (mostly arborvitae trees). While many areas are characterized by a single vegetation type, there are some areas with mixed vegetation, which justifies the use of linear spectral mixture analysis (LSMA) [38,43]—where the spectrum of a pixel is considered a linear combination of spectra of pure endmembers within the pixel weighted by their fractional abundance. To this end, a wide variety of features, such as spectral features (spectral reflectance and spectral indices), textural features (calculated by the gray level co-occurrence matrix), and vegetation abundances (the abundances of coniferous forest, broad-leaved forest, and low vegetation, obtained by LSMA) were derived from the Sentinel-2A image data and combined with topographical features (DEM—digital elevation model, and slope and aspect derived from DEM) to classify urban vegetation classification using the support vector machine (SVM) method. SVM is a machine learning algorithm used for image classification [44,45] and can achieve high accuracy. We compared SVM with other classifiers, namely random forest (RF), artificial neural network (ANN), and quick unbiased efficient statistical tree (QUEST), and found that the SVM produced the best result when vegetation abundances were added for classification. For a detailed description of the classification procedure, please refer to our previous research [2]. The produced classification map helps to identify the dominant vegetation type of each pixel so the biomass of each vegetated pixel can be estimated with the models constructed later.

3.3. Candidate Variables for Modeling

A total of 116 variables (features) on spectral reflectance, vegetation indices, topographical features, and vegetation abundances were selected as candidate variables (features) for biomass estimation. They are given in Table 2 (see Table A1 for their description and calculation formulas).

Table 2. Candidate variables for biomass estimation.

Category	Variable	Number
Spectral reflectance	Blue, Green, Red, VRE1, VRE2, VRE3, NIR, N_NIR, SWIR1, SWIR2	10
Vegetation abundance	Low, BLF, CLF	3
Topographical features	DEM, Slope, Aspect	3
Vegetation indices	SAVI, MSAVI2, OSAVI, DVI, SR1–SR7, RVI, NDVIre1n, NDVIre1, NDVI, gNDVI, GI, Chlogreen, EVI2, NDII	20
Textural features	Mean (*), Var (*), Homo (*), Cont (*), Diss (*), Entr (*), Sec_M (*), Cor (*)	80
Total		116

Note: VRE1–VRE3 represent the spectral reflectance in the three red-edge bands of Sentinel-2A image data and N_NIR represents the narrow near-infrared band. Low, BLF, and CLF represent the abundances of low vegetation, broadleaved forest, and coniferous forest. The description and formulas for the vegetation indices are detailed in Table A1. Mean (*), Var (*), Homo (*), Cont (*), Diss (*), Entr (*), Sec_M (*), and Cor (*) refer to the eight textural features obtained by the gray level co-occurrence matrix using the 10 original image bands, namely mean, variance, homogeneity, contrast, difference, entropy, second moment, and correlation.

3.4. Field Measurements

Biomass sampling is necessary for vegetation biomass modeling. Usually, quadrat biomass is the sum of the dry weight of every single plant in the quadrat [12,13]. Despite high accuracy, this method requires the vegetation being investigated to be cut. As such, it is applicable to primeval forest or experimental plots but not desirable for urban green land. As a frequently used indirect biomass estimation method [46], the allometric biomass equations, where the quantitative relationships between the biomass and the growth variables of a plant are established [11], however, provide an alternative biomass sampling approach in an urban context. As they are reliable for determining tree biomass, a growing number of biomass equations have been proposed for various vegetation species across the world [47–54]. In this study, the allometric biomass equations were considered for calculating the biomass of each quadrat.

From extensive literature, the allometric biomass equations for various types of trees and shrubs in Xuzhou were summarized (Tables A2 and A3). For grass, a different estimation approach was adopted in this study: the average unit grassland biomass of Xuzhou is the spatially weighted biomass of Jiangsu, Anhui, Henan, and Shandong provinces [55] since Xuzhou is located at the junction of these four provinces (Table 3). Through the calculation, the average unit biomass of Xuzhou's grassland is 61.89 g/m^2.

Table 3. The calculation of the average unit (aboveground) biomass of grassland of Xuzhou [55].

Area	Grassland ($\times 10^4$ km^2)	(Aboveground) Biomass of Grassland (Tg)	Average Unit (Aboveground) Biomass of Grassland (g/m^2)
Jiangsu	0.31	0.17	54.48
Anhui	1.08	0.69	63.89
Henan	1.80	1.14	63.33
Shandong	1.35	0.81	60.00
Total	4.54	2.81	61.89

The growth variables of plants required in the allometric biomass equations were measured in the field investigations conducted from October to December 2017. The general investigation procedure is as follows: (1) a total of 192 urban vegetation quadrats were randomly pre-selected over the false-color Sentinel-2A imagery of the study area and their central coordinates were retrieved; (2) 10 m × 10 m quadrats were determined (matching the spatial resolution of Sentinel-2A imagery) by navigation in the field with hand-held GPS (Global Positioning System) devices to these coordinates; (3) the growth variables of each single plant (shrubs and trees only) in each quadrat were recorded and the biomass of each single plant using the plant-specific allometric biomass equations was calculated; and (4) the biomass of the all the plants in a quadrat were summed to obtain the total biomass of that quadrat and this was repeated for each quadrat.

Note that our records varied with vegetation type. Within each quadrat, we documented the name, tree height (from the base to the crown), and DBH (diameter at breast height, i.e., ~1.3 m) for trees, the name, basal diameter, height, and crown width for shrubs, and the name, height, and coverage area for grass. Different measuring tools were used in accordance with the plants to be investigated and the parameters to be recorded. The DBHs and basal diameters were measured by a 2-m tape measure with a minimum scale of 1 mm while shrub heights were measured by a 5-m tape measure with a minimum scale of 1 mm. For tree heights, we used a telescopic height measuring rod with a maximum range of 20 m and a minimal scale of 1 mm. Photos illustrating the fieldwork are shown in Figure 2.

Figure 2. Photos taken in the field illustrating the measurements.

Although 192 vegetation quadrats were initially selected, only 140 quadrats of them (shown in Figure 1) were visited and investigated in practice—because some of the pre-selected quadrats were not accessible for various reasons (e.g., physical barriers and refusal to access). Among the 140 quadrats were 35 dominated by coniferous forest, 73 by broadleaved forest, and 32 by low vegetation. The results of quadrat biomass calculated mainly by using the allometric biomass equations are detailed in Table A4.

3.5. Modeling

3.5.1. Correlation Analysis

Prior to modeling, the relationship between the candidate variables (Table 2) and the vegetation biomass was examined through correlation analysis. The biomass of the quadrats dominated by low vegetation, broadleaved forest, and coniferous forest is hereinafter referred to as low vegetation biomass, broadleaved forest biomass, and coniferous forest biomass, respectively. The correlation coefficients were computed with and without vegetation types discriminated.

3.5.2. Stepwise Regression Modeling

Stepwise regression (SR) is essentially a multiple linear regression method, but it is different from the general multiple linear regression in the selection of variables. In a stepwise regression analysis, the most significant or least significant variable is added to or removed with iteration from the multiple linear regression model based on its statistical significance [56,57]. At each iteration of adding or removing a potential independent variable, resultant models are assessed by means of the p-value of an F-statistic (p-value < 0.05 for statistical significance) [56,57]. Stepwise regression has proved effective in selecting variables for modeling and has been widely used in different fields [58,59], including forest biomass estimation [60]. As such, it was considered more suitable for constructing the urban vegetation biomass estimation models in this study.

As it is likely that collinearity exists in the predictive variables, the variance inflation factor (*VIF*) [57,61] is used to examine it in this study:

$$VIF = 1/\left(1 - R_i^2\right) \tag{1}$$

where R_i is the correlation coefficient between the ith predictive variable and the remaining predictive variables. There is no multicollinearity if VIF ranges between 0 and 10. If VIF $\geq$ 10, high multicollinearity exists between variables and some of them should be removed from the model [62].

3.5.3. Boruta Based Multiple Linear Regression Modeling

In addition to the SR modeling, the general multiple linear regression (MLR) is also considered in this study for comparative analysis. It is too complicated to include all the 116 candidate variables (Table 2) in the MLR modeling as it would decrease accuracy, cause overfitting, and slow computation. It is advisable to reduce the dimensionality of data when there are a large number of variables [63]. To this end, a group of important variables is then selected, which is done in this study by using the Boruta algorithm. Boruta is a feature selection wrapper built around the random forest classification algorithm and helps to determine important variables [64,65]. A detailed description of this feature selection technique can be found in [65,66]. The Boruta algorithm can be performed in the statistical software of R, where important variables are confirmed for modeling and unimportant one are rejected, and some artificial variables called shadow variables are generated from the original variables [65]).

Despite the capability to locate important variables, the Boruta algorithm does not consider the collinearity among these variables. Like the SR modeling, closely correlated variables are removed if VIF $\geq$ 10. The final MLR biomass estimation models are finally determined until the VIF of each remaining variable is less than 10.

3.5.4. Accuracy Assessment

While 70% of the calculated quadrat biomass were used for modeling, the remaining 30% were reserved for assessing the models using two measures, namely the coefficient of determination (R_{yz}^2) and the root-mean-square-error ($RMSE_{yz}$):

$$R_{yz}^2 = \sum_{i=1}^{n}\left(B_{modeled,i} - \overline{B}\right)^2 / \sum_{i=1}^{n}\left(B_{calculated,i} - \overline{B}\right)^2 \tag{2}$$

$$RMSE_{yz} = \sqrt{\frac{1}{n}\sum_{i=1}^{n}\left(B_{calculated,i} - B_{modeled,i}\right)^2} \tag{3}$$

where $B_{measured,i}$ is the calculated quadrat biomass, $B_{modeled,i}$ is the modeled quadrat biomass, $\overline{B}$ is the average of calculated biomass of all quadrats, and n is the number of quadrats.

3.6. Seasonal Variation of Urban Vegetation Biomass

After the accuracy assessment, the superior models can be determined and used for exploring the seasonal vegetation biomass variation of the study area. With the variables required by the determined models derived from the Sentinel-2A image data (Table 1), the biomass of low vegetation, broadleaved forest, and coniferous forest can be estimated for January, March, May, July, September, and December of 2017, respectively. The total urban vegetation biomass of the study area is then calculated by summing the estimated type-specific biomass. The change rate (*CR*) is defined by the following equation:

$$CR = \frac{Bio_{max} - Bio_{min}}{Bio_{min}} \tag{4}$$

where Bio_{max} and Bio_{max} are the maximum and minimum biomass of the year 2017.

4. Results and Analysis

4.1. Urban Vegetation Classification

By the SVM classifier, the urban vegetation of the study area was classified into three types, namely low vegetation, broadleaved forest, and coniferous forest (Figure 3) in the 24-July-2107 image; the overall accuracy of this classification was 89.86% with a Kappa coefficient of 0.83. While the central part of the study area had limited vegetation, vegetated areas were mostly covered by low vegetation, followed by coniferous forest.

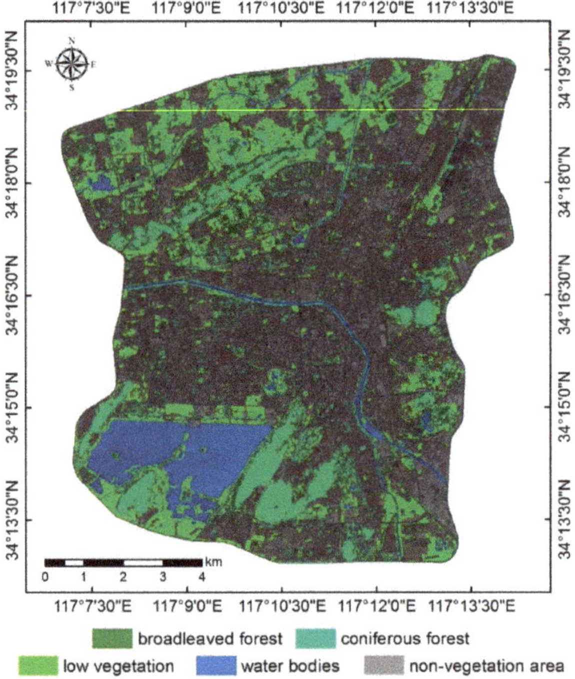

Figure 3. Urban vegetation classification by support vector machine.

4.2. Correlations between Candidate Variables and Urban Vegetation Biomass

4.2.1. For Low Vegetation

There were 14 candidate variables significantly correlated with low vegetation biomass (Table 4). Eight spectral reflectance variables had negative correlations with all-vegetation biomass, coefficients ranging from −0.364 to −0.553. It was negatively associated with low vegetation abundance and positively with coniferous forest abundance. Low vegetation biomass is generally lower than the biomass of broadleaved and coniferous forests, and more low vegetation in the quadrat means lower quadrat biomass. The correlation of low vegetation biomass with topographic features was not significant because low vegetation is usually scattered in the study area. Low vegetation biomass was negatively correlated with two vegetation indices and two textural features.

Table 4. Variables significantly correlated with low vegetation biomass.

Variable	Correlation (*p*-Value)	Variable	Correlation (*p*-Value)
Blue	−0.397 (0.025)	SWIR2	−0.364 (0.041)
Green	−0.473 (0.006)	Low	−0.564(0.001)
VRE2	−0.370 (0.037)	CLF	0.356 (0.046)
VRE3	−0.397 (0.024)	DVI	−0.399 (0.024)
NIR	−0.553 (0.001)	SR6	−0.455 (0.009)
N_NIR	−0.460 (0.008)	Cor (VRE2)	−0.411(0.019)
SWIR1	−0.431 (0.014)	Cor (VRE3)	−0.423 (0.016)

4.2.2. For Broadleaved Forest

A total of 54 variables were significantly correlated with broadleaved forest biomass (Table 5). Four spectral reflectance variables were negatively correlated with broadleaved forest biomass. Regarding vegetation abundance variables, only low vegetation abundance was negatively correlated with broadleaved forest biomass, but the coefficient was low. As for topographic features, broadleaved forest grows in relatively flat areas (e.g., parks and residential land) and low-elevated hills in the study area and, therefore, no significant correlation exists between topography and broadleaved forest biomass. The biomass was also correlated with seven vegetation indices, higher correlation coefficients with *DVI* and *SR4*. Textural features had close, mostly positive, correlations with broadleaved forest biomass, although the highest correlation (−0.72), with *Cor (VRE2)*, was negative.

Table 5. Variables significantly correlated with broadleaved forest biomass.

Variable	Correlation (*p*-Value)	Variable	Correlation (*p*-Value)	Variable	Correlation (*p*-Value)
Green	−0.424 (0.000)	Homo (Red)	0.255 (0.030)	Cont (NIR)	0.357 (0.002)
VRE1	−0.297 (0.011)	Entr (Red)	0.245 (0.037)	Diss (NIR)	0.339 (0.003)
NIR	−0.412 (0.000)	Sec_M (Red)	0.231 (0.049)	Entr (NIR)	0.322 (0.005)
SWIR1	−0.272 (0.020)	Homo (VRE1)	0.252 (0.031)	Cor (NIR)	−0.379 (0.001)
Low	−0.281 (0.016)	Diss (VRE1)	0.232 (0.048)	Mean (N_NIR)	0.310 (0.008)
MSAVI2	−0.341 (0.003)	Entr (VRE1)	0.265 (0.023)	Var (N_NIR)	0.332 (0.004)
OSAVI	−0.272 (0.020)	Mean (VRE2)	0.296 (0.011)	Cont (N_NIR)	0.527 (0.000)
DVI	−0.382 (0.001)	Var (VRE2)	0.268 (0.022)	Diss (N_NIR)	0.482 (0.000)
SR4	0.388 (0.001)	Cont (VRE2)	0.490 (0.000)	Entr (N_NIR)	0.250 (0.033)
gNDVI	0.366 (0.001)	Diss (VRE2)	0.433 (0.000)	Mean (SWIR1)	0.273 (0.019)
Chlogreen	0.276 (0.018)	Entr (VRE2)	0.399 (0.000)	Cont (SWIR1)	0.302 (0.009)
EVI2	−0.352 (0.002)	Cor (VRE2)	−0.720 (0.000)	Diss (SWIR1)	0.327 (0.005)
Homo (Blue)	0.275 (0.018)	Mean (VRE3)	0.300 (0.010)	Entr (SWIR1)	0.369 (0.001)
Entr (Blue)	0.231 (0.049)	Cont (VRE3)	0.353 (0.002)	Mean (SWIR2)	0.267 (0.023)
Sec_M (Blue)	0.288 (0.014)	Diss (VRE3)	0.358 (0.002)	Cont (SWIR2)	0.441 (0.000)
Homo (Green)	0.254 (0.030)	Entr (VRE3)	0.286 (0.014)	Diss (SWIR2)	0.406 (0.000)
Diss (Green)	0.259 (0.027)	Mean (NIR)	0.289 (0.013)	Entr (SWIR2)	0.373 (0.001)
Entr (Green)	0.294 (0.011)	Var (NIR)	0.254 (0.030)	Cor (SWIR2)	−0.324 (0.005)

4.2.3. For Coniferous Forest

Among the 116 candidate variables, 16 were significantly correlated with coniferous forest biomass (Table 6). Seven spectral reflectance variables were all negatively correlated with coniferous forest biomass, with correlation coefficients mostly higher than 0.5. Not surprisingly, only coniferous forest abundance (*CLF*) was highly positively correlated with coniferous forest biomass. *DEM* was the only topographic feature significantly correlated with coniferous forest biomass, and the negative correlation is probably linked to the fact that coniferous forest grows in hills and its biomass decreases with elevation. Coniferous forest biomass was highly significantly correlated with several vegetation indices but, interestingly, no correlation was found with textural features. The *Var* (variance), *Cont* (contrast), *Diss* (difference), *Entr* (entropy) values were all zero while *Mean* (mean), *Homo* (homogeneity), *Sec_M* (second moment), and *Cor* (correlation) values were all one—coniferous forest is densely distributed in the study area, thus no clear textural characteristics.

Table 6. Variables significantly correlated with coniferous forest biomass.

Variable	Correlation (*p*-Value)	Variable	Correlation (*p*-Value)
VRE1	−0.335 (0.049)	BLF	−0.371 (0.028)
VRE2	−0.637 (0.000)	CLF	0.531 (0.001)
VRE3	−0.588 (0.000)	DEM	−0.337 (0.047)
NIR	−0.551 (0.001)	SAVI	−0.559 (0.000)
N_NIR	−0.560 (0.000)	MSAVI2	−0.567 (0.000)
SWIR1	−0.636 (0.000)	OSAVI	−0.514 (0.002)
SWIR2	−0.541 (0.001)	DVI	−0.562 (0.000)
Low	−0.580 (0.000)	EVI2	−0.558 (0.000)

4.2.4. For All-Type Vegetation

Results show that 39 variables were significantly correlated with all-type vegetation biomass (Table 7). In total, ten spectral reflectance variables had negative correlations with all-type vegetation biomass, coefficients ranging from −0.308 (Red) to −0.496 (Green). It was negatively associated with low vegetation abundance but positively with broadleaved and coniferous forest abundances. Low vegetation has lower biomass than coniferous and broadleaved forest and, in a given area (e.g., a pixel size), the all-type vegetation biomass would be lower if low vegetation abundance is larger than the other two vegetation abundances. While it had no significant correlation with topographic features, all-type vegetation biomass was correlated with half of the vegetation indices. The highest positive correlation coefficient was found with SR4 (0.390) while the highest negative with DVI (−0.396) (Table A1). In addition, only 14 (17.50% of the total) textural features were significantly correlated with all-type vegetation biomass and coefficients were generally low.

Table 7. Variables significantly correlated with all-type vegetation biomass.

Variable	Correlation (*p*-Value)	Variable	Correlation (*p*-Value)	Variable	Correlation (*p*-Value)
VRE1	−0.335 (0.049)	BLF	−0.371 (0.028)	VRE1	−0.335 (0.049)
VRE2	−0.637 (0.000)	CLF	0.531 (0.001)	VRE2	−0.637 (0.000)
VRE3	−0.588 (0.000)	DEM	−0.337 (0.047)	VRE3	−0.588 (0.000)
NIR	−0.551 (0.001)	SAVI	−0.559 (0.000)	NIR	−0.551 (0.001)
N_NIR	−0.560 (0.000)	MSAVI2	−0.567 (0.000)	N_NIR	−0.560 (0.000)
SWIR1	−0.636 (0.000)	OSAVI	−0.514 (0.002)	SWIR1	−0.636 (0.000)
SWIR2	−0.541 (0.001)	DVI	−0.562 (0.000)	SWIR2	−0.541 (0.001)
Low	−0.580 (0.000)	EVI2	−0.558 (0.000)	Low	−0.580 (0.000)
VRE1	−0.335 (0.049)	BLF	−0.371 (0.028)	VRE1	−0.335 (0.049)
VRE2	−0.637 (0.000)	CLF	0.531 (0.001)	VRE2	−0.637 (0.000)
VRE3	−0.588 (0.000)	DEM	−0.337 (0.047)	VRE3	−0.588 (0.000)
NIR	−0.551 (0.001)	SAVI	−0.559 (0.000)	NIR	−0.551 (0.001)

4.3. Urban Vegetation Biomass Estimation Models

4.3.1. Stepwise Regression Models

The results of performing SR for constructing vegetation biomass estimation models are presented in Table A5. All the (adjusted) coefficients of determination (R_{nh}^2 and adj-R_{nh}^2) were higher than 0.70, and the fitting was generally good. The variables in the models were less than those (highly) significantly correlated with vegetation biomass (Tables 4–7). The type-specific and all-vegetation biomass estimation models are given below.

The SR biomass estimation model for low vegetation:

$$B = 10 \times [-171.896 - 49.335 \times Low + 76.406 \times CLF + 316.404 \times gNDVI - 13.710 \\ \times SR2 - 0.365 \times Cor(VRE2) + 1.087 \times DEM] \tag{5}$$

The SR biomass estimation model for broadleaved forest:

$$B = 10 \times [660.327 \quad -16.739 \times Cor(VRE2) - 3601.606 \times Green \\ +9.944 \times Cor(SWIR1) - 695.210 \times OSAVI \\ -196.861 \times Var(VRE2) + 98.126 \times Cont(SWIR1)] \tag{6}$$

The SR biomass estimation model for coniferous forest:

$$B = 10 \times [183.909 \quad -473.034 \times SWIR1 - 0.016 \times SR3 - 0.232 \times DEM + 0.299 \times GI \\ +14.747 \times Cor(VRE2)] \tag{7}$$

The SR biomass estimation model for all-type vegetation:

$$B = 10 \times [213.811 \quad -4566.311 \times Green - 5.370 \times Cor(VRE2) + 2655.001 \times Red \\ +237.815 \times Cont(SWIR2) - 108.805 \times Cont(VRE1) \\ +0.366 \times Cor(N\_NIR) - 273.149 \times Var(SWIR1) - 395.915 \times Var(Blue) \\ +157.094 \times Var(VRE1) - 49.701 \times Cont(Red) + 163.695 \times Entr(Green) \\ -203.368 \times Sec\_M(VRE2)] \tag{8}$$

4.3.2. Multiple Linear Regression Models

The results of performing the Boruta algorithm in the statistical software of R are shown in Figure 4. Important variables were labeled as Confirmed in blue, unimportant ones as Rejected in red, and shadow ones as Shadow in grey.

Using the same biomass data as the SR modeling, the MLR biomass estimation models for low vegetation, broadleaved forest, coniferous forest, and all-type vegetation were built with the important variables identified through the Boruta algorithm and the use of VIF.

The MLR biomass estimation model for low vegetation biomass:

$$= 10 \times [110.92 - 77.401 \times Low - 199.972 \times SR6 + 70.94 \times CLF] \tag{9}$$

The MLR biomass estimation model for broadleaved forest:

$$B = 10 \times [409.043- \quad 12.234 \times Cor(VRE2) - 2222.677 \times Green \\ -696.378 \times NIR - 124.43 \times Var(N_{NIR}) \\ +27.297 \times Cont(VRE2)] \tag{10}$$

The MLR biomass estimation model for coniferous forest:

$$BB = 10 \times [170.234 - 301.27 \times VER2 - 0.712 \times Slope] \tag{11}$$

The MLR biomass estimation model for all-type vegetation:

$$B = 10 \times [156.94 \quad -4011.984 \times Green + 37.17 \times Cont(VRE2) \\ +2201.306 \times Red + 4.449 \times SR4] \tag{12}$$

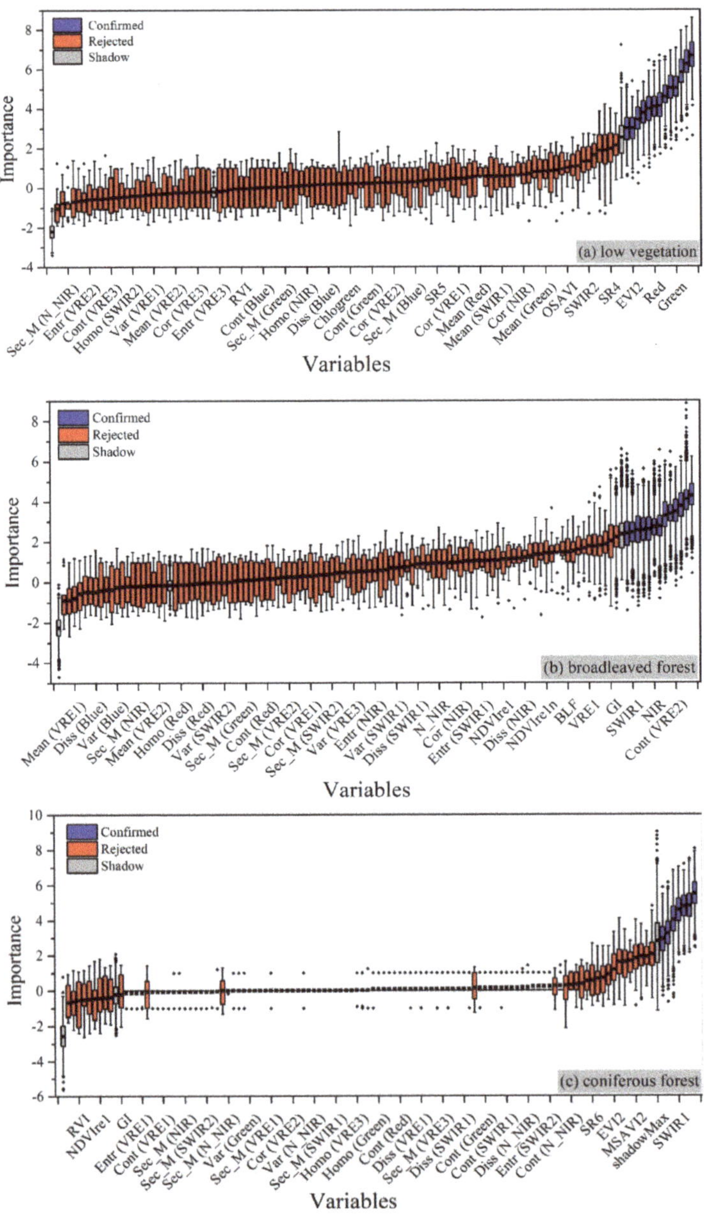

Figure 4. *Cont.*

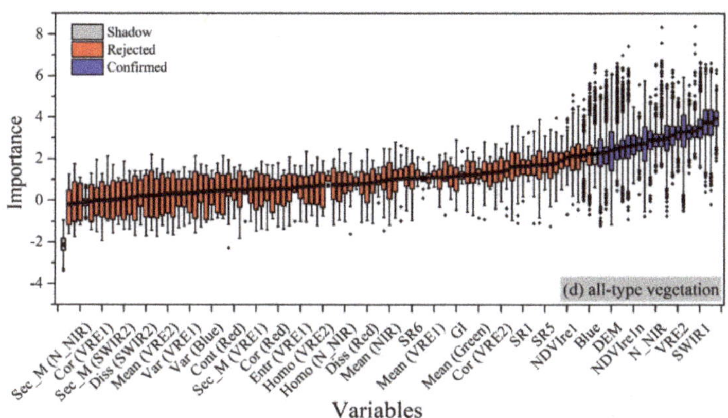

Figure 4. Importance of candidate variables: (**a**) low vegetation; (**b**) broadleaved forest; (**c**) coniferous forest; and (**d**) all-type vegetation. Important variables are labeled as Confirmed in blue, unimportant ones as Rejected in red, and shadow ones as Shadow in grey.

4.3.3. Accuracy Assessment

Figure 5 illustrates the results of assessing the SR biomass estimation models for low vegetation, broadleaved forest, coniferous forest, and all-type vegetation. It shows that $R_{yz}{}^2$ values of the models for specific vegetation types (viz. the models for low vegetation, broadleaved forest, and coniferous forest) were all higher than 0.7. The coniferous model had the highest $R_{yz}{}^2$ (0.786) and the lowest $RMSE_{yz}$ (6.89 t/hm^2). The all-type model had a larger $RMSE$ than the type-specific models.

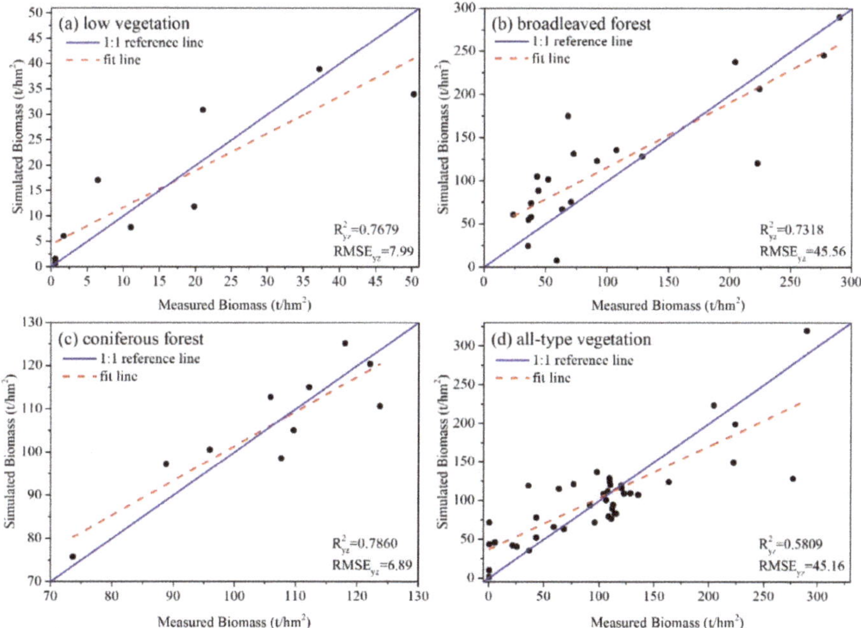

Figure 5. Accuracy assessment of the SR biomass estimation models: (**a**) low vegetation; (**b**) broadleaved forest; (**c**) coniferous forest; and (**d**) all-type vegetation.

Similarly, the remaining 30% of field observation data are used to assess the accuracy of the *MLR* biomass estimation models. After this, the two types of models are compared in terms of accuracy measured by the coefficient of determination ($R_{yz}{}^2$) and root-mean-square-error ($RMSE_{yz}$) (Table 8).

Table 8. Comparing the accuracies of the SR and MLR biomass estimation models (unit for *RMSE*: t/hm^2).

Vegetation Type	Low Vegetation		Broadleaved Forest		Coniferous Forest		All-Type Vegetation	
	$R_{yz}{}^2$	$RMSE_{yz}$	$R_{yz}{}^2$	$RMSE_{yz}$	$R_{yz}{}^2$	$RMSE_{yz}$	$R_{yz}{}^2$	$RMSE_{yz}$
SR	0.77	7.99	0.73	45.66	0.79	6.89	0.58	45.16
MLR	0.70	10.89	0.62	57.06	0.64	9.67	0.49	60.19

4.4. Seasonal Variation

As the SR models produced better estimates, they were used to calculate the biomass of each urban vegetation type in January, March, May, July, September, and December of 2017. The type-specific vegetation biomass and total vegetation biomass are shown in Figure 6.

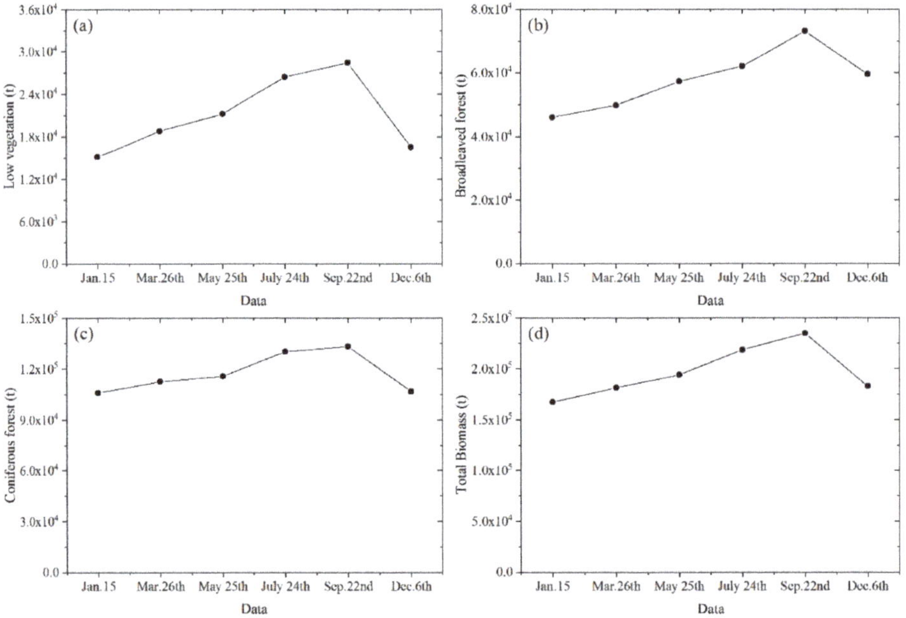

Figure 6. Type-specific biomass and the total vegetation biomass in the selected months of 2017: (**a**) low vegetation; (**b**) broadleaved forest; (**c**) coniferous forest; and (**d**) all vegetation.

Overall, vegetation biomass increased over time and decreased after peaking in autumn. The highest biomass of low vegetation was in September (28,423 t) and lowest in January and December (~15,000 t) with a maximal change rate of 87.60%. Despite an increase of 27,150 t biomass from January to September, the change rate of broadleaved forest was 58.93%, much lower than low vegetation (Figure 7). The biomass change rate of coniferous forest (25.58%) was the lowest in the three vegetation types. The total vegetation biomass change was 67,524 t with a change rate of 40.39%.

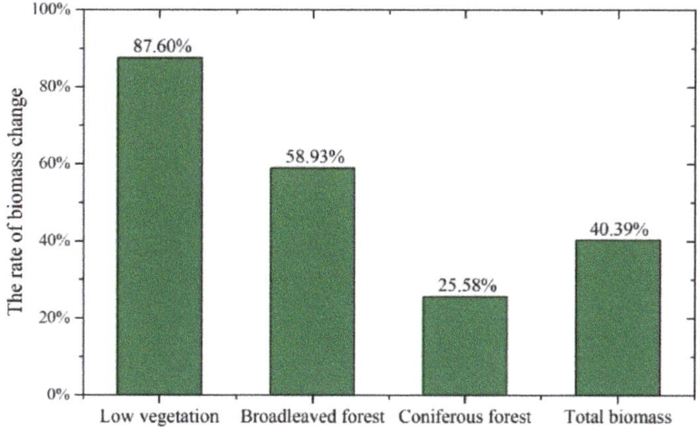

Figure 7. The seasonal change rate of vegetation biomass of the study area.

5. Discussion

Correlation analysis is useful to identify what variables are related to the dependent variable [59]. While the biomass of low vegetation and broadleaved forest is correlated mostly with spectral reflectance, broadleaved biomass is correlated mostly with textural features. Although there might be close correlations among some of the candidate variables (e.g., NDVI and RVI in the category of vegetation indices), we here did not provide a full correlation matrix for this because the number of variables was so large and would take substantial space of the publication. In addition, the use of stepwise regression and variance inflation factor can avoid the models with correlated variables [57].

Our modeling results show that for both individual vegetation types and all-type vegetation, the SR models have higher coefficients of determination and lower root-mean-square-errors than the MLR models. This clearly suggests that the SR modeling outperforms the MLR modeling in the estimation of urban vegetation biomass. The superiority of SR modeling is also noted in the study of Xu et al., where degraded grassland biomass was estimated using machine learning methods from terrestrial laser scanning data [27]. By comparing SR, random forest, and artificial neural network, they claimed that SR produced the highest accuracy (R^2 = 0.84, $RMSE$ = 48.89g/m^2). However, it might be controversial to conclude that SR is best for vegetation biomass modeling as some researchers favor machine learning algorithms. For example, Lu et al. report that RF (R^2 = 0.78, $RMSE$ = 1.34 t/ha) performs better than SR (R^2 = 0.75, $RMSE$ = 1.46 t/ha) in wheat biomass estimation with unmanned aerial vehicle data [67]. We here do not attempt to compare the results of our models with those of others because the data for modeling and the contexts (various vegetation types in an urban area vs. a single type of vegetation in (semi-) environments) were different.

Although some researchers estimated vegetation biomass from remote sensing without discriminating types [29], our study revealed that vegetation biomass should be modeled for specific vegetation types for higher modeling accuracy. This is often done for different contexts by other researchers, e.g., Gao et al. who discriminated broadleaved, coniferous, mixed, and bamboo forest in China's Zhejiang province [68], and González-Jaramillo et al. who divided vegetation of the San Francisco watershed (south Ecuador) into tropical mountain forest, subpáramo, and pastures [23]. In fact, the finding of correlation analysis that variables significantly correlated with vegetation biomass varies largely with vegetation type implies that type-specific biomass estimations models should be constructed. Similarly, non-species-specific allometric growth models yielded larger errors than species-specific ones [69]. Urban vegetation cannot be regarded as a single vegetation type as it varies largely in biophysical characteristics and thus biomass. Such variations, which might be minimized in plantations, should be considered for urban green areas. As such, it is important to discriminate urban vegetation types through image classification before modeling urban vegetation biomass from remote sensing image data.

Regarding the seasonal variation of vegetation biomass, coniferous forest has much lower biomass loss than low vegetation and broadleaved forest, which is because coniferous forest consists mainly of evergreen arborvitae trees that do not lose their leaves through the year. This suggests that more coniferous trees should be planted if the biomass loss of low vegetation and broadleaved forest needs to be compensated. In this multi-season analysis, the same type-specific estimation models were used for estimating vegetation biomass from remote sensing data imaged in different months. For a plant species in an area, there is only one allometric growth equation, which is often built with measurements acquired, e.g., when plants are luxuriant with maximal biomass in a year. The biomass estimation models constructed with quadrat biomass calculated using these equations should best reflect that time. If these models are used for other dates, estimation biomass would be less accurate (e.g., due to less leaves in winter). Remote sensing variables derived from remote sensing images can however characterize the vegetative status of the plants and compensate the impact.

In addition, there are some other limitations that might undermine the results. Firstly, the allometric biomass equations for a variety of plant species with high reported accuracies were borrowed from previous studies, but we were not able to individually verify these equations as this work is out of the scope of the present study. Secondly, tree biomass could be, to some extent, underestimated from remote sensing image data. While it is likely that under large coniferous and broadleaved and coniferous trees grow some low vegetation like grass and bushes, this cannot be recognized in pixels, notwithstanding the application of linear spectral mixture analysis. Despite these limitations, our study proves the capability of free optical sensor data like Sentinel-2A to estimate urban vegetation biomass. It would be interesting if urban vegetation biomass could be regularly monitored; however, this seems currently challenging as Sentinel-2A data now remains scarce and does not allow a retrospective assessment.

6. Conclusions

This study demonstrates how Sentinel-2A image data can be used for vegetation biomass in an urban context. The main findings and conclusions of this study are as follows:

- Freely available multispectral Sentinel-2A satellite data has proven its capability in urban vegetation biomass estimation. The measured biomass of each vegetation type is closely correlated with different remote sensing derived variables, mostly spectral reflectance for low vegetation and coniferous forest and mostly textural features for broadleaved forest.
- The vegetation biomass estimation models built by the stepwise regression (SR) outperform those with the multiple linear regression. It is necessary to discriminate vegetation types in biomass modeling and the highest accuracy is obtained by the SR model for coniferous forest.
- Highest vegetation biomass occurs in autumn (September) while lowest in winter (January and December). Low vegetation and broadleaved forest have larger seasonal change rates than coniferous forest that consists mostly of evergreen trees.

Urban green areas are a key component of urban eco-environment and make a vital contribution to improving the quality of life and moderating climate. In general, trees have a stronger carbon sequestration capability and produce more biomass than low vegetation. More coniferous trees can maintain less biomass loss in winter. However, tree species should be diversified to reduce ecological vulnerability and guarantee a more robust urban ecosystem and more sustainable urban development.

Author Contributions: Conceptualization, X.Z. and L.L.; methodology, X.Z., L.L., and L.C. (Longqian Chen); software, X.Z., Y.L., and Y.Z.; validation, Y.L. and Y.Z.; formal analysis, L.L., and X.Z; investigation, X.Z., Y.L., and L.L.; resources, L.C (Longqian Chen). and L.C. (Longgao Chen); data curation, X.Z. and L.L.; writing—original draft preparation, L.L. and X.Z.; writing—review and editing, L.L., L.C. (Longqian Chen), and L.C. (Longgao Chen); visualization, Y.Z. and Y.L.; supervision, L.C. (Longqian Chen); project administration, L.L. and L.C. (Longqian Chen); funding acquisition, L.L. All authors have read and agreed to the published version of the manuscript.

Funding: This research was supported by "the Fundamental Research Funds for the Central Universities" (Grant No.: 2018QNB06).

Acknowledgments: The authors would like to thank the United States Geological Survey (USGS) for freely providing satellite remote sensing image data required in this study, Guanghui Hong for his assistance in the fieldwork, and Jinyu Zang, Ruiyang Liu, Zhiqiang Wang, and Ziqi Yu for their help in preparing an early version of this manuscript. Comments and suggestions from three anonymous reviewers are greatly appreciated for improving the study.

Conflicts of Interest: The authors declare no conflict of interest.

Appendix A

Table A1. Formulas used for calculating spectral indices [70].

Spectral Index	Formula
Green index (GI)	$GI = Green/Red$
Green normalized different vegetation index (gNDVI)	$gNDVI = (N\_NIR - Green)/(N\_NIR + Green)$
Normalized difference vegetation index (NDVI)	$NDVI = (NIR - Red)/(NIR + Red)$
Ratio vegetation index (RVI)	$RVI = NIR/Red$
Difference vegetation index (DVI)	$DVI = NIR - Red$
Enhanced vegetation index 2 (EVI2)	$EVI2 = (NIR - Red)/(1 + NIR + 2.4 \times Red)$
Chlorophyll green index (Chlogreen)	$Chlogreen = N\_NIR/(Green + VER1)$
Normalized difference vegetation index ($NDVI_{re1}$)	$NDVI_{re1} = (NIR - VER1)/(NIR + VER1)$
Normalized difference vegetation index ($NDVI_{re1n}$)	$NDVI_{re1n} = (N\_NIR - VER1)/(N\_NIR + VER1)$
Simple ratio 1 (SR1)	$SR1 = NIR/VER1$
Simple ratio 2 (SR2)	$SR2 = N\_NIR/VER1$
Simple ratio 3 (SR3)	$SR3 = N\_NIR/Red$
Simple ratio 4 (SR4)	$SR4 = N\_NIR/Green$
Simple ratio 5 (SR5)	$SR5 = N\_NIR/Blue$
Simple ratio 6 (SR6)	$SR6 = Blue/VER1$
Simple ratio 7 (SR7)	$SR7 = NIR/Red$
Normalized difference infrared index (NDII)	$NDII = (NIR - SWIR1)/(NIR + SWIR1)$
Soil-adjusted vegetation index (SAVI)	$SAVI = \frac{N\_NIR - Red}{N\_NIR + Red + L} \times 0.5$
Modified soil-adjusted vegetation index 2 (MSAVI2)	$MSAVI2 = 0.5 \times [(2 \times NIR + 1) - \sqrt{(2 \times NIR + 1)^2 - 8 \times (NIR - Red)}]$
Optimized soil-adjusted vegetation index (OSAVI)	$OSAVI = (NIR - Red)/(NIR + Red + 0.16)$

Note: VRE1–VRE3 represent the three red-edge bands; N_NIR represents the narrow near-infrared bands.

Table A2. Allometric biomass equations for trees, used for calculating quadrat biomass.

Tree species	Model	R^2	Reference
Platycladus orientalis	$W_S = 0.0573 (D^2H)^{0.8657}$	0.97	[71]
	$W_B = 0.0043 (D^2H)^{1.1085}$	0.89	
	$W_L = 0.0038 (D^2H)^{1.0385}$	0.84	
	$W_R = 0.0485 (D^2H)^{0.6886}$	0.80	
Robinia pseudoacacia	$W_S = 0.0681 (D^2H)^{0.9865}$	0.9545	[72]
	$W_B = 12020 + 0.009 (D^2H)$	0.8862	
	$W_L = -0.549 + 0.007 (D^2H)$	0.9174	
	$W_R = 0.0087 (D^2H)^{1.0513}$	0.9472	
Metasequoia glyptostroboides	$W_S = 0.0146 (D^2H)^{0.9835}$	0.993	[73]
	$W_B = 0.0243 (D^2H)^{0.7359}$	0.993	
	$W_L = 0.0949 (D^2H)^{0.4795}$	0.982	
	$W_R = 0.0102 (D^2H)^{0.8745}$	0.975	

Table A2. *Cont.*

Tree species	Model	R^2	Reference
Populus euramevicana	$W_S = 0.006 \, (D^2H)^{1.098}$	0.995	[74]
	$W_B = 0.001 \, (D^2H)^{1.157}$	0.984	
	$W_L = 0.012 \, (D^2H)^{0.685}$	0.955	
	$W_R = 0.083 \, (D^2H)^{0.636}$	0.915	
Cinnamomum camphora	$W_S = 0.0914 \, (D^2H)^{0.7755}$	0.944	[73]
	$W_B = 0.0099 \, (D^2H)^{1.0256}$	0.946	
	$W_L = 0.0011 \, (D^2H)^{1.1713}$	0.941	
	$W_R = 0.0298 \, (D^2H)^{0.8740}$	0.935	
Ginkgo biloba	$\ln W_S = -3.84 + 0.95 \ln (D^2H)$	0.98	[75]
	$\ln W_B = -9.38 + 1.46 \ln (D^2H)$	0.852	
	$\ln W_L = -6.95 + 1.03 \ln (D^2H)$	0.853	
	$\ln W_R = -5.60 + 1.07 \ln (D^2H)$	0.967	
Platanus acerifolia	$W_T = 0.0690 (D^2H)^{0.9133}$	/	[76]
Larix gmelinii	$\ln W_S = -2.8319 + 0.8379 \ln (D^2H)$	0.9996	[77]
	$\ln W_B = -3.9021 + 0.8822 \ln (D^2H)$	0.9015	
	$\ln W_L = -4.0174 + 0.7659 \ln (D^2H)$	0.9007	
	$\ln W_R = -3.6497 + 0.8247 \ln (D^2H)$	0.9994	
Broussonetia papyrifera	$W_T = 0.07112 \, (D^2H)^{0.910358078}$	/	[78]
Ligustrum lucidum	$W_S = 0.03939 \, (D^2H)^{0.95679}$	0.97	[79]
	$W_B = 0.03357 \, (D^2H)^{0.77809}$	0.84	
	$W_L = 0.11613 \, (D^2H)^{0.45871}$	0.61	
	$W_T = 0.11394 \, (D^2H)^{0.84957}$	0.97	
Koelreuteria bipinnata	$W_S = 0.08259 \, (D^2H)^{0.80831}$	0.97	[79]
	$W_B = 0.00053 \, (D^2H)^{1.29104}$	0.94	
	$W_L = 0.01286 \, (D^2H)^{0.69408}$	0.81	
	$W_T = 0.12238 \, (D^2H)^{0.84468}$	0.98	
Magnolia grandiflora	$W_S = 0.0649 \, (D^2H)^{0.8131}$	0.969	[73]
	$W_B = 0.0431 \, (D^2H)^{0.6697}$	0.904	
	$W_L = 0.0254 \, (D^2H)^{0.8701}$	0.837	
	$W_R = 0.0885 \, (D^2H)^{0.6713}$	0.883	
Liriodendron chinense	$W_S = 0.02426 \, (D^2H)^{0.942303}$	0.99537	[80]
	$W_B = 0.000349 \, (D^2H)^{1.268207}$	0.962865	
	$W_L = 0.000419 \, (D^2H)^{1.048786}$	0.834806	
	$W_R = 0.023475 \, (D^2H)^{0.770233}$	0.918072	
Paulownia fortunei	$W_S = 0.021158 D^{2.43244}$	0.9978	[81]
	$W_B = 0.057869 D^{2.06599}$	0.9959	
	$W_L = 0.060045 D^{1.54688}$	0.9891	
	$W_R = 0.030740 D^{2.10612}$	0.8387	

Note: D is DBH (diameter at breast height); H is tree height; W_S, W_B, W_L, refer to the biomass of stem, branch, and leaves; and W_T and W_R to the total aboveground biomass and root biomass.

Table A3. Allometric biomass equation of shrubs, used for calculating quadrat biomass [82,83].

Species	Model	R^2	Species	Model	R^2
Ligustrum quihoui	$W_B = 26.332 \, (CH)^{0.666}$	0.950	*Buxus bodinieri*	$W_B = 262.879 \, (CH)^{1.546}$	0.895
	$W_L = 14.646 C^{1.164}$	0.972		$W_L = 224.662 \, (CH)^{1.364}$	0.890
	$W_R = 18.721 \, (V_C)^{0.421}$	0.965		$W_R = 294.262 \, (CH)^{1.639}$	0.889
	$W_T = 52.388 \, (CH)^{0.654}$	0.959		$W_T = 756.343 \, (CH)^{1.497}$	0.913
Berberis thunbergii	$W_B = 73.468 \, (A_C)^{0.766}$	0.927	*Buxus megistophylla*	$W_B = 15.572 D^{1.325}$	0.979
	$W_L = 3.340 \, (A_C)^{0.465}$	0.601		$W_L = 20.649 + 9.047 \ln (CH)$	0.902
	$W_R = 29.029 \, (A_C)^{0.721}$	0.785		$W_R = 9.654 D^{1.308}$	0.975
	$W_T = 104.637 \, (A_C)^{0.734}$	0.903		$W_T = 35.982 D^{1.212}$	0.980

Table A3. *Cont.*

Species	Model	R^2	Species	Model	R^2
Photinia serrulata	$W_B = 0.310\,(D^2H)^{1.097}$	0.985	*Pittosporum tobira*	$W_B = 765.073\,(V_C)^{0.824}$	0.991
	$W_L = 0.264\,(D^2H)^{0.916}$	0.986		$W_L = 2.958\,(D^2H)^{0.607}$	0.911
	$W_R = 0.259\,(D^2H)^{1.053}$	0.988		$W_R = 445.103\,(V_C)^{0.742}$	0.972
	$W_T = 0.805\,(D^2H)^{1.051}$	0.988		$W_T = 1411.387\,(V_C)^{0.742}$	0.979
Hibiscus syriacus	$W_B = 108.688\,(V_C)^{1.693}$	0.984	*Nandina domestica*	$W_B = 75.700\,(CH)^{1.110}$	0.980
	$W_L = 18.925\,(CH)^{1.565}$	0.969		$W_L = 11.109 + 17.911\ln H$	0.971
	$W_R = 69.564\,(V_C)^{1.563}$	0.985		$W_R = 57.553\,(CH)^{1.187}$	0.939
	$W_T = 206.627\,(V_C)^{1.589}$	0.986		$W_T = 167.114\,(CH)^{1.174}$	0.960
Lagerstroemia indica	$W_B = 30.213H^{6.318}$	0.987	*Syringa oblata*	$W_B = 0.876\,(D^2H)^{0.894}$	0.988
	$W_L = 6.656H^{5.065}$	0.994		$W_L = 0.683\,(D^2H)^{0.715}$	0.988
	$W_R = 20.934H^{5.905}$	0.989		$W_R = 0.603\,(D^2H)^{0.877}$	0.991
	$W_T = 58.305H^{6.065}$	0.989		$W_T = 2.011\,(D^2H)^{0.863}$	0.991
Forsythia suspensa	$W_B = 0.385\,(D^2H)^{1.025}$	0.997			
	$W_L = 0.187\,(D^2H)^{0.868}$	0.985			
	$W_R = 0.176\,(D^2H)^{0.954}$	0.990			
	$W_T = 0.716\,(D^2H)^{0.989}$	0.996			

Note: D is basal diameter; H is height; C is crown width (which is the average of south-north crown diameter C_1 and east-west crown diameter C_2; $C = (C_1 + C_2)/2$); A_C is the area of crown ($A_C = \pi \times C_1 \times C_2$); V_C is the volume of crown ($V_C = A_C \times H$); W_S, W_B, W_L, refer to the biomass of stem, branch, and leaves; and W_T and W_R to the total biomass and root biomass.

Table A4. The calculated biomass for each quadrat. As only 140 of the 192 pre-selected quadrats were visited and investigated, the quadrat ID ranges from 1 to 192.

ID	Biomass (kg)	ID	Biomass (kg)	ID	Biomass (kg)	ID	Biomass (kg)
1	1005.74	2	654.00	3	1192.37	5	1372.71
6	1711.81	7	11,250.00	8	972.12	9	1286.95
10	2118.96	11	1043.43	13	1258.78	14	1114.78
15	502.87	16	431.85	17	638.50	21	985.40
22	918.67	23	989.10	24	1212.90	25	349.37
26	732.72	27	838.38	28	1580.27	29	383.54
30	110.76	31	100.03	33	766.57	34	1556.81
35	917.00	36	56.07	37	383.64	38	1171.94
39	759.66	40	519.31	41	1383.84	42	1300.94
43	711.91	45	1158.74	46	831.58	47	447.30
48	906.36	50	607.56	51	362.14	52	325.44
53	734.24	54	634.06	55	2152.83	56	965.29
59	240.68	60	777.64	63	2042.87	65	1237.21
66	1573.60	67	901.01	68	1641.06	69	805.89
70	612.32	71	1658.79	72	433.56	74	2225.07
75	257.17	76	893.24	77	1209.58	80	8.23
82	706.14	83	989.41	84	1105.56	86	551.12
87	38.43	88	222.73	90	879.47	91	1285.60
92	17.95	94	442.08	95	822.52	96	680.24
97	1085.41	99	1153.80	100	188.35	101	11,085.08
106	2771.95	107	590.58	110	884.22	111	1531.30
116	984.55	117	3452.27	118	525.76	119	120.16
121	371.84	123	663.12	124	559.32	126	1610.53
127	866.05	129	3437.13	132	384.20	134	2179.22
135	216.23	137	2915.05	139	610.32	140	6.19

Table A4. *Cont.*

ID	Biomass (kg)	ID	Biomass (kg)	ID	Biomass (kg)	ID	Biomass (kg)
142	1100.06	143	1634.99	145	973.77	146	364.50
147	1421.20	148	1841.15	149	2707.82	151	969.84
152	2899.26	153	6.19	154	1176.14	156	1100.67
157	3279.12	158	6.19	159	1265.47	162	1007.43
163	2240.88	164	1189.30	166	6.19	168	6.19
169	6.19	170	698.32	172	198.72	173	6.19
174	1772.22	175	2304.31	176	6.19	177	6.19
178	1045.89	179	131.80	180	78.79	181	888.34
182	64.71	183	320.74	185	210.05	187	724.18
188	308.44	189	1002.98	190	6.19	192	623.21

Table A5. The results of the *SR* modeling.

Vegetation Type	R_{nh}^2	Adj-R_{nh}^2	Variable	Coefficient	VIF
Low vegetation	0.853	0.818	Constant	−171.896	
			Low	−49.335	1.382
			CLF	76.406	3.254
			gNDVI	316.404	3.181
			SR2	−13.710	4.274
			Cor (VRE2)	−0.365	1.311
			DEM	1.087	1.207
Broadleaved forest	0.821	0.805	Constant	660.327	
			Cor (VRE2)	−16.739	2.095
			Green	−3601.606	1.066
			Cor (SWIR1)	9.944	2.317
			OSAVI	−695.210	1.375
			Var (VRE2)	−196.861	5.043
			Cont (SWIR1)	98.126	5.674
Coniferous forest	0.838	0.810	Constant	183.909	
			SWIR1	−473.034	1.151
			SR3	−0.016	1.346
			DEM	−0.232	1.109
			GI	0.299	1.461
			Cor (VRE2)	14.747	1.079
All vegetation	0.754	0.721	Constant	213.811	
			Green	−4566.311	4.279
			Cor (VRE2)	−5.370	1.889
			Red	2655.001	4.530
			Cont (SWIR2)	237.815	9.833
			Cont (VRE1)	−108.805	6.695
			Cor (N_NIR)	0.366	1.905
			Var (SWIR1)	−273.149	3.947
			Var (Blue)	−395.915	3.295
			Var (VRE1)	157.094	4.905
			Cont (Red)	−49.701	3.396
			Entr (Green)	163.695	9.353
			Sec_M (VRE2)	−203.368	4.150

References

1. United Nations. *World Urbanization Prospects: The 2080 Revision*; United Nations: New York, NY, USA, 2018.
2. Zhou, X.; Li, L.; Chen, L.; Liu, Y.; Cui, Y.; Zhang, Y.; Zhang, T. Discriminating urban forest types from Sentinel-2A image data through linear spectral mixture analysis: A case study of Xuzhou, East China. *Forests* **2019**, *10*, 478. [CrossRef]

3. Miller, R.W.; Hauer, R.J.; Werner, L.P. *Urban Forestry: Planning and Managing Urban Greenspaces*, 3rd ed.; Waveland Press, Inc.: Long Grove, IL, USA, 2015; ISBN 9781478606376.

4. Zhao, S.; Tang, Y.; Chen, A. Carbon storage and sequestration of urban street trees in Beijing, China. *Front. Ecol. Evol.* **2016**, *4*, 53. [CrossRef]

5. Cohen-Cline, H.; Turkheimer, E.; Duncan, G.E. Access to green space, physical activity and mental health: A twin study. *J. Epidemiol. Community Health* **2015**, *69*, 523–529. [CrossRef] [PubMed]

6. White, M.P.; Alcock, I.; Wheeler, B.W.; Depledge, M.H. Would you be happier living in a greener urban area? A fixed-effects analysis of Panel Data. *Psychol. Sci.* **2013**, *24*, 920–928. [CrossRef] [PubMed]

7. Wilkes, P.; Disney, M.; Vicari, M.B.; Calders, K.; Burt, A. Estimating urban above ground biomass with multi-scale LiDAR. *Carbon Balance Manag.* **2018**, *13*, 10. [CrossRef]

8. Reis, C.; Lopes, A. Evaluating the cooling potential of urban green spaces to tackle urban climate change in Lisbon. *Sustainability* **2019**, *11*, 2480. [CrossRef]

9. Pérez, G.; Perini, K. (Eds.) *Nature Based Strategies for Urban and Building Sustainability*; Elsevier: Oxford, UK, 2018; ISBN 9780128121504.

10. He, M.; Zhao, B.; Ouyang, Z.; Yan, Y.; Li, B. Linear spectral mixture analysis of Landsat TM data for monitoring invasive exotic plants in estuarine wetlands. *Int. J. Remote Sens.* **2010**, *31*, 4319–4333. [CrossRef]

11. He, H.; Zhang, C.; Zhao, X.; Fousseni, F.; Wang, J.; Dai, H.; Yang, S.; Zuo, Q. Allometric biomass equations for 12 tree species in coniferous and broadleaved mixed forests, Northeastern China. *PLoS ONE* **2018**, *13*, e0186226. [CrossRef]

12. Weaver, T.; Collins, D. Measuring vegetation biomass and production. *Am. Biol. Teach.* **1988**, *50*, 164–166. [CrossRef]

13. Launchbaugh, K. Direct Measures of Biomass. Available online: https://www.webpages.uidaho.edu/veg_measure/ Modules/Lessons/Module7(Biomass&Utilization)/7_3_DirectMethods.htm (accessed on 1 October 2019).

14. Wu, J. Developing general equations for urban tree biomass estimation with high-resolution satellite imagery. *Sustainability* **2019**, *11*, 4347. [CrossRef]

15. Galidaki, G.; Zianis, D.; Gitas, I.; Radoglou, K.; Karathanassi, V.; Tsakiri-Strati, M.; Woodhouse, I.; Mallinis, G. Vegetation biomass estimation with remote sensing: Focus on forest and other wooded land over the Mediterranean ecosystem. *Int. J. Remote Sens.* **2017**, *38*, 1940–1966. [CrossRef]

16. Lu, D. The potential and challenge of remote sensing-based biomass estimation. *Int. J. Remote Sens.* **2006**, *27*, 1297–1328. [CrossRef]

17. Kumar, L.; Mutanga, O. Remote sensing of above-ground biomass. *Remote Sens.* **2017**, *9*, 935. [CrossRef]

18. Sousa, A.M.O.; Gonçalves, A.C.; Mesquita, P.; Marques da Silva, J.R. Biomass estimation with high resolution satellite images: A case study of Quercus rotundifolia. *ISPRS J. Photogramm. Remote Sens.* **2015**, *101*, 69–79. [CrossRef]

19. Simonson, W.; Ruiz-Benito, P.; Valladares, F.; Coomes, D. Modelling above-ground carbon dynamics using multi-temporal airborne lidar: Insights from a Mediterranean woodland. *Biogeosciences* **2016**, *13*, 961–973. [CrossRef]

20. Gonzalez, P.; Asner, G.P.; Battles, J.J.; Lefsky, M.A.; Waring, K.M.; Palace, M. Forest carbon densities and uncertainties from Lidar, QuickBird, and field measurements in California. *Remote Sens. Environ.* **2010**, *114*, 1561–1575. [CrossRef]

21. Nijland, W.; Addink, E.A.; De Jong, S.M.; Van der Meer, F.D. Optimizing spatial image support for quantitative mapping of natural vegetation. *Remote Sens. Environ.* **2009**, *113*, 771–780. [CrossRef]

22. Vaglio Laurin, G.; Chen, Q.; Lindsell, J.A.; Coomes, D.A.; Del Frate, F.; Guerriero, L.; Pirotti, F.; Valentini, R. Above ground biomass estimation in an African tropical forest with lidar and hyperspectral data. *ISPRS J. Photogramm. Remote Sens.* **2014**, *89*, 49–58. [CrossRef]

23. González-Jaramillo, V.; Fries, A.; Zeilinger, J.; Homeier, J.; Paladines-Benitez, J.; Bendix, J. Estimation of above ground biomass in a tropical mountain forest in Southern Ecuador using airborne LiDAR data. *Remote Sens.* **2018**, *10*, 660. [CrossRef]

24. Vafaei, S.; Soosani, J.; Adeli, K.; Fadaei, H.; Naghavi, H.; Pham, T.; Tien Bui, D. Improving accuracy estimation of forest aboveground biomass based on incorporation of ALOS-2 PALSAR-2 and Sentinel-2A imagery and machine learning: A case study of the Hyrcanian forest area (Iran). *Remote Sens.* **2018**, *10*, 172. [CrossRef]

25. Ren, H.; Zhou, G. Estimating green biomass ratio with remote sensing in arid grasslands. *Ecol. Indic.* **2019**, *98*, 568–574. [CrossRef]

26. Jia, W.; Liu, M.; Yang, Y.; He, H.; Zhu, X.; Yang, F.; Yin, C.; Xiang, W. Estimation and uncertainty analyses of grassland biomass in Northern China: Comparison of multiple remote sensing data sources and modeling approaches. *Ecol. Indic.* **2016**, *60*, 1031–1040. [CrossRef]

27. Xu, K.; Su, Y.; Liu, J.; Hu, T.; Jin, S.; Ma, Q.; Zhai, Q.; Wang, R.; Zhang, J.; Li, Y.; et al. Estimation of degraded grassland aboveground biomass using machine learning methods from terrestrial laser scanning data. *Ecol. Indic.* **2020**, *108*, 105747. [CrossRef]

28. Klemas, V. Remote Sensing of Coastal Wetland Biomass: An Overview. *J. Coast. Res.* **2013**, *290*, 1016–1028. [CrossRef]

29. Han, M.; Pan, B.; Liu, Y.B.; Yu, H.Z.; Liu, Y.R. Wetland biomass inversion and space differentiation: A case study of the Yellow River Delta Nature Reserve. *PLoS ONE* **2019**, *14*, e0210774. [CrossRef]

30. Zhang, C.; Lu, D.; Chen, X.; Zhang, Y.; Maisupova, B.; Tao, Y. The spatiotemporal patterns of vegetation coverage and biomass of the temperate deserts in Central Asia and their relationships with climate controls. *Remote Sens. Environ.* **2016**, *175*, 271–281. [CrossRef]

31. Mueed Choudhury, M.A.; Costanzini, S.; Despini, F.; Rossi, P.; Galli, A.; Marcheggiani, E.; Teggi, S. Photogrammetry and remote sensing for the identification and characterization of trees in urban areas. *J. Phys. Conf. Ser.* **2019**, *1249*, 12008. [CrossRef]

32. SUHET Sentinel-2 User Handbook. Available online: https://sentinel.esa.int/documents/247904/685211/Sentinel-2_User_Handbook (accessed on 1 July 2016).

33. Fernández-Manso, A.; Fernández-Manso, O.; Quintano, C. SENTINEL-2A red-edge spectral indices suitability for discriminating burn severity. *Int. J. Appl. Earth Obs. Geoinf.* **2016**, *50*, 170–175. [CrossRef]

34. Navarro, G.; Caballero, I.; Silva, G.; Parra, P.C.; Vázquez, Á.; Caldeira, R. Evaluation of forest fire on Madeira Island using Sentinel-2A MSI imagery. *Int. J. Appl. Earth Obs. Geoinf.* **2017**, *58*, 97–106. [CrossRef]

35. Chrysafis, I.; Mallinis, G.; Siachalou, S.; Patias, P. Assessing the relationships between growing stock volume and Sentinel-2 imagery in a Mediterranean forest ecosystem. *Remote Sens. Lett.* **2017**, *8*, 508–517. [CrossRef]

36. Puliti, S.; Saarela, S.; Gobakken, T.; Ståhl, G.; Næsset, E. Combining UAV and Sentinel-2 auxiliary data for forest growing stock volume estimation through hierarchical model-based inference. *Remote Sens. Environ.* **2018**, *204*, 485–497. [CrossRef]

37. Korhonen, L.; Packalen, P.; Rautiainen, M. Comparison of Sentinel-2 and Landsat 8 in the estimation of boreal forest canopy cover and leaf area index. *Remote Sens. Environ.* **2017**, *195*, 259–274. [CrossRef]

38. Li, H.; Li, L.; Chen, L.; Zhou, X.; Cui, Y.; Liu, Y.; Liu, W. Mapping and characterizing spatiotemporal dynamics of impervious surfaces using Landsat images: A case study of Xuzhou, East China from 1995 to 2018. *Sustainability* **2019**, *11*, 1224. [CrossRef]

39. Zhang, Y.; Li, L.; Qin, K.; Wang, Y.; Chen, L.; Yang, X. Remote sensing estimation of urban surface evapotranspiration based on a modified Penman–Monteith model. *J. Appl. Remote Sens.* **2018**, *12*, 046006. [CrossRef]

40. Xinhua China's Xuzhou City Wins UN-Habitat Scroll of Honor for Promoting Urban Renewal. Available online: http://www.xinhuanet.com/english/2018-10/01/c_137506123.htm (accessed on 1 July 2019).

41. Zhou, W. Study on Carbon Stock of Vegetation and Its Affecting Factors in Xuzhou. Ph.D. Thesis, Nanjing Forestry University, Nanjing, China, 2012.

42. ESA SNAP. Available online: http://step.esa.int/main/toolboxes/snap/ (accessed on 1 October 2019).

43. Li, L.; Canters, F.; Solana, C.; Ma, W.; Chen, L.; Kervyn, M. Discriminating lava flows of different age within Nyamuragira's volcanic field using spectral mixture analysis. *Int. J. Appl. Earth Obs. Geoinf.* **2015**, *40*, 1–10. [CrossRef]

44. Ahmad, M.; Khan, A.; Mazzara, M.; Distefano, S. Multi-layer extreme learning machine-based autoencoder for hyperspectral image classification. In Proceedings of the 14th International Joint Conference on Computer Vision, Imaging and Computer Graphics Theory and Applications, Prague, Czech Republic, 25–27 February 2019; SCITEPRESS—Science and Technology Publications: Setúbal, Portugal, 2019; pp. 75–82.

45. Ahmad, M.; Khan, A.; Khan, A.M.; Mazzara, M.; Distefano, S.; Sohaib, A.; Nibouche, O. Spatial prior fuzziness pool-based interactive classification of hyperspectral images. *Remote Sens.* **2019**, *11*, 1136. [CrossRef]

46. Somogyi, Z.; Cienciala, E.; Mäkipää, R.; Muukkonen, P.; Lehtonen, A.; Weiss, P. Indirect methods of large-scale forest biomass estimation. *Eur. J. For. Res.* **2007**, *126*, 197–207. [CrossRef]

47. Chave, J.; Andalo, C.; Brown, S.; Cairns, M.A.; Chambers, J.Q.; Eamus, D.; Fölster, H.; Fromard, F.; Higuchi, N.; Kira, T.; et al. Tree allometry and improved estimation of carbon stocks and balance in tropical forests. *Oecologia* **2005**, *145*, 87–99. [CrossRef]

48. Pastor, J.; Aber, J.D.; Melillo, J.M. Biomass prediction using generalized allometric regressions for some northeast tree species. *For. Ecol. Manag.* **1984**, *7*, 265–274. [CrossRef]

49. Haase, R.; Haase, P. Above-ground biomass estimates for invasive trees and shrubs in the Pantanal of Mato Grosso, Brazil. *For. Ecol. Manag.* **1995**, *73*, 29–35. [CrossRef]

50. Kuyah, S.; Dietz, J.; Muthuri, C.; van Noordwijk, M.; Neufeldt, H. Allometry and partitioning of above- and below-ground biomass in farmed eucalyptus species dominant in Western Kenyan agricultural landscapes. *Biomass Bioenergy* **2013**, *55*, 276–284. [CrossRef]

51. Cushman, K.C.; Muller-Landau, H.C.; Condit, R.S.; Hubbell, S.P. Improving estimates of biomass change in buttressed trees using tree taper models. *Methods Ecol. Evol.* **2014**, *5*, 573–582. [CrossRef]

52. Mosseler, A.; Major, J.E.; Labrecque, M.; Larocque, G.R. Allometric relationships in coppice biomass production for two North American willows (*Salix* spp.) across three different sites. *For. Ecol. Manag.* **2014**, *20*, 190–196. [CrossRef]

53. Zhao, H.; Li, Z.; Zhou, G.; Qiu, Z.; Wu, Z. Site-specific allometric models for prediction of above-and belowground biomass of subtropical forests in Guangzhou, southern China. *Forests* **2019**, *10*, 862. [CrossRef]

54. Zianis, D.; Muukkonen, P.; Mäkipää, R.; Mencuccini, M. *Biomass and Stem Volume Equations for Tree Species in Europe*; Tammer-Paino Oy: Tampere, Finland, 2005.

55. Piao, S.; Fang, J.; He, J.; Xiao, Y. Spatial distribution of grassland biomass in China. *Acta Phytocol. Sin.* **2004**, *28*, 491–498.

56. Draper, N.R.; Smith, H. *Applied Regression Analysis*, 3rd ed.; John Wiley & Sons, Inc.: Hoboken, NJ, USA, 1998; ISBN 9781118625590.

57. Li, L.; Bakelants, L.; Solana, C.; Canters, F.; Kervyn, M. Dating lava flows of tropical volcanoes by means of spatial modeling of vegetation recovery. *Earth Surf. Process. Landf.* **2018**, *43*, 840–856. [CrossRef]

58. Whittingham, M.J.; Stephens, P.A.; Bradbury, R.B.; Freckleton, R.P. Why do we still use stepwise modelling in ecology and behaviour? *J. Anim. Ecol.* **2006**, *75*, 1182–1189. [CrossRef]

59. Yang, X.; Li, L.; Chen, L.; Chen, L.; Shen, Z. Improving ASTER GDEM accuracy using land use-based linear regression methods: A case study of Lianyungang, East China. *ISPRS Int. J. Geo-Inf.* **2018**, *7*, 145. [CrossRef]

60. Li, Y.; Li, C.; Li, M.; Liu, Z. Influence of variable selection and forest type on forest aboveground biomass estimation using machine learning algorithms. *Forests* **2019**, *10*, 1073. [CrossRef]

61. Cheng, L.; Li, L.; Chen, L.; Hu, S.; Yuan, L.; Liu, Y.; Cui, Y.; Zhang, T. Spatiotemporal variability and influencing factors of aerosol optical depth over the Pan Yangtze River Delta during the 2014–2017 period. *Int. J. Environ. Res. Public Health* **2019**, *16*, 3522. [CrossRef]

62. Hair, J.F.; Black, W.C.; Babin, B.J.; Anderson, R.E. *Multivariate Data Analysis*, 7th ed.; Pearson: New Jersey, NJ, USA, 2009; ISBN 9780138132637.

63. Sauerbrei, W.; Royston, P.; Binder, H. Selection of important variables and determination of functional form for continuous predictors in multivariable model building. *Stat. Med.* **2007**, *26*, 5512–5528. [CrossRef] [PubMed]

64. Shaheen, A.; Iqbal, J. Spatial distribution and mobility assessment of carcinogenic heavy metals in soil profiles using geostatistics and random forest, Boruta algorithm. *Sustainability* **2018**, *10*, 799. [CrossRef]

65. Kursa, M.B.; Rudnicki, W.R. Feature selection with the Boruta package. *J. Stat. Softw.* **2010**, *36*, 1–13. [CrossRef]

66. Liaw, A.; Wiener, M. Classification and Regression by randomForest. *R News* **2002**, *2*, 18–22.

67. Lu, N.; Zhou, J.; Han, Z.; Li, D.; Cao, Q.; Yao, X.; Tian, Y.; Zhu, Y.; Cao, W.; Cheng, T. Improved estimation of aboveground biomass in wheat from RGB imagery and point cloud data acquired with a low-cost unmanned aerial vehicle system. *Plant Methods* **2019**, *15*, 17. [CrossRef]

68. Gao, Y.; Lu, D.; Li, G.; Wang, G.; Chen, Q.; Liu, L.; Li, D. Comparative analysis of modeling algorithms for forest aboveground biomass estimation in a subtropical region. *Remote Sens.* **2018**, *10*, 627. [CrossRef]

69. Rocha de Souza Pereira, F.; Kampel, M.; Gomes Soares, M.; Estrada, G.; Bentz, C.; Vincent, G. Reducing uncertainty in mapping of mangrove aboveground biomass using airborne discrete return Lidar data. *Remote Sens.* **2018**, *10*, 637. [CrossRef]

70. Huete, A. Vegetation indices. In *Encyclopedia of Remote Sensing*; Njoku, E.G., Ed.; Springer: New York, NY, USA, 2014; pp. 883–886.

71. Li, C.; Zhou, W.; Guan, Q.; Wei, W.; Dong, P.; Zhang, H. Biomass and its influencing factors of Platyclatdus orientalis plantation in the limestone mountains. *J. Anhui Agric. Univ.* **2010**, *37*, 669–674.

72. Lu, Y.; Liang, Z.; Wu, Z.; Cai, Y.; Zhou, K.; Yang, G.; Yin, Z. Biomass and productivity of main afforestation tree species on the seawall in Northern Jiangsu. *J. Jiangsu For. Sci. Technol.* **2000**, *27*, 12–15.

73. Zhu, Y. Characteristics of Structure and Carbon Storage of Greening on the Campus of Anhui Agricultural University. Master's Thesis, Anhui Agricultural University, Hefei, China, 2016.

74. Li, J.; Li, C.; Peng, S. Study on the biomass expansion factor of poplar plantation. *J. Nanjing For. Univ.* **2007**, *31*, 37–40.

75. Kun, L.; Cao, L.; Wang, G.; Cao, F. Biomass allocation patterns and allometric models of Ginkgo biloba. *J. Beijing For. Univ.* **2017**, *39*, 12–20.

76. Wen, J. Effects of Urbanization on Carbon Storage and Sequestration in the Built-Up Area. Master's Thesis, Zhejiang University, Hangzhou, China, 2010.

77. Che, R. Study on single tree biomass model for Larix Principis-rupprechtii. *Shanxi For. Sci. Technol.* **2017**, *46*, 35–36.

78. State Forestry Administration of China. *Carbon Accounting and Monitoring Guide for Afforestation Projects*; China Forestry Press: Beijing, China, 2014; ISBN 9787503873676.

79. Zhang, X.; Leng, H.; Zhao, G.; Jing, J.; Tu, A.; Song, K.; Da, L. Allometric models for estimating aboveground biomass for four common greening tree species in Shanghai City, China. *J. Nanjing For. Univ. (Nat. Sci. Ed.)* **2018**, *42*, 141–146.

80. Huang, T.; Zhong, Q.; Peng, X. Sutdy on biomass and productivity of Liriodendron chinense plantation. *For. Sci. Technol.* **2000**, *9*, 12–15.

81. Yang, X.; Wu, G.; Huang, D.; Yang, C. Quantitative study on biomass accumulation of Paulownia. *Chin. J. Appl. Ecol.* **1999**, *10*, 143–146.

82. Yao, Z.; Liu, J.; Zhao, X.; Long, D.; Wang, L. Spatial dynamics of aboveground carbon stock in urban green space: A case study of Xi'an, China. *J. Arid Land* **2015**, *7*, 350–360. [CrossRef]

83. Yao, Z.; Liu, J. Models for biomass estimation of four shrub species planted in urban area of Xi'an City, Northwest China. *Chin. J. Appl. Ecol.* **2014**, *25*, 111–116. [CrossRef]

 *forests*

Article

Estimation of Forest Biomass in Beijing (China) Using Multisource Remote Sensing and Forest Inventory Data

Yan Zhu [1], Zhongke Feng [1,*], Jing Lu [1] and Jincheng Liu [2]

1   Precision Forestry Key Laboratory of Beijing, Beijing Forestry University, Beijing 100083, China;
    zybjlydx@126.com (Y.Z.); lujing95831@163.com (J.L.)
2   College of Natural Resources and Environment, Northwest A&F University, Yangling 712100, Shanxi, China;
    jinchengl@nwafu.edu.cn
*   Correspondence: zhongkefeng@bjfu.edu.cn; Tel.: +86-138-1030-5579

Received: 2 December 2019; Accepted: 28 January 2020; Published: 31 January 2020

Abstract: Forest biomass reflects the material cycle of forest ecosystems and is an important index to measure changes in forest structure and function. The accurate estimation of forest biomass is the research basis for measuring carbon storage in forest systems, and it is important to better understand the carbon cycle and improve the efficiency of forest policy and management activities. In this study, to achieve an accurate estimation of meso-scale (regional) forest biomass, we used Ninth Beijing Forest Inventory data (FID), Landsat 8 OLI Image data and ALOS-2 PALSAR-2 data to establish different forest types (coniferous forest, mixed forest, and broadleaf forest) of biomass models in Beijing. We assessed the potential of forest inventory, optical (Landsat 8 OLI) and radar (ALOS-2 PALSAR-2) data in estimating and mapping forest biomass. From these data, a wide range of parameters related to forest structure were obtained. Random forest (RF) models were established using these parameters and compared with traditional multiple linear regression (MLR) models. Forest biomass in Beijing was then estimated. The results showed the following: (1) forest inventory data combined with multisource remote sensing data can better fit forest biomass than forest inventory data alone. Among the three forest types, mixed forest has the best fitting model. Forest inventory variables and multisource remote sensing variables can match each other in time and space, capturing almost all spatial variability. (2) The 2016 forest biomass density in Beijing was estimated to be 52.26 Mg ha^{-1} and ranged from 19.1381–195.66 Mg ha^{-1}. The areas with high biomass were mainly distributed in the north and southwest of Beijing, while the areas with low biomass were mainly distributed in the southeast and central areas of Beijing. (3) The estimates from the RF model are better than those from the MLR model, showing a high R^2 and a low root mean square error (RMSE). The R^2 values of the MLR models of three forest types were greater than 0.5, and RMSEs were less than 15.5 Mg ha^{-1}, The R^2 values of the RF models were higher than 0.6, and the RMSEs were lower than 13.5 Mg ha^{-1}. We conclude that the methods in this paper can help improve the accurate estimation of regional biomass and provide a basis for the planning of relevant forestry decision-making departments.

Keywords: forest biomass estimation; forest inventory data; multisource remote sensing; random forest; biomass density

1. Introduction

Forest ecosystems are an important component of the terrestrial ecosystem. Forests store 76%~98% of the organic carbon in terrestrial ecosystems [1] and play an irreplaceable role in mitigating global warming caused by the increase in atmospheric carbon dioxide [2]. Forest biomass reflects the material cycle of forest ecosystems and is an important indicator for measuring changes in forest structure and function. Additionally, forest biomass is closely related to the carbon sources and sinks in forest ecosystems [3]. Because the monitoring of forest biomass resources is expensive and time consuming, most countries do not have effective monitoring systems. Therefore, accurate estimations of forest biomass can effectively replace forest monitoring systems and are an important basis for assessing ecosystem processes, the carbon balance of ecosystems and climate change [4].

Meso-scale forest biomass estimations are usually obtained from forest inventory data [5]. In many countries, the use of large-scale forest inventories is considered an effective method for estimating biomass accurately [6]. China conducts a large-scale forest resource survey every five years to provide good data for statistical forest resources. Using these inventory data, forest biomass can be estimated at provincial or national scales [7]. However, with the continuous change in the forest resource structure of China, there have been some problems with these inventory data in regional biomass estimation [8]. To obtain the total volume or biomass of forest, the volume or biomass of one tree is calculated, and then the volumes or biomasses of all the trees in the sample plot are added together. Obviously, this method requires a high amount of manpower and material resources [9]. Moreover, inventory data cannot fully reflect forest information [6,10]. Therefore, we need data that cover a wide area and contain a high amount of vegetation information to supplement forest inventory data.

With developments in technology, remote sensing has increased the possibilities for forest biomass research [11,12]. The use of remote sensing data in the research of meso-scale biomass is an important technical method. Various remote sensing indicators based on optical sensors, such as the normalized difference vegetation index (NDVI) and other factors obtained by image transformations, have been shown to be well correlated with the ground vegetation, providing reliable information for forest biomass estimation [13–15]. However, applications with optical data are often limited due to the complexity of biomass in time and space and limitations in the spatial and spectral characteristics of satellite data [16]. More abundant remote sensing data are needed to depict detailed forest information. Lidar can penetrate dense forests, provide accurate three-dimensional information of trees, and then be used to obtain forest biomass [17]. However, because of its limited coverage, high cost and inconvenience to transport, Lidar is not suitable for forest biomass estimation at the meso-scale [18]. Synthetic Aperture Radar (SAR) data, such as L-band Advanced Land Observing Satellite/Phased Array L-band Synthetic Aperture Radar (ALOS/PALSAR) [19] and X-band TerraSAR-X data, are widely used in the estimation of forest biomass [20,21]. SAR is not affected by illumination and climate conditions and it can penetrate vegetation to obtain information, covering relatively large areas in a short period of time [20].

At the meso-scale, many studies have demonstrated the potential of optical and radar remote sensing-derived indicators to estimate forest biomass [22,23]. However, there is a large range and many uncertainties of remote sensing. For example, the resolution of remote sensing images might be insufficient, and the vertical structure information of forest canopies cannot be obtained, which has certain limitations in high biomass areas. Therefore, at the regional scale, the accuracy of forest biomass estimation using remote sensing data is low [24].

Optical and radar remote sensing data can match forest inventory data in time and space [25]. In addition, these data can provide forest attributes and structural information that are missing from inventory data. Therefore, combining multisource remote sensing data with forest inventory data for regional forest biomass research provides a more consistent spatial and temporal analyses than forest inventory data alone. Generally, the uncertainty in the estimation can be reduced by this combination [25]. Furthermore, this combination can promote the application of forest biomass estimation and other ecological research at the meso-scale. However, there is little information on the potential of the combination of sample plot survey data with multisource remote sensing data to estimate and map biomass. Most forest biomass estimation studies focus on the impact of environmental variables on forest biomass [26]. Therefore, it is necessary to better assess and understand the modeling potential of sample survey factors and remote sensing factors to provide decision makers with information on forest resources.

In this study, by synthesizing the existing technical means, we combined forest inventory data with multisource remote sensing data to estimate forest biomass and improve the accuracy of biomass estimation at the regional scale. There are three objectives of the present study: the primary objective is to assess the potential of forest inventory data combined with multisource remote sensing data in modeling and mapping forest biomass. The second objective is to estimate the biomass of different forest types in Beijing in 2016 and provide data support for regional biomass estimation. The third objective is to estimate biomass using multiple linear regression (MLR) and the random forest (RF) model and compare the performances of the two models.

2. Materials and Methods

2.1. Data Collection

2.1.1. Forest Inventory Data

The National Forest Resources Continuous Inventory system is a method of forest resource investigation that aims to understand the status and dynamics of macroforest resources and periodically reviews both with fixed sample plots. It is an important part of the comprehensive monitoring system for forest resources and ecological conditions in China. China's Ministry of Forestry has carried out eight consecutive national surveys and inventories of forest resources [27]. According to the technical regulations of the national forest resources continuous inventory, systematic sampling is used to lay out fixed sample plots, the size of which is 4 km × 4 km, and the sample plots are laid out at the intersection point of the kilometer network of the newly compiled 50,000 or 100,000 topographic map of the country. To ensure that the sample points are not repeated and missed, computer technology such as GIS, is used as far as possible [28]. In recent years, the collection of forest inventory data depends primarily on manual work, and it is supported by high-tech survey instruments that can automatically collect data to improve the accuracy of inventory results [29].

In this study, we used the Ninth Beijing Forest Inventory data of 2016, which involve 1431 sample plots located in all districts and counties of Beijing, as shown in Figure 1, covering coniferous, broadleaf and mixed coniferous-broadleaf forest types. The dataset describes in detail plot locations, measurement dates and forest compositions. For each plot, multiple attributes were collected, including the mean diameter at breast height (DBH), mean tree height, mean age, crown density, volume, land use and cover, and ecological conditions. The biomass, mean DBH, mean tree height, mean age and crown density were used as the inventory variables to establish the model, as shown in Table 1. The real biomass value was calculated using the equation for the biomass-volume relationship of the stand type and age group [30]. The stem volume of each tree was provided by FID, and the stand volume of each fixed sample plot was the sum of all tree volumes. The area of each plot was 0.0667 hectares.

Table 1. Statistics of the main forest inventory dataset (plot number and biomass of the three forest types in Beijing, China).

Forest Types	Coniferous Forest (n_1 = 663)			Mixed Forest (n_2 = 272)			Broadleaf Forest (n_3 = 496)		
	Max	Min	Mean	Max	Min	Mean	Max	Min	Mean
Biomass (Mg ha^{-1})	260.45	5.08	51.83	170.45	21.63	51.78	138.23	11.82	51.71
Mean H(m)	32.40	1.00	7.42	27.7	1.50	7.41	22.80	1.50	7.41
Mean DBH (cm)	51.00	1.00	12.8	40.60	3.50	12.8	42.50	3.00	12.8
Mean age	115	1	25	100	3	24	80	3	24
Crown density	90	20	48	90	20	48	90	20	48

There are significant differences in topography and administrative functions among different districts in Beijing. From the topographic point of view, the northwest is a mountainous area with a higher terrain, and the southeast is a plain; from the administrative function point of view, the central part of Beijing is the capital and core functional area, the Northwest Mountainous Area and the southwest are ecological conservation functional areas, and the plain is a densely populated scientific and technological innovation and economic development area, which also leads to differences in forest biomass distribution. More than 80% of Beijing's forest resources are distributed in mountainous counties in the west and north of the city. The forest coverage in mountainous areas of Beijing has reached more than 50%, but the forest area in the plain of southeast Beijing accounts for less than 20% of the whole city.

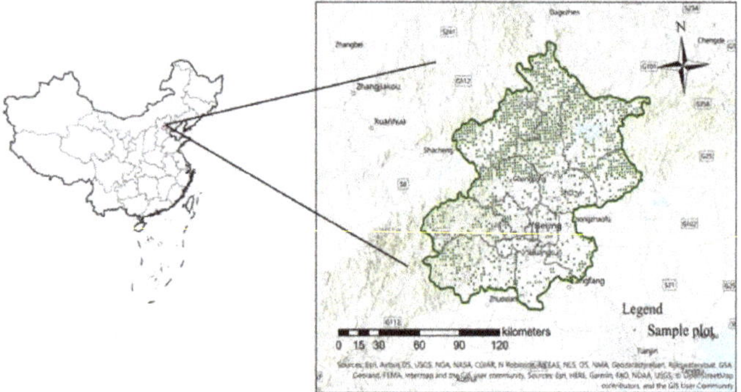

Figure 1. Spatial distribution map of the forest sample plots in Beijing.

The forest ecosystem contains arbors, shrubs and herbs, but the amount of biomass from shrubs and herbs is less than the amount of arboreal biomass [31]. Therefore, this study considers only arboreal biomass and does not consider shrub and herb biomass.

2.1.2. Remote Sensing Data and Preprocessing

We used optical (Landsat 8 OLI) and radar (ALOS-2 PALSAR-2) remote sensing sources.

Landsat 8

Landsat 8 Operational Land Imager (OLI, which developed by Bauer Aerospace and Technology Corp, Colorado, USA) images included a 15 m panchromatic band, with a spectral range from 0.500 to 0.680 μm and eight 30 m multispectral bands, with a spectral range from 0.433 to 2.300 μm. They were selected for biomass estimation due to their suitability in terms of their resolution ratio; a spatial resolution of approximately 30 m by 30 m is adequate to assess information at the forest stand level. We selected eight Landsat 8 OLI scenes of Beijing with low cloud cover as the research images. The image range was 122 to 124 paths and 31 to 33 rows. The image acquisition time used in this study was June–August 2016, and the time phase was basically the same as the time phase of the Beijing forest inventory data.

The image preprocessing steps included geometric correction, radiation correction, atmospheric correction, and image clipping. Because the downloaded images were Level-1 data products, the geometric accuracy was high, so only radiation and atmospheric corrections were needed.

Based on the preprocessed Landsat 8 OLI data, we acquired the surface reflectance for 6 bands of the Landsat 8 OLI (Band 2-Band 7) and then acquired the vegetation indices, namely, normalized difference vegetation index (NDVI), difference vegetation index (DVI) and ratio vegetation index (RVI) (Table 2), through band processing and the Landsat 8 OLI image calculation. The NDVI, DVI, and RVI are commonly used vegetation indices that are sensitive to vegetation, as shown in Equations (1)–(3). Another dataset was derived by image transformation from the original satellite band, which involved tasseled cap transformation (TCT) and texture features as shown in Table 2. Tasseled cap transformation, also known as a K-T transform, is an image enhancement method for vegetation information extraction. It can enhance vegetation information of images. After the K-T transform, the same number of components as the number of bands can be obtained, and the second component is the green index, which has a strong relationship with the vegetation coverage and biomass on the ground [32]. Therefore, based on the TCT coefficients of the OLI sensor onboard Landsat 8, we chose the second band generated from the TCT, which was marked as the greenness [32].

Table 2. Remote sensing factors calculated from the Landsat 8 and ALOS-2/PALSAR-2 images.

Factor Type	Remote Sensing Factors	Data Source
Band value	Band 2, Band 3, Band 4, Band 5, Band 7	Landsat 8 OLI
Vegetation index	NDVI, DVI, RVI	Landsat 8 OLI
Tasseled cap transformation	Greenness	Landsat 8 OLI
Texture analysis	Mean, Variance, Contrast, Correlation, Second moment	Landsat 8 OLI
Backscattering coefficients	$\Gamma^0_{HH}, \Gamma^0_{Hv}, \Gamma^0_{HH}+\Gamma^0_{Hv}, \Gamma^0_{HH}-\Gamma^0_{Hv}, \Gamma^0_{HH}/\Gamma^0_{Hv}$	ALOS-2/PALSAR-2

In addition, we extracted the texture factor of the image. Texture is an important feature of remote sensing images and can be extracted by using the gray-level cooccurrence matrix (GLCM). Previous research has shown that Band 2 of a Landsat 8 OLI image contains much information about the image; thus, we extracted the texture feature of Band 2. The larger the selected window is, the greater the information content will be [33]. According to the sample area, five texture eigenvalues were extracted from the 15×15 window, namely, the mean, variance, contrast, correlation and second moment [34], as shown in Equations (4)–(8):

Normalized Difference Vegetation Index (NDVI):

$$\text{NDVI} = \frac{\text{NIR1} - R}{\text{NIR1} + R} \tag{1}$$

Difference Vegetation Index (DVI):

$$\text{RVI} = \frac{\text{NIR1}}{R} \tag{2}$$

Ratio Vegetation Index (RVI):

$$\text{DVI} = \text{NIR1} - R \tag{3}$$

Mean(ME):

$$\text{ME} = \sum_{i,j=0}^{N-1} iP_{ij} \tag{4}$$

Variance(VA):

$$\text{VA} = \sum_{i,j=0}^{N-1} P_{ij}(i - \text{ME})^2 \tag{5}$$

Contrast(CO):

$$\text{CO} = \sum_{i,j=0}^{N-1} iP_{ij}(i - j)^2 \tag{6}$$

Correlation (CC):

$$\text{CC} = \sum_{i,j=0}^{N-1} iP_{ij} \left| \frac{(i - \text{ME})(j - \text{ME})}{\sqrt{VA_i VA_j}} \right| \tag{7}$$

Second Moment(SM):

$$\text{SM} = \sum_{i,j=0}^{N-1} iP_{ij}^2 \tag{8}$$

Using bilinear interpolation, the average values of the remote sensing factors at and near sampling points can be extracted. This method effectively solves the problem that occurs when sampling sites do not match the image completely and can cover areas that the inventory data cannot fully cover, thus improving the estimation accuracy.

ALOS-2/PALSAR-2

We downloaded six images of ALOS-2/PALSAR-2 (L-band) taken in 2016 from Japan Aerospace Exploration Agency (JAXA (http://www.eorc.jaxa.jp/ALOS/en/index.htm)). The PALSAR data had a 25-m spatial resolution and contain two polarized bands, HH and HV. The preprocessing of PALSAR data was completed by the JAXA. The digital numbers (DN) of the PALSAR signal amplitude were extracted and converted to gamma naught backscattering coefficients (dB) in decimal units using the following equation [35,36]:

$$\Gamma^0 = 10 \times log_{10}DN^2 - CF \tag{9}$$

where Γ^0 is the backscattering coefficient, DN is the digital number value of pixels, and CF is the calibration factor, which equals -83 [36]. Then, we calculated the sum, difference and ratio values using the backscattering coefficients of HH and HV, as shown in Equations (10)–(12):

$$sum = \Gamma^0_{HH} + \Gamma^0_{Hv} \tag{10}$$

$$difference = \Gamma^0_{HH} - \Gamma^0_{Hv} \tag{11}$$

$$ratio = \Gamma^0_{HH}/\Gamma^0_{Hv'} \tag{12}$$

where Γ^0_{HH} and Γ^0_{Hv} are the backscattering coefficients of HH and HV in decibels.

2.2. Multiple Regression Model

The allometric growth equation is the most widely used model for estimating forest biomass. Many studies have confirmed the advantages of the allometric growth equation for estimating forest biomass [37–39]. This model regresses a correlated variable (biomass) based on one or more independent variables. The DBH and tree height, as the two most relevant factors of biomass, are often used in biomass prediction in the form of single or compound variables. Based on the allometric model and previous research results, we introduced new variables (Landsat 8 data and backscattering coefficients) into the model to explore its ability to estimate forest biomass, as shown in Equation (13).

$$\ln(B) = \beta_0 + a\ln(d^2H) + \beta_1x_1 + \beta_2x_2 + \ldots + \beta_jx_j \tag{13}$$

where B is the biomass of the sample plot, each x_j is an independent variable ($j = 1, 2, 3\ldots$), β_j is the regression coefficient of x_j, β_0 is a constant, and a is the regression coefficient of the model.

However, the remote sensing variables were highly collinear. To overcome this problem, we used a stepwise regression analysis method, which gradually screens variables and leaves highly correlated variables that are not collinear in the model, to retain a model that was not very complex and to reduce the number of calculations. The basic idea of stepwise regression is to introduce variables into the model one by one. After introducing an explanatory variable, we need to conduct F-test and t-test for the selected explanatory variables one by one. When the original explanatory variables are no longer significant due to the introduction of later explanatory variables, they will be deleted. To ensure that only significant variables are included in the regression equation before each new variable is introduced. This is a repeated process until neither significant explanatory variables are selected into the regression equation nor insignificant explanatory variables are removed from the regression equation. To ensure that the final set of explanatory variables is optimal.

2.3. Feature Selection and Random Forest Model

We used R to establish an RF model to estimate forest biomass. The RF model was a classification and regression algorithm based on decision trees [40]. By establishing and combining multiple decision tree predictions (1000 trees in our study), the average value of all the decision tree prediction results was taken as the final prediction result [41]. The RF model can effectively alleviate the problem of overfitting and is insensitive to the collinearity between variables, so it is suitable for establishing a nonlinear model [42]. RF is increasingly used to perform biomass regression and estimate forest biomass [43,44].

First, subsets of variables were selected as input for the RF prediction using feature selection to ensure that the input variables were highly correlated with biomass. Feature selection refers to the selection of subsets from the original feature set to optimize a certain evaluation criterion so that the model established with the optimal feature subset can achieve a prediction accuracy similar to or better than that of the model established without feature selection. RF provides an increase in the mean-squared error (percentage of IncMSE, where IncMSE indicates the increase in MSE) for each independent indicator, quantifying the increase in the MSE when the indicator is randomly permuted. This error measures the relative importance of each indicator, where a high IncMSE implies that the indicator has a high weight in the model prediction and vice versa [23]. Then, we used the data after feature selection as the independent variable, forest biomass as the dependent variable, and the random forest software package in R to establish an RF model.

2.4. Model Accuracy Evaluation

To test these models, we assessed the prediction accuracy on randomly selected subsets (20%) of the original dataset retained before the model was developed. To evaluate the advantage of the use of an advanced regression tree model versus more traditional approaches, the performance of the RF model was computed and compared with that of a stepwise multiple linear regression model.

We used the proportion of variance explained (R^2) and the root mean square error (RMSE) to evaluate the model performance on the complete datasets. In addition, we computed the relative RMSE (RMSE%), the bias and the relative bias (bias%). Bias was calculated as the difference between a population mean of the measurements or test results and an accepted reference or true value, R^2 values were used to judge the model, and RMSE, Bias%, RMSE% reflect the precision of the model [45].

These statistics were calculated as follows:

$$R^2 = 1 - \frac{\Sigma(y_i - \hat{y}_i)^2}{\Sigma\left(y_i - \overline{y}_i\right)^2} \tag{14}$$

$$RMSE = \sqrt{\frac{\Sigma(y_i - \hat{y}_i)^2}{n-1}} \tag{15}$$

$$RMSE\% = \frac{RMSE}{\overline{y}_i} \times 100\% \tag{16}$$

$$Biasc = \frac{1}{n}\sum_{i=1}^{n}(y_i - \hat{y}_i) \tag{17}$$

$$Bias\% = \frac{BIAS}{\overline{y}_i} \times 100\% \tag{18}$$

where y_i is the observed biomass of the plot, $\hat{y}_i$ is the predicted biomass of the plot, and $\overline{y}_i$ is the mean biomass of n plots.

3. Results

3.1. Univariate Correlation Analysis

Previous studies have typically analyzed the relationship between a single remote sensing variable and the forest biomass or have used the original band and variables transformed from images for feature selection [46,47]. Through the Pearson correlation coefficient (r), we analyzed the ability of each variable to estimate biomass and obtained the correlation between each variable and the biomass, as shown in Table 3.

Table 3. Coefficients of correlation between forest biomass and variables.

	Variables	Code	Correlation (r)		
			Coniferous Forest	**Mixed Forest**	**Broadleaf Forest**
FID	d^2H	N1	0.492	0.510	0.489
	Crown density	N2	0.373	0.451	0.401
	Mean age	N3	0.455	0.412	0.359
Original bands	B2	X1	0.359	0.290	0.367
	B3	X2	0.118	0.257	0.213
	B4	X3	0.219	0.197	0.156
	B5	X4	0.237	0.211	0.181
	B7	X5	0.235	0.243	0.221
Vegetation index	NDVI	X6	0.336	0.332	0.317
	DVI	X7	0.105	0.116	0.132
	RVI	X8	0.089	0.124	0.098
Tasseled cap	Greenness	X9	0.239	0.159	0.224
Texture (15 × 15)	Mean	X10	0.196	0.195	0.174
	Variance	X11	0.089	0.077	0.103
	Contrast	X12	0.130	0.114	0.082
	Correlation	X13	0.138	0.211	0.243
	Second moment	X14	0.145	0.056	0.097
Backscattering coefficients	Γ^0_{HH}	X15	0.135	0.125	0.126
	Γ^0_{Hv}	X16	0.165	0.187	0.173
	$\Gamma^0_{HH} + \Gamma^0_{Hv}$	X17	0.148	0.099	0.106
	$\Gamma^0_{HH} - \Gamma^0_{Hv}$	X18	0.132	0.071	0.121
	$\Gamma^0_{HH} / \Gamma^0_{Hv}$	X19	0.127	0.126	0.187

Among all variables, forest inventory variables were highly correlated with biomass in three forest types. Different remote sensing variables (OLI data and PALSAR data) showed different degrees of correlation. The shortwave infrared (SWIR) optical band (Band 7) showed the greatest relevant biomass among the Landsat data because it allowed an effective separation between high- and low-biomass data. The importance of the SWIR wavelengths in biomass prediction is consistent with previous studies [40]. In addition, Band 5, Band 4, Band 3, Band 2, the NDVI, and the greenness were also highly correlated with biomass, and the most relevant texture factors were the mean, correlation and second moment. The other Landsat variables had little correlation with biomass. SAR data can penetrate dense forests and obtain the vertical structure information of forests, so the PALSAR HH and HV backscatter coefficients and their derivative variables (sum, difference, ratio) were correlated with forest biomass. In addition, correlations between the forest biomass and HV backscatter coefficients of different forest types were higher than those between the forest biomass and HH backscatter coefficients, which is in line with previous research results [22,48]. All these factors can be considered potential variables for forest biomass estimation.

3.2. Results of Forest Biomass Model Establishment

3.2.1. Multiple Stepwise Regression Model

To avoid overfitting, the multiple stepwise regression method was used to screen variables and establish a multiple linear model. The results are as follows:

The multiple stepwise regression model of coniferous forests was:

$$\ln(B_C) = 3.821 + 0.226 \times N_1 + 0.111N_2 + 0.139N_3 - 0.120X_1 - 0.246X_6 - 0.089X_{10} - 0.656X_{16} + 0.538X_{17} - 0.308X_{19} \qquad (19)$$

The multiple stepwise regression model of mixed forest was:

$$\ln(B_M) = 3.776 + 0.291 \times N_1 + 0.155N_2 + 0.108N_3 - 0.153X_1 + 0.302X_2 - 0.132X_3 - 0.05X_4 + 0.052X_6 + 0.037X_{13} \qquad (20)$$

The multiple stepwise regression model of broadleaf forests was:

$$\ln(B_B) = 3.810 + 0.127 \times N_1 + 0.137N_2 + 0.181N_3 + 0.110X_1 + 0.039X_6 - 0.088X_9 + 0.024X_{13} + 0.043X_{15} - 0.057X_{18} \quad (21)$$

Then, the biomass estimation model was obtained as follows:

$$B_C = e^{\begin{array}{l} 3.821 + 0.226 \times N_1 + 0.111N_2 + 0.139N_3 - 0.120X_1 \\ -0.246X_6 - 0.089X_{10} - 0.656X_{16} + 0.538X_{17} - 0.308X_{19} \end{array}} \quad (22)$$

$$B_M = e^{\begin{array}{l} 3.776 + 0.291 \times N_1 + 0.155N_2 + 0.108N_3 - 0.153X_1 \\ +0.302X_2 - 0.132X_3 - 0.05X_4 + 0.052X_6 + 0.037X_{13} \end{array}} \quad (23)$$

$$B_B = e^{\begin{array}{l} 3.810 + 0.127 \times N_1 + 0.137N_2 + 0.181N_3 + 0.110X_1 \\ +0.039X_6 - 0.088X_9 + 0.024X_{13} + 0.043X_{15} - 0.057X_{18} \end{array}} \quad (24)$$

where X_i and N_i in each formula correspond to the variables in Table 3.

P values represent the probability that the sample results differ from the original hypothesis. The smaller the P value is, the more significant the results are. Generally speaking, $p < 0.05$ indicates a significant difference, and $p < 0.01$ indicates a very significant difference. Table 4 shows that the P values of the model coefficients of different forest types, which are less than 0.05, some of which are less than 0.01. These results show that the differences in the selected variables are significant.

Table 4. P values of the coefficients of models for different forest types.

Forest Type	Variables	*p*-Value
Coniferous forest	N1	0.001
	N2	0.004
	N3	0.010
	X1	0.011
	X6	0.020
	X10	0.010
	X16	0.015
	X17	0.009
	X19	0.022
Mixed forest	N1	0.002
	N2	0.001
	N3	0.004
	X1	0.016
	X2	0.011
	X3	0.002
	X4	0.015
	X6	0.008
	X13	0.037
Broadleaf forest	N1	0.005
	N2	0.003
	N3	0.010
	X1	0.033
	X6	0.02
	X9	0.01
	X13	0.025
	X15	0.019
	X18	0.002

3.2.2. Random Forest Model

First, feature variables were selected for the variables involved in the modeling. The importance of the variables was ranked according to IncMSE%, and the unimportant variables were eliminated. Generally, the number of final variables is 1/3 of the total number of input variables [49]. Table 5 shows that d^2H and the mean age are two very important variables in the RF model. IncMSE% was more than 20% in the different forest types. NDVI, Band 2 and Band 7 were also important to the model with regard to optical data. The most influential backscattering coefficient factors were Γ^0_{Hv} and $\Gamma^0_{HH} - \Gamma^0_{Hv}$. The IncMSE% of these factors were all higher than 10%.

Table 5. The IncMSE% of the top five most important variables in the biomass fitting of different forest types in the random forest model.

Forest Type	Variables	IncMSE%
	d^2H	40.82
	The mean age	19.83
Coniferous forest	Band 2	15.39
	NDVI	12.36
	Γ^0_{Hv}	11.15
	The mean age	32.28
	d^2H	25.52
Mixed forest	$\Gamma^0_{HH} - \Gamma^0_{Hv}$	18.48
	$Band2_{mean}$	14.23
	Crown density	11.88
	d^2H	43.46
	Crown density	23.02
Broadleaf forest	Band 7	12.77
	Γ^0_{Hv}	11.70
	NDVI	10.08

3.3. Model Precision Evaluation and Comparison of Two Models

To test the goodness of fit of the model, 20% of the samples were used for validation. We analyzed the scatter diagrams of different forest types in Figure 2 and obtained the accuracy of the linear regression and RF models in Table 6.

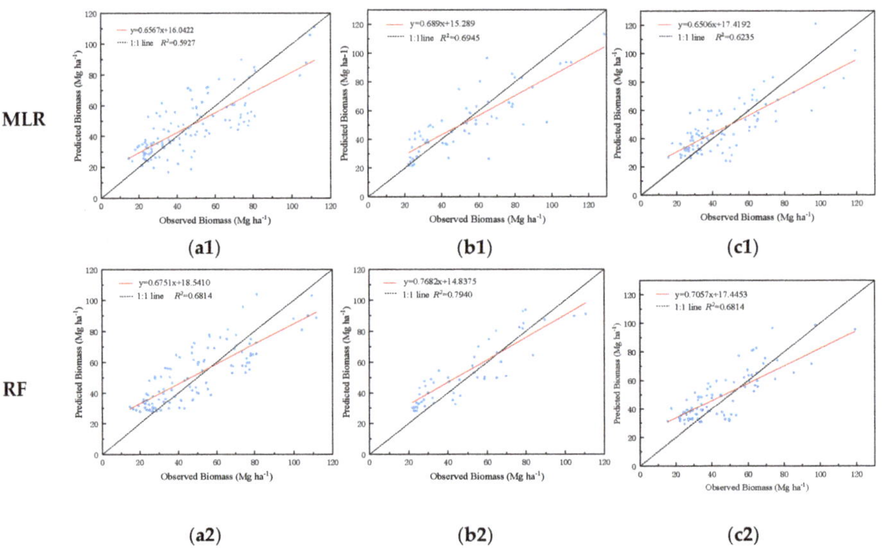

Figure 2. Graphs of the predicted versus observed values of three forest types for two models. (**a1**) and (**a2**) Coniferous forest; (**b1**) and (**b2**) Mixed forest; (**c1**) and (**c2**): Broadleaf forest.

Figure 2 shows that correlations between the estimated biomass and observed biomass in the coniferous forests, mixed forests, and broadleaf forests were all better for the RF models than for linear regression. In the linear model, as the biomass value increased, the performance of the model decreased, and most of the high-value biomass was underestimated. In particular, the high biomass values of mixed forests were greatly underestimated. RF can improve the performance of the model. When the biomass was less than 100 Mg ha^{-1}, the difference between predicted and observed values is lower than that of the linear model; the error under higher biomass values was slightly larger, and some of the higher values were underestimated.

Table 6. Estimation accuracy of different models.

Model	Forest Type	R^2	RMSE (Mg ha^{-1})	RMSE%	Bias (Mg ha^{-1})	Bias%
	Coniferous forest	0.59	14.15	29.65	0.34	0.71
MLR	Mixed forest	0.7	14.54	27.92	0.9	1.73
	Broadleaf forest	0.53	15.26	32.48	0.16	0.33
	Coniferous forest	0.66	13.23	27.24	−2.67	−5.50
RF	Mixed forest	0.77	11.09	22.89	−3.34	−6.89
	Broadleaf forest	0.64	11.98	27.02	−4.02	−9.08

Table 6 shows that the Bias% values were all near 0, and the RMSE% ranged from 27.92% to 32.48% for the MLR model. For the RF model, the Bias ranged from −2.67 to −4.02 and the RMSE% ranged from 22.89%–27.24%. These results showed that the two types of models were relatively stable and could be used to estimate biomass. However, an improvement in performance was found in the RF models for coniferous forest ($R^2 = 0.66$, RMSE = 13.23 Mg ha^{-1}), mixed forest ($R^2 = 0.77$, RMSE = 11.09 Mg ha^{-1}), and broadleaf forest ($R^2 = 0.64$, RMSE = 11.98 Mg ha^{-1}), in comparison to the linear regression models for coniferous forest ($R^2 = 0.59$, RMSE = 14.15 Mg ha^{-1}), mixed forest $R^2 = 0.70$, RMSE = 14.54 Mg ha^{-1}), and broadleaf forest ($R^2 = 0.53$, RMSE = 15.26 Mg ha^{-1}). Generally, the RF model was characterized by a high R^2 and a low RMSE, indicating a good fitting result.

For the same model with different forest types, the fit of the mixed-forest models was better than that of the models with the other two forest types. The R^2 of the linear regression model based on mixed forests was 0.11 and 0.17 higher than the R^2 of the linear regression models based on coniferous forests and broadleaf forests, respectively. The R^2 of the RF model based on mixed forests was 0.11 and 0.13 higher than the R^2 of the RF models based on coniferous forests and broadleaf forests, respectively, and the RMSE was 4.35 and 4.13 Mg ha^{-1} lower, respectively. Overall, the model for mixed forests had a high estimation accuracy.

3.4. Results of Biomass for Different Forest Types and Spatial Distribution of the Forest Biomass Density in Beijing

Based on the model estimation results, the forest biomass and biomass density of coniferous forests, broadleaf forests and mixed forests were estimated, the kriging interpolation was used and biomass density distribution maps of three forest types were obtained. Biomass of different forest types are shown in Table 7 and biomass density distribution in Figure 3.

Table 7. Biomass and biomass density of each forest type.

Forest Type	Area (ha^{-1})	Biomass (Mg)	Biomass Density (Mg ha^{-1})
Coniferous forest	663	35,622.99	53.73
Mixed forest	272	13,926.40	51.20
Broadleaf forest	496	25,196.80	50.80

The total forest biomass obtained from the survey data was 74,746.10Mg, and the biomass density was 19.14–195.66 Mg ha^{-1}, with an average biomass density of 52.26 Mg ha^{-1}. Among these values, the total biomass of coniferous forest was 35,622.99 Mg, the average biomass density was 53.73 Mg ha^{-1}; the total biomass of mixed forest was 13,926.40 Mg, the average biomass density was 51.20 Mg ha^{-1}; the total biomass of broadleaf forest was 25,196.80Mg, and the average biomass density was 50.80 Mg ha^{-1}.

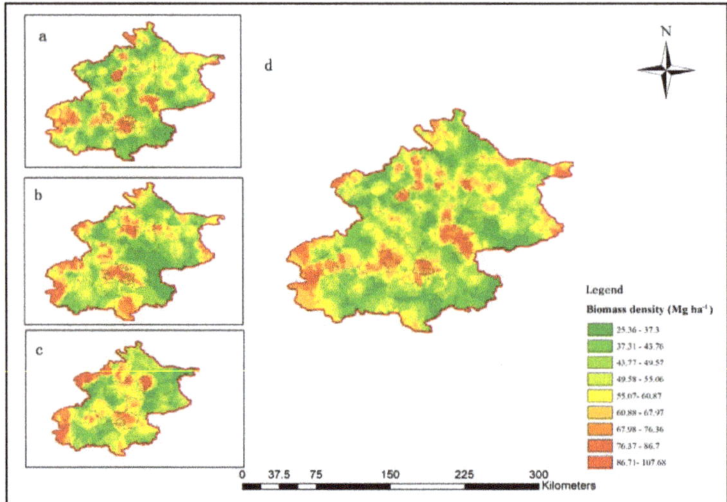

Figure 3. Biomass density distribution in Beijing, China (**a**) Coniferous forest; (**b**) Mixed forest; (**c**) Broadleaf forest; (**d**) all forest sampling plots.

As shown in Figure 3, the biomass distribution of arbor forests was basically consistent with the distribution of forestland in Beijing. The high-biomass area corresponded to dense forestland, and most of these forests were mature and overmature and were mainly distributed in the north and southwestern part of Beijing. The low-biomass area was mainly located in the southeast and central parts of Beijing. Because this area is urban with mostly developed land, the biomass in this area is low. The biomass density in most areas of Beijing was less than 70 Mg ha^{-1}.

The distribution of the three types of forest was obviously different. Coniferous forests were mainly distributed to the west and south of Beijing, mixed forests were mainly distributed in the west, and broadleaf forests were mainly distributed in the north and southwest.

This study did not consider shrubs and herbs, so the estimation of biomass can be considered relatively conservative but can also represent the basic situation of biomass in Beijing. At present, China's biomass estimation system is still not perfect. This study provides a feasible method for regional biomass estimation.

4. Discussion

4.1. Forest Biomass Estimation Model Based on Forest Inventory and Multisource Remote Sensing Data

In this paper, we propose a novel approach to modeling and mapping the biomass of forests at the regional scale that provides more detailed and accurate information than other approaches, such as estimating using only a single remote sensing data source or forest inventory data.

We combined forest inventory data with multisource remote sensing data (OLI and PALSAR) to estimate forest biomass, capturing almost all forest biomass spatial variability, and producing spatially explicit biomass estimates over regions.

According to the biomass characteristics of different forest types, it is very important to select variables with a high importance to the model [48]. The forest inventory factors selected in this study included not only the DBH and height, which are the two most relevant factors to biomass [38,39,50] but also the mean age and canopy density, which have received increasing attention in recent studies. Many previous studies have demonstrated that these two factors show a good correlation with biomass [40,51], which is consistent with our results. Landsat optical data are sensitive to forest vegetation, and their spatial resolution is suitable for the sample plot size. Zheng et al. confirmed that the red and NIR bands (Bands 4 and 5, respectively) are effective predictors of biomass [13]. Zheng et al. found that the SWIR band (Band 7) showed a satisfactory estimation ability in forests with a high canopy density [52], Foody et al. found that the NDVI and other vegetation indices are strongly correlated with biomass [53]. These results are consistent with our results. The models established by these factors fit the data relatively well. Multiple-variable PALSAR data have a higher correlation with forest biomass than individual-variable PALSAR HH and HV data because of their ability to detect canopy structure and retrieve forest biomass [54,55]. In addition, the biomass estimation model based on multisource remote sensing data combined with forest inventory data had a higher accuracy than that based on single-source data. Zhao used Landsat TM and ALOS PALSAR data to establish forest biomass models in Zhejiang Province. The R^2 of each forest type was below 0.5 [56]. Urbazaev used SAR backscatter, Landsat images and topographic factors to obtain the best R^2, a value of only 0.62 [57], which was lower than the accuracy in this study. This indicates that our model is suitable for estimating forest biomass in Beijing.

Optical and radar data are an effective supplement to inventory data, provide spatial information for estimating regional forest biomass, and can continuously estimate forest biomass [51]. In the results, the R^2 and RMSE of the three forest types were all greater than 0.5 and less than 20 Mg ha^{-1}. The models are reliable, but the model accuracy differed among different forest types.

In our results, mixed forests had the highest estimation accuracy, followed by coniferous forests and broadleaf forests, which is inconsistent with previous research.

Previous studies have shown that coniferous forest biomass estimation models have a high accuracy [58]. This inconsistency may be because the three types of modeling factors included in this study are highly sensitive to the structure of mixed forests. Moreover, differences in the study area location, tree species and forest types lead to different model estimation accuracies for different forest types.

However, the R^2 values of the model were all less than 0.8, which indicates the present model has less precision than the model established in a previous study [25]. A possible reason is that previous studies have mostly focused on small-scale areas. Our research mainly focuses on meso-scale areas, including a variety of terrain and environmental conditions, causing different environmental factors to have a certain impact on the modeling accuracy.

4.2. Estimation and Spatial Distribution of Forest Biomass in Beijing

The results of this study suggest that it is possible to produce spatially explicit biomass estimates over regions if adequate inventory data and remote sensing data are available. This meso-scale study was based on a relatively large sample size.

Because the data set used in this research did not contain all sample plot data from the forest inventory in Beijing, it can represent the distribution but not the total amount of biomass, while biomass density can represent the state of forest resources in Beijing well [9]. The 2016 forest biomass density range of Beijing was estimated by this model to be 19.14–195.66 Mg ha^{-1}, and the average biomass density was 52.26 Mg ha^{-1}. The forest biomass density in Beijing increased compared with previous studies [24].

Overall, forest resources show a pattern of more forests in mountainous areas, less forests in plains, as well as more forests outside urban areas and less forests in urban areas [59]. This is consistent with the distribution map of biomass density obtained in this study, as shown in Figure 3. However, this study shows that the biomass density in the central city is high, which is inconsistent with the distribution of forest resources in Beijing. This result may be caused by the fact that in recent years, the central city has insisted on the construction of ecological cities and increases in the area of green space, so the forest biomass has increased.

China is vigorously promoting the construction of eco-cities. At the meso-scale, forestry biomass estimation and biomass density mapping can provide decision makers with detailed information on forest resources to strengthen the management of forest resources.

4.3. Comparing Performance between MLR and RF Models

We also compared the performance of the MLR model and the RF model. We found that the prediction effect of the linear model for extreme biomass values (extremely high or very low) was not completely ideal. Low values were overestimated, and high values were underestimated, which is also common in previous studies [25,60]. The range in biomass for the multiple regression prediction is larger than that for the RF model.

This finding indicates that this model may be applicable only in the estimation of biomass values within a certain range. The use of the RF model had a positive impact on the estimation accuracy of extreme values [23]. The results showed that the RF model captured the complex nonlinear relationship between the optical and SAR data and biomass and compensated for the lack of inventory data, capitalizing on the strengths of both the forest inventory and remote sensing data. Therefore, the fit of the RF model is better than that of the linear regression model.

Linear regression and RF model contain different independent variable factors. Because of the collinearity between variables, some variables will be eliminated in linear model, while RF model can fully consider the fitting problem. The two model types include forest investigation factors such as d^2H, crown density and the mean age, indicating that the FID variables has a greater impact on the estimation of biomass and is not affected by collinearity. The RF models contain more PALSAR variables, and the model accuracy is higher, which indicates that the PALSAR variables are more sensitive to forest biomass.

Certainly, there are still some limitations in our research. Two thirds of Beijing is plain, and one third is mountainous area. Terrain correction is a part of remote sensing image correction under rugged surface, which can offset the influence of terrain to a certain extent, and is helpful to improve the accuracy of biomass estimation. Therefore, the research of terrain correction will be strengthened in future research. In the image preprocessing stage, the inaccuracy of the biomass models based on forest types and age classes and the lack of a consideration for the impact of environmental factors such as topography, soil and hydrology on biomass, which will be strengthened in the future research. Despite these problems, this study aimed to improve the performance of the regional forest biomass model and can provide a reference and support for future plans of relevant forestry departments, which has certain practical significance.

5. Conclusions

This paper proposed an approach for establishing the forest biomass of different forest type models and calculating forest biomass in Beijing by combining forest inventory data with multisource remote sensing data. The approach can capture all spatial variability and provide a reliable method for estimating forest biomass at a meso-scale with a high efficiency and low cost. In addition, we used this model to predict the forest biomass in Beijing in 2016. Among the three studied forest types, coniferous forest had the highest biomass density. According to the distribution of forest biomass in Beijing, the northern and southwestern parts of Beijing had a high biomass, while the central and eastern parts have a low biomass density. At present, there is no perfect biomass estimation system in China. Therefore, this method can provide a basis for meso-level biomass estimation and a reference for the planning of relevant forestry decision-making departments.

Author Contributions: Conceptualization, Z.F. and Y.Z.; Methodology, Z.F.; Validation, Z.F. and Y.Z.; Formal Analysis, Y.Z.; Investigation, Z.F. and Y.Z.; Resources, Z.F.; Data Curation, Z.F. and J.L. (Jincheng Liu); Writing-Original Draft Preparation, Y.Z.; Writing-Review & Editing, Y.Z. and J.L. (Jing Lu); Visualization, Y.Z.; Supervision, Z.F.; Project Administration, Z.F.; and Funding Acquisition, Z.F. All authors have read and agreed to the published version of the manuscript.

Funding: This research was jointly supported by the medium long-term project of "Precision Forestry Key Technology and Equipment Research" (No. 2015ZCQ-LX-01) and the National Natural Science Foundation of China (No.U1710123).

Acknowledgments: We would like to acknowledge support from Beijing Key Laboratory for Precision Forestry, Beijing Forestry University, as well as all the people who have contributed to this paper.

Conflicts of Interest: The authors declare no conflict of interest.

References

1. Cheng, W.; Yang, C.; Zhou, J.; Zhou, W.; Liu, Y. Research Summary of Forest Volume Quantitative Estimation Based on Remote Sensing Technology. *J. Anhui Agric. Sci.* **2009**, *37*, 7746–7750.
2. Wang, X.; Shao, G.; Chen, H.; Lewis, B.J. An Application of Remote Sensing Data in Mapping Landscape-Level Forest Biomass for Monitoring the Effectiveness of Forest Policies in Northeastern China. *Environ. Manag.* **2013**, *52*, 612–620. [CrossRef]
3. Roy, P.S.; Ravan, S.A. Biomass estimation using satellite remote sensing data—An investigation on possible approaches for natural forest. *J. Biosci.* **1996**, *21*, 535–561. [CrossRef]
4. Li, D.; Wang, C.; Hu, Y.; Liu, S. General review on remote sensing-based biomass estimation. *Geomat. Inf. Sci. Wuhan Univ.* **2012**, *37*, 631–635.
5. Tang, X.G.; Liu, D.W.; Wang, Z.M.; Jia, M.M.; Xu, W.M. Estimation of forest aboveground biomass based on remote sensing data: A review. *Chin. J. Ecol.* **2012**, *31*, 1311–1318.
6. Fang, J. Changes in Forest Biomass Carbon Storage in China between 1949 and 1998. *Science* **2001**, *292*, 2320–2322. [CrossRef]
7. Zhang, H.; Feng, Z.; Chen, P.; Chen, X. Development of a Tree Growth Difference Equation and Its Application in Forecasting the Biomass Carbon Stocks of Chinese Forests in 2050. *Forests* **2019**, *10*, 582. [CrossRef]
8. Liu, J.; Feng, Z.; Mannan, A.; Khan, T.; Cheng, Z. Comparing Non-Destructive Methods to Estimate Volume of Three Tree Taxa in Beijing, China. *Forests* **2019**, *10*, 92. [CrossRef]
9. Lu, J.; Feng, Z.; Zhu, Y. Estimation of Forest Biomass and Carbon Storage in China Based on Forest Resources Inventory Data. *Forests* **2019**, *10*, 650. [CrossRef]
10. Chave, J.; Réjou-Méchain, M.; Búrquez, A.; Chidumayo, E.; Colgan, M.S.; Delitti, W.B.; Duque, A.; Eid, T.; Fearnside, P.M.; Goodman, R.C.; et al. Improved allometric models to estimate the aboveground biomass of tropical trees. *Glob. Chang. Biol.* **2014**, *20*, 3177–3190. [CrossRef]
11. Gu, Y.; Brown, J.F.; Verdin, J.P.; Wardlow, B. A five-year analysis of MODIS NDVI and NDWI for grassland drought assessment over the central Great Plains of the United States. *Geophys. Res. Lett.* **2007**, *34*, 6. [CrossRef]
12. Gu, Y.; Wylie, B.K. Detecting Ecosystem Performance Anomalies for Land Management in the Upper Colorado River Basin Using Satellite Observations, Climate Data, and Ecosystem Models. *Remote Sens.-Basel* **2010**, *2*, 1880–1891. [CrossRef]
13. Zheng, D.; Rademacher, J.; Chen, J.; Crow, T.; Bresee, M.; Le Moine, J.; Ryu, S. Estimating aboveground biomass using Landsat 7 ETM+ data across a managed landscape in northern Wisconsin, USA. *Remote Sens. Environ.* **2004**, *93*, 402–411. [CrossRef]
14. Running, S.W.; Nemani, R.R.; Peterson, D.L.; Band, L.E.; Potts, D.F.; Pierce, L.L.; Spanner, M.A. Mapping Regional Forest Evapotranspiration and Photosynthesis by Coupling Satellite Data with Ecosystem Simulation. *Ecology* **1989**, *70*, 1090–1101. [CrossRef]
15. Xiao, X.; Hollinger, D.; Aber, J.; Goltz, M.; Davidson, E.A.; Zhang, Q.; Moore, B., III. Satellite-based modeling of gross primary production in an evergreen needleleaf forest. *Remote Sens. Environ.* **2004**, *89*, 519–534. [CrossRef]
16. Nichol, J.E.; Sarker, M.L. Improved Biomass Estimation Using the Texture Parameters of Two High-Resolution Optical Sensors. *IEEE Trans. Geosci. Remote Sens.* **2011**, *49*, 930–948. [CrossRef]
17. Næsset, E. Predicting forest stand characteristics with airborne scanning laser using a practical two-stage procedure and field data. *Remote Sens. Environ.* **2002**, *80*, 88–99. [CrossRef]
18. Hanes, J.; Zhang, X.; Nimeister, W. *Biophysical Applications of Satellite Remote Sensing*; Springer: Berlin/Heidelberg, Germany, 2016.

19. Zhang, H.; Zhu, J.; Wang, C.; Lin, H.; Long, J.; Zhao, L.; Fu, H.; Liu, Z. Forest Growing Stock Volume Estimation in Subtropical Mountain Areas Using PALSAR-2 L-Band PolSAR Data. *Forests* **2019**, *10*, 276. [CrossRef]

20. Karjalainen, M.; Kankare, V.; Vastaranta, M.; Holopainen, M.; Hyypp, J. Prediction of plot-level forest variables using TerraSAR-X stereo SAR data. *Remote Sens. Environ.* **2012**, *117*, 338–347. [CrossRef]

21. Vastaranta, M.; Holopainen, M.; Karjalainen, M.; Kankare, V.; Hyyppa, J.; Kaasalainen, S. TerraSAR-X Stereo Radar grammetry and Airborne Scanning LiDAR Height Metrics in Imputation of Forest Aboveground Biomass and Stem Volume. *IEEE Trans. Geosci. Remote. Sens.* **2013**, *52*, 1197–1204. [CrossRef]

22. Sinha, S.; Jeganathan, C.; Sharma, L.K.; Nathawat, M.S.; Das, A.K.; Mohan, S. Developing synergy regression models with space-borne ALOS PALSAR and Landsat TM sensors for retrieving tropical forest biomass. *J. Earth Syst. Sci.* **2016**, *125*, 725–735. [CrossRef]

23. Bourgoin, C.; Blanc, L.; Bailly, J.; Cornu, G.; Berenguer, E.; Oszwald, J.; Tritsch, I.; Laurent, F.; Hasan, A.; Sist, P.; et al. The Potential of Multisource Remote Sensing for Mapping the Biomass of a Degraded Amazonian Forest. *Forests* **2018**, *9*, 303. [CrossRef]

24. Wang, G.; Liu, Q. Estimating Carbon Sequestration of Beijing's Forests Based on TM Images and Forest Inventory Data. *J. Basic Sci. Eng.* **2013**, *21*, 224–235. [CrossRef]

25. Zheng, D.; Heath, L.S.; Ducey, M.J. Forest biomass estimated from MODIS and FIA data in the Lake States: MN, WI and MI, USA. *Forestry* **2007**, *80*, 265–278. [CrossRef]

26. Longo, M.; Keller, M.; Dos-Santos, M.N.; Leitold, V.; Pinagé, E.R.; Baccini, A.; Saatchi, S.; Nogueira, E.M.; Batistella, M.; Morton, D.C. Aboveground biomass variability across intact and degraded forests in the Brazilian Amazon. *Glob. Biogeochem. Cycles* **2016**, *30*, 1639–1660. [CrossRef]

27. Shao, W.; Cai, J.; Wu, H.; Liu, J.; Zhang, H.; Huang, H. An Assessment of Carbon Storage in China's Arboreal Forests. *Forests* **2017**, *8*, 110. [CrossRef]

28. Qiu, Z.; Feng, Z.; Song, Y.; Li, M.; Zhang, P. Carbon sequestration potential of forest vegetation in China from 2003 to 2050: Predicting forest vegetation growth based on climate and the environment. *J. Clean. Prod.* **2020**, *252*, 119715. [CrossRef]

29. Zhiguo, H.E.; Yang, Z. The problems and countermeasures of continuous forest inventory in China. *Hunan For. Sci. Technol.* **2012**, *39*, 80–83.

30. Xinliang, X.U.; Mingkui, C.; Kerang, L.I. Temporal- Spatial Dynamics of Carbon Storage of Forest Vegetation in China. *Prog. Geogr.* **2007**, *26*, 1–10.

31. Wang, H.; Niu, S.K.; Xiao, S.; Zhang, C. Study on biomass estimation methods of understory shrubs and herbs in forest ecosystem. *Acta Prataculturae Sin.* **2014**, *23*, 20–29.

32. Crist, E.P.; Cicone, R.C. A Physically-Based Transformation of Thematic Mapper Data—The TM Tasseled Cap. *IEEE Trans. Geosci. Remote Sens.* **1984**, *GE-22*, 256–263. [CrossRef]

33. You, J. The research on High Resolution Texture Information of Remote Sensing in Forest Volume Estimate. Master's Thesis, Xi'an University of Science and Technology, Xi'an, China, 2014.

34. Gao, Y. Research on the Ratio Band of Remote Sensing Image in Forest Stock Volume and Choice of Texture Information. Master's Thesis, Xi'an University of Science and Technology, Xi'an, China, 2014.

35. Qin, Y.; Xiao, X.; Dong, J.; Zhang, G.; Shimada, M.; Liu, J.; Li, C.; Kou, W.; Moore, B. Forest cover maps of China in 2010 from multiple approaches and data sources: PALSAR, Landsat, MODIS, FRA, and NFI. *ISPRS J. Photogramm.* **2015**, *109*, 1–16. [CrossRef]

36. Shimada, M.; Isoguchi, O.; Tadono, T.; Isono, K. PALSAR radiometric and geometric calibration. *IEEE Trans. Geosci. Remote Sens.* **2009**, *47*, 3915–3932. [CrossRef]

37. Vieilledent, G.; Vaudry, R.; Andriamanohisoa, S.F.; Rakotonarivo, O.S.; Randrianasolo, H.Z.; Razafindrabe, H.N.; Rakotoarivony, C.B.; Ebeling, J.; Rasamoelina, M. A universal approach to estimate biomass and carbon stock in tropical forests using generic allometric models. *Ecol. Appl.* **2012**, *22*, 572–583. [CrossRef]

38. Fehrmann, L.; Kleinn, C. General considerations about the use of allometric equations for biomass estimation on the example of Norway spruce in central Europe. *For. Ecol. Manag.* **2006**, *236*, 412–421. [CrossRef]

39. Rutishauser, E.; Noor An, F.; Laumonier, Y.; Halperin, J.; Rufi, I.; Hergoualc, H.K.; Verchot, L. Generic allometric models including height best estimate forest biomass and carbon stocks in Indonesia. *For. Ecol. Manag.* **2013**, *307*, 219–225. [CrossRef]

40. Baccini, A.; Friedl, M.A.; Woodcock, C.E.; Warbington, R. Forest biomass estimation over regional scales using multisource data. *Geophys. Res. Lett.* **2004**, *31*. [CrossRef]
41. Fang, K.; Wu, J.; Zhu, J.; Xie, B. A Review of Technologies on Random Forests. *Stat. Inf. Forum* **2011**, *26*, 32–38.
42. Zhao, T.; Yang, D.; Cai, X.; Cao, Y. Predict seasonal low flows in the upper Yangtze River using random forests model. *J. Hydroelectr. Eng.* **2012**, *31*, 18–24.
43. Asner, G.P.; Mascaro, J.; Muller-Landau, H.C.; Vieilledent, G.; Vaudry, R.; Rasamoelina, M.; Hall, J.S.; van Breugel, M. A universal airborne LiDAR approach for tropical forest carbon mapping. *Oecologia* **2012**, *168*, 1147–1160. [CrossRef]
44. Goetz, S.J.; Baccini, A.; Laporte, N.T.; Johns, T.; Walker, W.; Kellndorfer, J.; Houghton, R.A.; Sun, M. Mapping and monitoring carbon stocks with satellite observations: A comparison of methods. *Carbon Balance Manag.* **2009**, *4*, 2. [CrossRef] [PubMed]
45. Saud, P.; Lynch, T.B.; KC, A.; Guldin, J.M. Using quadratic mean diameter and relative spacing index to enhance height–diameter and crown ratio models fitted to longitudinal data. *Forestry* **2016**, *89*, 215–229. [CrossRef]
46. Gjertsen, A. Accuracy of forest mapping based on Landsat TM data and a kNN-based method. *Remote Sens. Environ.* **2007**, *110*, 420–430. [CrossRef]
47. Katila, M.; Tomppo, E. Selecting estimation parameters for the Finnish multisource National Forest Inventory. *Remote Sens. Environ.* **2001**, *76*, 16–32. [CrossRef]
48. Ma, J.; Xiao, X.; Qin, Y.; Chen, B.; Hu, Y.; Li, X.; Zhao, B. Estimating aboveground biomass of broadleaf, needleleaf, and mixed forests in Northeastern China through analysis of 25-m ALOS/PALSAR mosaic data. *For. Ecol. Manag.* **2017**, *389*, 199–210. [CrossRef]
49. Li, X.-H. Using "random forest"for classification and regression. *Chin. J. Appl. Entomol.* **2013**, *50*, 1190–1197.
50. Fu, W.; Wu, Y. Estimation of aboveground biomass of different mangrove trees based on canopy diameter and tree height. *Procedia Environ. Sci.* **2011**, *10*, 2189–2194. [CrossRef]
51. Avitabile, V.; Baccini, A.; Friedl, M.A.; Schmullius, C. Capabilities and limitations of Landsat and land cover data for aboveground woody biomass estimation of Uganda. *Remote Sens. Environ.* **2012**, *117*, 366–380. [CrossRef]
52. Zheng, G.; Chen, J.M.; Tian, Q.J.; Ju, W.M.; Xia, X.Q. Combining remote sensing imagery and forest age inventory for biomass mapping. *J. Environ. Manag.* **2007**, *85*, 616–623. [CrossRef]
53. Foody, G.M.; Cutler, M.E.; McMorrow, J.; Pelz, D.; Tangki, H.; Boyd, D.S.; Douglas, I. Mapping the Biomass of Bornean Tropical Rain Forest from Remotely Sensed Data. *Glob. Ecol. Biogeogr.* **2001**, *10*, 379–387. [CrossRef]
54. Saatchi, S.; Halligan, K.; Despain, D.G.; Crabtree, R.L. Estimation of Forest Fuel Load from Radar Remote Sensing. *IEEE Trans. Geosci. Remote Sens.* **2007**, *45*, 1726–1740. [CrossRef]
55. Carreiras, J.M.B.; Vasconcelos, M.J.; Lucas, R.M. Understanding the relationship between aboveground biomass and ALOS PALSAR data in the forests of Guinea-Bissau (West Africa). *Remote Sens. Environ.* **2012**, *121*, 426–442. [CrossRef]
56. Zhao, P.; Lu, D.; Wang, G.; Liu, L.; Li, D.; Zhu, J.; Yu, S. Forest aboveground biomass estimation in Zhejiang Province using the integration of Landsat TM and ALOS PALSAR data. *Int. J. Appl. Earth Obs.* **2016**, *53*, 1–15. [CrossRef]
57. Urbazaev, M.; Thiel, C.; Migliavacca, M.; Reichstein, M.; Rodriguez-Veiga, P.; Schmullius, C. Improved Multi-Sensor Satellite-Based Aboveground Biomass Estimation by Selecting Temporally Stable Forest Inventory Plots Using NDVI Time Series. *Forests* **2016**, *7*, 169. [CrossRef]
58. Ma, J.; Bu, R.; Miao, L.; Yu, C.; Han, F.; Qin, Q.; Hu, Y. Recovery of understory vegetation biomass and biodiversity in burned larch boreal forests in Northeastern China. *Scand. J. For. Res.* **2015**, *31*, 1–12. [CrossRef]
59. Chen, H.Y.; Li, W. Analysis on Forest Distribution and Structure in Beijing. *For. Resour. Manag.* **2011**, *2*, 32–35.
60. Fuchs, H.; Magdon, P.; Kleinn, C.; Flessa, H. Estimating aboveground carbon in a catchment of the Siberian forest tundra: Combining satellite imagery and field inventory. *Remote Sens. Environ.* **2009**, *113*, 518–531. [CrossRef]

Article

Spatially Explicit Analysis of Trade-Offs and Synergies among Multiple Ecosystem Services in Shaanxi Valley Basins

Yijie Sun [1], Jing Li [1,*], Xianfeng Liu [1], Zhiyuan Ren [1], Zixiang Zhou [2] and Yifang Duan [1]

[1] School of Geography and Tourism, Shaanxi Normal University, Xi'an 710119, China; sunjy1018@126.com (Y.S.); liuxianfeng7987@163.com (X.L.); renzhy@snnu.edu.cn (Z.R.); duanyifang@snnu.edu.cn (Y.D.)

[2] College of Geomatics, Xi'an University of Science and Technology, Xi'an 710054, China; zhouzixiang@xust.edu.cn

* Correspondence: lijing@snnu.edu.cn

Received: 12 January 2020; Accepted: 7 February 2020; Published: 12 February 2020

Abstract: Understanding the spatiotemporal characteristics of trade-offs and synergies among multiple ecosystem services (ESs) is the basis of sustainable ecosystem management. The ecological environment of valley basins is very fragile, while bearing the enormous pressure of economic development and population growth, which has damaged the balance of the ecosystem structure and ecosystem services. In this study, we selected two typical valley basins—Guanzhong Basin and Hanzhong Basin—as study areas. The spatial heterogeneity of trade-offs and synergies among multiple ESs (net primary production (NPP), habitat quality (HQ), soil conservation (SC), water conservation (WC), and food supply (FS)) were quantified using the correlation analysis and spatial overlay based on the gird scale to quantitatively analyze and compare the interaction among ESs in two basins. Our results found that: (1) Trade-offs between FS and other four services NPP, HQ, SC, and WC were discovered in two basins, and there were synergistic relationships between NPP, HQ, SC, and WC. (2) From 2000 to 2018, the conflicted relationships between paired ESs gradually increased, and the synergistic relationship became weaker. Furthermore, the rate of change in Guanzhong Basin was stronger than that in Hanzhong Basin. (3) The spatial synergies and trade-offs between NPP and HQ, WC and NPP, FS and HQ, SC and FS were widespread in two basins. The strong trade-offs between pair ESs were widly distributed in the central and southwest of Guanzhong Basin and the southeast of Hanzhong Basin. (4) Multiple ecosystem service interactions were concentrated in the north of Qinling Mountain, the central of Guanzhong Basins, and the east of Hanzhong Basin. Our research highlights the importance of taking spatial perspective and accounting for multiple ecosystem service interactions, and provide a reliable basis for achieving ecological sustainable development of the valley basin.

Keywords: ecosystem services; trade-off; synergy; multiple ES interactions; valley basin

---

1. Introduction

Ecosystem services (ESs) are the benefits that people derive from ecosystems [1–3] and include four categories (supporting, regulating, provisioning, and cultural services). According to the Millennium Ecosystem Assessment (MEA) reported, 60% of worldwide ecosystem services have degraded or been in an unsustainable state because of the rapid economic development and global population growth. Therefore, it is urgent to improve the capacity of ESs by improving the eco-management measures to maintain social and economic sustainable development [4].

Due to the diversity of ecosystem services, the heterogeneity of the spatial distribution and the selectivity of human use, the multiple relationships between ecosystem services show the dynamic variation under the influence of natural factors and human activities, which are characterized by different patterns such as trade-offs and synergies [5]. Trade-offs are the situations where one service increases at the cost of another services [6–8]. Such as, in an agricultural system, increasing fertilizer use to improve crop yields may have significant negative effects on water purification, and indirectly decrease fishery and recreational values [9]. Synergies are the reverse of trade-offs, which can be defined as situations in which both services either increase or decrease [6]. For instance, increasing net primary productivity simultaneously increases the values of water yield and soil conservation [10]. In addition, the ecosystem has diverse functions and, thus, provide multi-level services to humans. The multiple relationships of the ecosystem service is a challenge for local ecological management [11]. Moreover, trade-offs and synergies between ESs can differ in different regions because of landscape heterogeneity across the region, and the interactions between ESs would behave in diverse ways during different periods [12]. At the same time, the distinct ecosystem management strategies of the local region may also cause various interactions among multiple ESs. Therefore, identifying trade-offs and synergies between the ecosystem service could provide a powerful message to policy makers, and better inform management choices to achieve a "win-win" situation [13,14].

The identification of trade-offs and synergies between ecosystem services can be conducted through the methods: statistical analysis [15–17], mapping comparison [18,19], model simulation [20,21], and scenario analysis [22,23]. In this, correlation statistical analysis is a common method used in trade-off and synergy analysis, which can usually be used in combination with other methods. By spatial correlation analysis and calculating the changes of the relationship between ESs, which can quantitatively reveal the relationship between ESs within a certain period. However, there are still some limitations in previous studies, such as trade-off and synergy analysis. These are mostly based on quantitative statistical analysis, lack of dynamic trend changes of relationships between ES for long time series, and mostly consider the pairwise interactions between ES [10,24] while neglecting the study of multiple ecosystem service interactions. Furthermore, to local ecological management, policymakers need to know the location of trade-offs and synergies among multiple ESs. Therefore, spatial explicit analysis of trade-offs and synergies will be the core research in the future study of an ecosystem service.

With global population growth and rapid economic development, urbanization has brought a great threat to local ecological environments. The expansion of urban land, the influx of migrant populations, the reduction of carbon storage and soil degradation, which have occurred in the typical valley basins [4], Yanhe Watershed [25], Guanzhong-Tianshui economic region [26], and Grain-for-Green Programme region [27], are the relevant examples. Guanzhong Basin and Hanzhong Basin is located in the central and south of Shaanxi Province, respectively, as typical Shaanxi valley basins, which are sensitive to climate change, natural disaster, landscape fragmentation, and rapid degeneration of biodiversity [24]. In addition, Guanzhong Basin and Hanzhong Basin is located on either side of Qinling Mountains, which is the geographical boundary of northern and southern of China. Therefore, they have clear differences of the natural environment and social development.

Therefore, we used the Guanzhong Basin (as an economically developed region) and Hanzhong Basin (as the ecological environment region), as case study areas to explore the temporal and spatial variations of trade-offs and synergies among multiple ecosystem services and compare the region difference. Based on five ESs (net primary production (NPP), habitat quality (HQ), soil conservation (SC), water conservation (WC), food supply (FS)) from 2000 to 2018, the identification of trade-offs, and synergies between paired ESs and correlation coefficients were calculated by spatial statistical analysis. Meanwhile the spatial distributions of multiple interactions among ESs were classified by spatial overlay analysis based on the gird cell. Then, we compare the difference of trade-offs and synergies between Guanzhong Basin and Hanzhong Basin, in order to provide a theoretical basis for ecological management decisions in Northwestern China.

2. Materials and Methods

2.1. Study Area

The Guanzhong Basin and Hanzhong Basin both belong to Shaanxi Province located in the middle part of inland China. Considering the unification of natural data and social statistical data, we used an administrative boundary to divide the Guanzhong Basin and Hanzhong Basin as study areas (Figure 1). Guanzhong Basin is located between the Loess Plateau and the Qinling Mountain. The region is between 33°35′ N and 35°51′ N and between 106°19′ E and 110°36′ E. The terrain is low in the west and high in the east. Meanwhile, Weihe River (a tributary of the Yellow River) runs through the central region, which forms a large area of alluvial plain. The Guanzhong Basin is a warm temperate zone with a semi-humid climate, distinct four seasons, hot and rainy summer, and cold and dry winter, which has diverse vegetation types and agrotypes. It is the important part of national and western economic and an ecological balanced development strategy base [28]. With the rapid development of economy, it has attracted much influx of external population until 2017. The total population was approximately 23.94 million. The GDP was 1409.20 billion and accounted for 64.35% of Shaanxi Province. The ecological environment is greatly damaged by human activities.

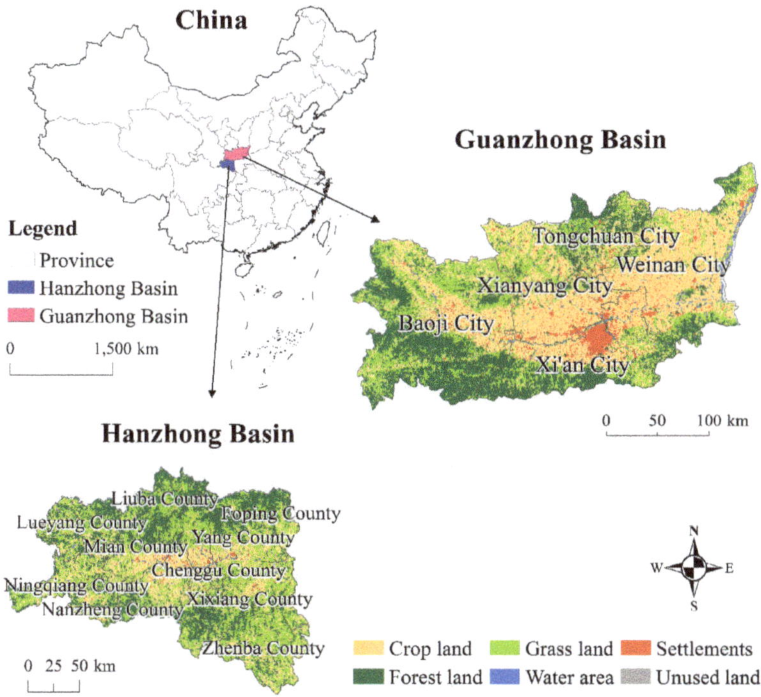

Figure 1. The geographical location and land use types of study areas in 2018.

Hanzhong Basin is located between Qinling Mountain and Daba Mountain (geographical coordinate is 32°08′54″ N~33°53′16″N, 105°30′50″E~108°16′45″E). While the Han River (a tributary of the Yangtze River) runs through the whole region, the terrain is gradually decreasing from northwest to southeast. The Hanzhong Basin is a typical north subtropical monsoon climate zone. The climate is often mild and humid, and there is no chilly winner and there is a hot summer. Species diversity is rich and it has a fine ecological environment. The forest coverage rate is 52%, the vegetation coverage rate of forest and grass is up to 60%, and it has the reputation of "Land of Fish and Rice" and "Land of

Abundance." By 2017, the GDP was 133.33 billion, the population was 3.44 million, the crop areas were 2102.10 km^2, and the grain yield reached 1.04 million tons. The region agriculture output contributes to more than 20% of the gross output. The level of economic development is low.

2.2. Data Sources

The data used in this study were obtained using the following sources.

(1) The land use/cover map in 2000 with a spatial resolution of 30 m × 30 m were applied from the Cold and Arid Region Sciences Data Center (http://westdc.westgis.ac.cn), 2005 and 2017 were provided by National Earth System Science Data Sharing Infrastructure (http://www.geodata.cn), 2010 and 2018 were down from Resource and Environment Data Cloud Platform (http://www.resdc.cn). The land use/cover data both covered six primary types and 25 secondary land use types. According to the actual land use settings in Shaanxi Province, and the need for quantitative evaluation of ecosystem services, the land cover types were classified into the following six categories: (1)Crop land, including plain dryland and irrigated land, mainly for agricultural cultivation. (2) Forest land, containing closed forest land, shrubbery, sparse wood land, and other forest land. (3) Grass land, referring to high, middle, and lower cover grassland. (4) Water area, including lake, river, reservoirs, and ponds, bottomland. (5) Settlements, containing urban land, rural residential area, industrial and mining, and other conservation land. (6) Unused land, which is currently unused and may be hard to use, including sand, bare land, swale land, saline land, and others.

(2) Digital elevation model with a resolution of 30 m was used to calculate the terrain factors in the Revised Universal Soil Loss Equation (RUSLE) model, which are available for download at Geospatial Data Could Platform (http://www.gsclooud.cn).

(3) The soil and vegetation type map of Shaanxi Province were extracted from 1:1,000,000 soil and vegetation database of China, respectively. Those were used to compute the soil conservation and net primary production, which were obtained from the National Earth System Science Data Sharing Infrastructure (http://www.geodata.cn).

(4) The Normalized Difference Vegetation Index (NDVI) was divided from the MOD13A2 product synthesized by the Maximum Value Composite (MVC) Method 16d and downloaded from the United States Geological Survey with a spatial resolution of 250 m (http://ladsweb.modaps.eosdis.nasa.gov).

(5) Meteorological data was obtained from the China Meteorological Science Data Sharing Service System (http://data.cma.cn), including average temperature, precipitation, wind speed, average air pressure, maximum temperature, minimum temperature, sunshine duration, solar radiation and more. Furthermore, via ArcGIS software using the Kriging interpolation method, we obtained the meteorological raster dataset.

(6) Major food productions, population, and gross domestic product were obtained from the Shaanxi statistical yearbooks and some statistical yearbooks from Hanzhong City and other cities from 2000 to 2018.

2.3. Quantifying Ecosystem Services

2.3.1. Net Primary Productivity (NPP)

Net Primary Productivity (NPP) is defined as the amount of organic energy produced by plant photosynthesis minus the energy consumed through autotrophic respiration [29]. This paper uses the Carnegie-Ames-Stanford approach (CASA) mode to estimate the value of NPP(net primary production) [30]. The formula is below.

$$NPP(x,t) = APAR(x,t) \times \varepsilon(x,t) \tag{1}$$

where NPP (x, t) represents the net primary productivity of pix x during month t (g C·m^{-2}·month^{-1}), $\varepsilon(x, t)$ describes the light utilization efficiency of pix x during month t (g C·MJ^{-1}), and APAR(x, t) is the absorbed photosynthetically active radiation(MJ·m^{-2}).

2.3.2. Habitat Quality (HQ)

Habitat quality refers to the ability of the ecosystem to provide conditions appropriate for individual and population persistence, and it depends on a habitat's proximity to human land uses and the intensity of these land uses [31,32]. InVEST models' habitat quality as a proxy for biodiversity, ultimately, estimates the extent of the habitat across the landscape, and their state of degradation [15]. The model integrates information on land use and threats to biodiversity to produce the habitat quality map [33]. The habitat quality of each grid is indicated by habitat suitability (value range from 0 to 1, 1 indicates the highest suitability of the habitat, while areas on the landscape that are not habitat get a quality score of 0) and habitat degradation. There are four factors in the function: each threat's relative impact, the relative sensitivity of each habitat type to each threat, the distance between habitats and sources of threats, and the degree to which the land is legally protected [34]. In this study, threats included urban land, rural residential areas, and industrial and mining construction land and cropland. Moreover, the impact of these four threats on habitat decreased as the distance from the degradation source increases. Now, we choose a linear distance-decay function to describe how a threat decays over space. The impact of threat r that originated in gird cell y, r_y, on habit in gird cell x is given by i_{rxy}, and the quality of the habitat in parcel x that was in LUCC j is given by Q_{xj} and was represented by the following equations.

$$i_{rxy} = 1 - \left(\frac{d_{xy}}{d_{r\max}}\right) \text{if linear} \tag{2}$$

$$i_{rxy} = \exp\left\{-\left(\frac{-2.99}{d_{r\max}}\right)d_{xy}\right\} \text{if exponential} \tag{3}$$

$$Q_{xj} = H_j\left(1 - \left(\frac{D^z_{xj}}{D^z_{xj} + k^2}\right)\right) \tag{4}$$

where r was the threat source of habitat, d_{xy} is the linear distance between grid cells x and y, $d_{r\max}$ is the maximum effective distance of threat r's reach across space, H_j indicates the habitat suitability of LULC type j, D_{xj} is the total threat level in grid cell x with LULC j, and z (we hard code $z = 2.5$) and k are scaling parameters, which are half of the maximum degradation. Furthermore, the sensitivity of different threat sources for land use is based on the InVEST 3.2.0 user's guide and other previous studies [26,35].

2.3.3. Water Conservation (WC)

Water conservation affects the ecosystem process and crop production through various land covers. Region rainfall, evapotranspiration, storage, and sorption are vegetation processes. Water conservation is integrated as the performance of water circulation and a different natural ecosystem, such as forest, vegetation coverage, and soil. We calculated water conservation through the summation of canopy interception, litter absorption, and soil storage.

$$Q = Q_1 + Q_2 + Q_3 \tag{5}$$

In the above equation, Q indicates the total amount of water retention capacity (t·year^{-1}). Q_1, Q_2, and Q_3 indicates the amount of vegetation canopy interception (t·year^{-1}), the amount of the litter retention capacity (t·year^{-1}), and the interception amount of the soil layer (t·year^{-1}).

$$Q_1 = \sum (\alpha_i \times \beta_i \times S_i) \tag{6}$$

$$Q_2 = \sum (\varepsilon_i \times \gamma_i \times S_i) \tag{7}$$

$$Q_3 = \sum (p_i \times h_i \times S_i) \tag{8}$$

where α_i is the annual rainfall (mm), β_i is the canopy retention (%), S_i is the area of vegetation type (ha), ε_i is the litter dry weight (t/ha), γ_i is the maximum water holding capacity (%), p_i is the soil non-capillary porosity (%), and h_i is soil thickness (mm) [36,37].

2.3.4. Soil Conservation (SC)

The Revised Universal Soil Loss Equation (RUSLE) is most widely used to calculate the average annual soil loss from each pixel of land. Based on the soil erosion theory and natural runoff observational data, the RUSLE model is applied using GIS software with some factors, including meteorological station dataset, the NDVI (normalized difference vegetation index) dataset, soil surveys, topographic maps, and land use data [38,39]. Therefore, the soil conservation is estimated from the difference between potential soil erosion and actual soil erosion [40].

$$A_m = R \times K \times C \times LS \times P \tag{9}$$

$$A_p = R \times K \times LS \tag{10}$$

$$A_c = A_p - A_m \tag{11}$$

In the above formula, A_m is the amount of actual soil erosion (t·ha^{-1}·year^{-1}), A_p is the potential soil erosion (t·ha^{-1}·year^{-1}), A_c is the amount of soil conservation (t· ha^{-1}··year^{-1}), R is the rainfall erosivity factor, and K is the soil erodibility factor, which indicates the physical and chemical properties of soil. C is a dimensionless crop management factor. LS includes the slope length factor (L) and the slope factor (S). P is the soil conservation measures factor, which reflects people using different protection measures to prevent soil erosion of various land use types.

2.3.5. Food Supply (FS)

Food supply services are one of the most important provisioning services in agricultural ecosystems [41]. Food is the most basic material that humans obtain from the natural ecosystem, which plays a decisive role in social development [42]. In this study, we use the land use dataset and region statistical yearbook data to estimate the total food supply of each land use in the study area, in order to realize the spatialization of the food supply. The equation is as follows.

$$G_i = A_i \times N_i \tag{12}$$

In the above equation, G_i is the amount of ith food for each pixel. A_i is the ith food area (km^2). The study area was divided into the unit grid of 1 km×1 km which is equal to 1km^2. N_i is the yield of ith food for the unit area (t/km^2).

$$N_i = \frac{F_i}{S_i} \tag{13}$$

where F_i is the total yield of food in the study area (t·year^{-1}). S_i is the total area of the ith food (km^2) in this research, which represents the area of each land use type. Among this, the grain, oil-bearing, and vegetables belong to the cropland. The output of meat and milk belong to the grassland. The aquatic products belong to the water area.

2.4. Spatial Correlation Analysis

The spatial statistical mapping method based on correlation coefficients on a pixel scale are used to quantify the relationship between ecosystem services. This method could explore the continuous temporal changes of various ecosystem services. Meanwhile, the relationship between ESs can be spatially expressed by quantitative mapping. In this paper, the correlation coefficient for each pair of ecosystem services is calculated by ArcGIS software (ArcGIS 10.2) at a pixel scale [25]. Furthermore, the correlation coefficients of two time series based on each pixel were calculated by Spearman's coefficient. Its expression is shown below.

$$r_{xy} = \frac{\sum\limits_{i=1}^{n}\sum\limits_{i=1}^{n}\left(x_{ij}-\bar{x}\right)\left(y_{ij}-\bar{y}\right)}{\sqrt{\sum\limits_{i=1}^{n}\left(x_{ij}-\bar{x}\right)^2}\sqrt{\sum\limits_{i=1}^{n}\left(y_{ij}-\bar{y}\right)^2}} \tag{14}$$

where r_{xy} is the spatial correlation coefficient, with values ranging from −1 to 1. If $r_{xy} > 0$, represents the positive correlation between two variables, which indicates that the two services are synergistic. If $r_{xy} < 0$, represents the negative correlation between two variables, which means there are trade-offs between two services. x_{ij}, y_{ij} indicates gird values for different types of ecosystem service spatial datasets.

3. Results

3.1. Spatial Distributions of Ecosystem Services

Figure 2 depicts spatial distribution of average ecosystem services in the two areas of study. In Guanzhong Basin, the average of NPP was 6.25 t·ha^{-1}·a^{-1}, the higher value of NPP was concentrated in the southwest, and the lowest was observed in the middle of the basin. While in Hanzhong Basin, the average of NPP is 8.21 t·ha^{-1}·a^{-1}, which is more than the Guanzhong Basin. The higher HQ was distributed in the northern Qinling mountain, which aggregated in forestland in the Guanzhong Basin. The average of HQ was 0.45 in the Guanzhong Basin, and 0.55 in the Hanzhong Basin. Moreover, the distribution of WC varied greatly in the Guanzhong Basin. Areas with high water conservation (WC) were concentrated in the north of Qinling Mountain and the southern of Xi'an City. It showed a clear trend from low in the north to high in the south in the Hanzhong Basin. Areas with high soil conservation (SC) were in the southwest of the Guanzhong Basin, but the central of the basin was mainly concentrated in crop land with low SC. Moreover, it could be seen the lower SC was in the central of the Hanzhong Basin, and lowest SC was just 0.01 t·ha^{-1}·a^{-1}. The highest value reached 1270 t·ha^{-1}·a^{-1}. Furthermore, the higher FS was mostly observed in the central region of the Guanzhong Basin, concentrated in farmland. The low values were in the southwest of Baoji City and most areas of the Tongchuan City. In the Hanzhong Basin, the food supply of crop land in the middle region was relatively higher, and those with low FS were distributed in the northern and southeastern basin, which are concentrated on forest land.

3.2. Spatial Correlations between Ecosystem Services

3.2.1. Trade-Offs and Synergies Analysis

We used the correlation analysis function to explore the trade-offs and synergies between each ES (Figure 3). In two basins, we could find that food supply (FS) with the other four services (soil conservation (SC), net primary productivity (NPP), habitat quality (HQ), and water conservation (WC)) both had negative relationships. While the negative relationships between FS and HQ were stronger than others, the correlation coefficient was −0.6333 and −0.5934 in the Guanzhong Basin and the Hanzhong Basin, respectively. In addition, NPP, HQ, SC, and WC presented positive relationships, and positive relationships between NPP and HQ were bigger in the Guanzhong Basin, where the correlation coefficient was up to 0.6173. The relationship between WC and SC was weak with only 0.03 and 0.002 in the Guanzhong Basin and the Hanzhong Basin, respectively. At the same time, in the scatterplot, as the FS changed, the other four ESs changed in opposite directions, while WC, HQ, NPP, and SC both had a consistent trend. Consequently, the trade-offs and synergies were identified by the correlation coefficient and the correlation diagram. The results indicated that trade-offs between FS and the other four services (NPP, HQ, SC, and WC) were discovered in two basins, and there were synergic relationships between NPP, HQ, SC, and WC. It could be explained that strong capacity of carbon storage and water retention was concentrated in forest land and grass land, but the food production was low. While the food production was bigger in cropland, the value of NPP and soil conservation was smaller. Moreover, because cropland is a core threat source of habitat quality, the habitat quality was low, which caused trade-offs between the food supply and habitat quality. When comparing the differences between the two basins, the trade-off between FS and NPP was clear in the Gunanzhong Basin. The correlation coefficient was −0.4790. While the synergy between WC and HQ was evident in theHanzhong Basin, the correlation coefficient was 0.3208. Consequently, the relationships between ecosystem services in Guanzhong Basin were more complex than those in the Hanzhong Basin. The trade-offs between FS and HQ, NPP, and SC were stronger in the Guanzhong Basin.

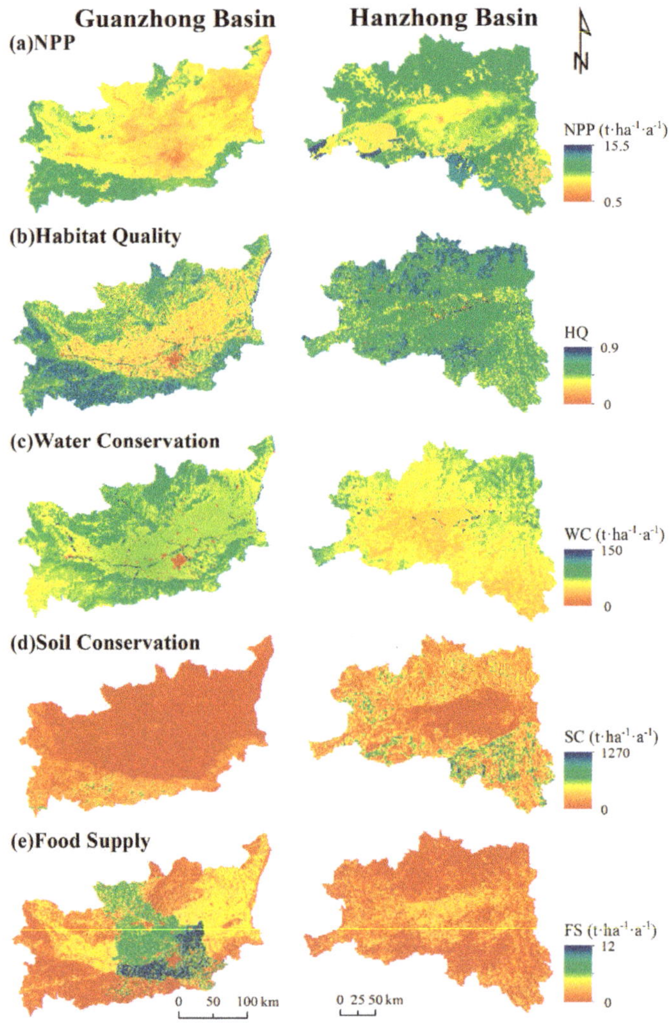

Figure 2. The spatial distribution of average ecosystem services in valley basins from 2000 to 2018.

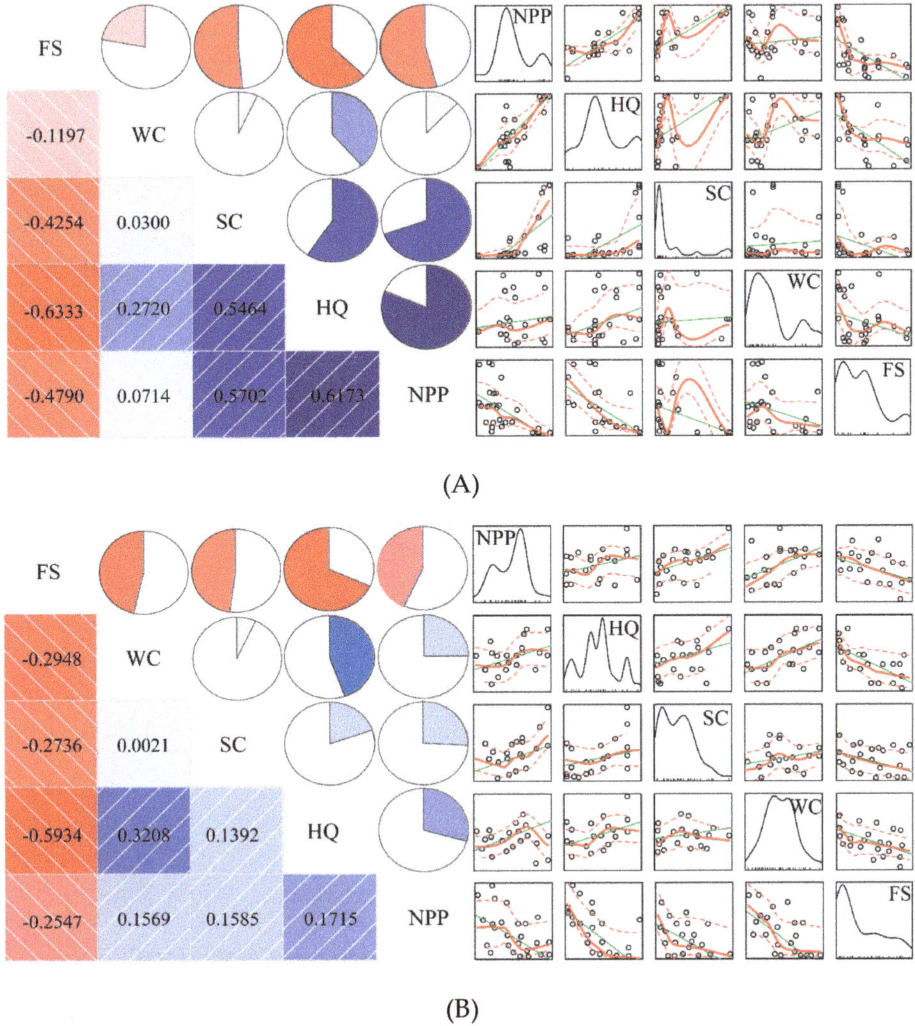

(A)

(B)

Figure 3. The relationships between average ecosystem services in the Guanzhong Basin (**A**) and in the Hanzhong Basin (**B**) from 2000 to 2018. Based on the R software, pie chart, and scatter plot can directly indicate the relationships between ecosystem services. For example, in the pie chart, the blue color represented positive values, the red color represented negative values, and the intensity of the color increased uniformly as the correlation value increased. The shading of the lower triangular in the figure had the same meanings in color as the circles of the upper triangular. Furthermore, the intensity of color scaled in proportion to the magnitude of correlation values, besides the circles were filled clockwise for positive values and anti-clockwise for negative values. In the scatter plot, all data may be considerably enhanced by the addition of linear regression lines, (loess) smoothed curves, and so forth. The key diagonal represented the kernel density curve and the lower horizontal axis was the shaft figure. Other figures contained linear and smooth fitting curves.

3.2.2. Temporal Analysis of Trade-Offs and Synergies

Trade-offs and synergies between ecosystem services at different periods varied greatly. We further explored the temporal changes of trade-offs and synergies from 2000 to 2018 (Figure 4). In the Guanzhong

Basin, there was a stable decrease of the synergistic relationship between NPP and HQ, NPP and SC, and HQ and SC. While the relationship between NPP and WC, HQ and WC, SC and WC presented a sudden drop in 2010, even between NPP and WC, SC and WC appeared to have negative relationships. From the temporal variations of correlation coefficients, it could be seen that WC increased with the decreasing trend of NPP in 2010. Some areas of the forest increased as the crop land and grassland decreased, which might increase the conflict between WC, NPP, and SC. The trade-offs between FS and other four ESs showed a fluctuating change during this period. Compared with 2000, the negative relationship between HQ and FS has decreased, whereas the trade-off between FS and WC became strong in 2010. In the Hanzhong Basin, the positive relationship between NPP and HQ presented an evident decreasing trend. It also discovered the relationship between NPP and WC, SC and WC appeared to have a decreasing trend in 2010. Moreover, the negative relationship between WC and FS was weaker. While the conflict relationship between NPP and FS, SC and FS decreased. Therefore, we found the synergistic relationship became weak as it changed and presented a decreasing trend in the Guanzhong Basin, which presented to be somewhat stronger than those in the Hanzhong Basin. On the other hand, the trade-offs between FS and SC, FS and WC were strong in the Guanzhong Basin and the conflicted relationship became more pronounced. On the whole, the change trend of the trade-off relationship in the Guanzhong Basin was slightly faster than that in the Hanzhong Basin.

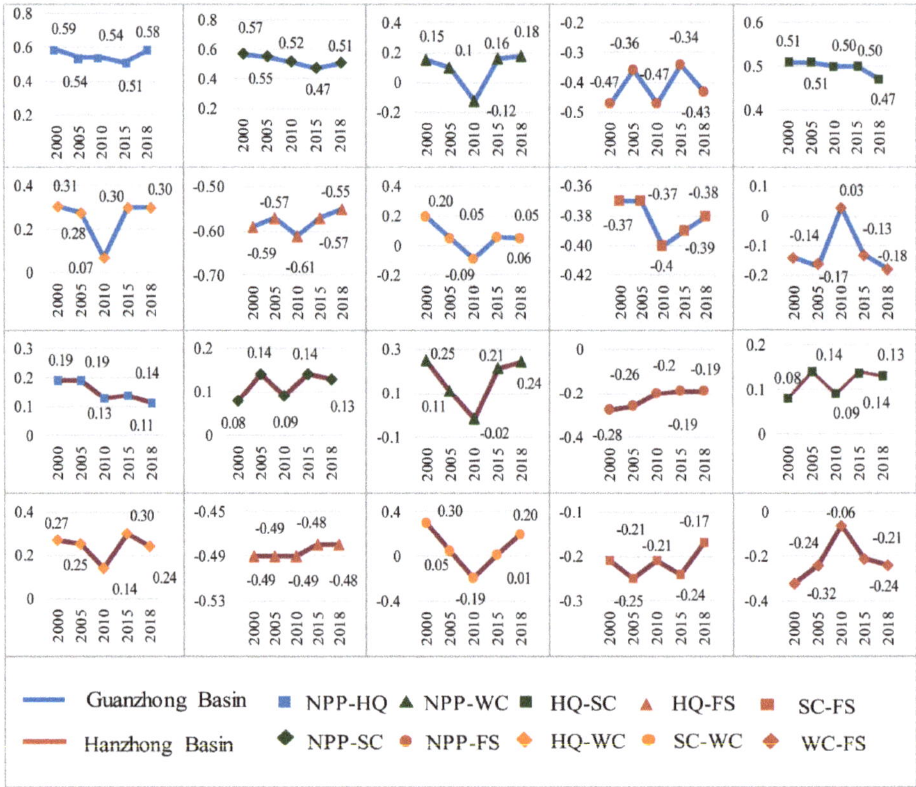

Figure 4. Spatial correlation coefficient for each pair of ecosystem services in two basins from 2000 to 2018.

3.3. Spatial Heterogeneity of Paired Ecosystem Service Interaction Based on the Grid Scale

Through the above study, we found that the multiple interactions among ES were different and had clear temporal and spatial characteristics. The spatial correlation coefficients based on the gird from 2000 to 2018 were shown in Figure 5, which were checked according to significance ($p < 0.05$). In this paper, we calculated spatial correlation coefficients between regulating and supporting services as well as provisioning and regulating services. Since this study involved five different services in two regions, we took four pairs services (HQ and NPP, WC and NPP, FS and HQ, FS and SC) with good correlation, as an example.

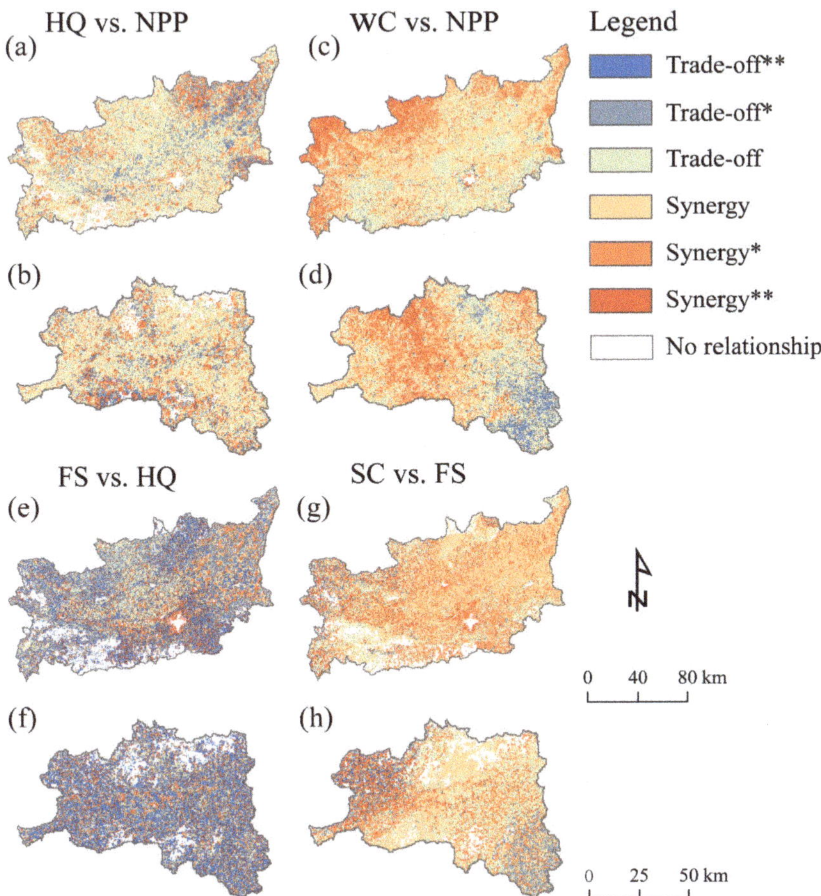

Figure 5. Spatial trade-offs and synergies of paired ecosystem services in two valley basins. ** Correlation were all significant at the 0.01 level. * Correlation were all significant at the 0.05 level. (**a,b**): Spatial trade-offs and synergies for HQ (habitat quality) and NPP (net primary production) in Guanzhong and Hanzhong Basin, respectively. (**c,d**): Spatial trade-offs and synergies for WC (water conservation) and NPP. (**e,f**): Spatial trade-offs and synergies for FS (food supply) and HQ. (**g,h**): Spatial trade-offs and synergies for FS and SC (soil conservation).

For HQ (habitat quality) and NPP (net primary production) (Figure 5a), the strong synergies ($p < 0.01$ and $p < 0.05$) were spatially aggregated in the north-east of Weinan City and the western region of the Guanzhong Basin, mostly concentrated in the forest land, which nearly accounted for 17.83%. Strong trade-offs ($p < 0.01$ and $p < 0.05$) account for 3.40%, which were in the south-east of Weinan City and the central region of the basin. At the same time, the strong synergies ($p < 0.01$ and $p < 0.05$) were mostly found in the edge of the Hanzhong Basin, which accounted for 20.93%, while the strong trade-offs accounted for 10.31% of the land use types that were in the central region of the basin and in the southeast of the Ningqiang County (Figure 5b).

For WC (water conservation) and NPP (Figure 5c), the average spatial correlation coefficient during 2000–2018 was 0.14 with a standard deviation of 0.34. Strong synergies ($p < 0.01$ and $p < 0.05$) were spatially aggregated in the northern edge of the Guanzhong Basin, which accounted for 26.44%, mostly discovered in the forest land. Strong trade-offs ($p < 0.01$ and $p < 0.05$) that accounted for 0.81% of the land use types were in the central region of the Xi'an City. Meanwhile, in the Hanzhong Basin, the average spatial correlation coefficient was 0.07 with a standard deviation of 0.44 from 2000-2018. Strong synergies ($p < 0.01$ and $p < 0.05$) were accounted for 28.91%, which were concentrated in the north-east and Mian County. The strong trade-offs ($p < 0.01$ and $p < 0.05$) were discovered in the west-south of the basin region and Zhenba County, which accounted for 0.82% (Figure 5d).

For FS (food supply) and HQ (Figure 5e), the trade-off relationships were widespread in two basins. In the Guanzhong Basin, the average spatial correlation coefficient during 2000–2018 was −0.30 with a standard deviation of 0.64. Strong trade-offs ($p < 0.01$ and $p < 0.05$) were spatially aggregated in the surrounding areas of the Guanzhong Basin and the edge of the bigger city (e.g., crop land and grass land) accounted for 46.94%. The strong synergies ($p < 0.01$ and $p < 0.05$) accounted for 21.20% of the land use types that were in the north-east of the basin and the central of Xian City. While the average spatial correlation coefficient during 2000–2018 was −0.40 with a standard deviation of 0.68 in the Hanzhong Basin (Figure 5f). It could be seen that the trade-offs widely exited in the edge of the basin and the strong trade-offs were approximately 57.46%. The strong synergies accounted for 20.17% of the land use types in the central region of the Hanzhong Basin and the settlements around the bigger county.

For SC (soil conservation) and FS (Figure 5g), the strong trade-offs ($p < 0.01$ and $p < 0.05$) were concentrated on the north-west region of the Guanzhong Basin and the north region of the Qinling Mountain (mostly concentrated on the forest land). While strong synergies ($p < 0.01$ and $p < 0.05$) accounted for 30.12%, which were in the central area of the basin and the north-east of Weinan City, which concentrated on the crop land. Furthermore, strong trade-offs were spatially aggregated in the south-west of the Hanzhong Basin and in grassland, which accounted for 9.56% of the whole basin. Strong synergies were aggregated in the northwest and the central region of the basin, which was concentrated in the grassland (Figure 5h).

3.4. Multiple Interactions among Ecosystem Services

3.4.1. Spatial Explicit Analysis of Multiple ESs Interactions

Figure 6 depicts the multiple interactions among ES from the perspective of the valley basin as a whole unit (the detailed method was seen in Appendix A). The relationship among multiple ESs in the Guanzhong Basin varied greatly. The complex relationships were discovered in the central region of the basin and the northern Qinling Mountain. In the Guanzhong Basin, the FS gradually decreased, while NPP, HQ, SC, and WC showed a simultaneous continuous increase from 2000 to 2018. This phenomenon meant that NPP, HQ, SC, and WC had synergistic interactions, and the four services had trade-off interactions with FS. It occurred in the west of Baoji City and the north of Xianyang City, which accounted for 8.31%. SC and FS simultaneously decreased, and NPP, HQ, and WC showed continuous increases. This result indicated the synergistic interaction occurred among NPP, HQ, and WC, and had trade-offs with SC and FS. It was greatly aggregated in the edge of the basin concentrated on grassland, which accounted for 4.21%. NPP, SC, and WC showed a simultaneous continuous increase, while HQ and FS showed a simultaneous decrease. This phenomenon declared that synergistic interactions occurred among NPP, SC, and WC, and these three services showed trade-offs with HQ and FS, which accounted for 4.88% in the north of Tongchuan City and the forest land around the bigger city. NPP and FS continuously increased, and HQ, SC, and WC simultaneously underwent a continuous decrease throughout these years. Synergies occurred among NPP and FS, and two services exhibited trade-offs with HQ, SC, and WC. This phenomenon accounted for 5.45% of the whole region was in the east of Xi'an City (Figure 6A) and the southeast of Baoji City. Furthermore, the NPP gradually increased, and HQ, SC, WC, and FS simultaneously underwent a continuous decrease from 2000 to 2018. Trade-offs occurred among NPP and those four services accounted for 4.98%, which were in the northern margin of the Qinling Mountain (Figure 6B).

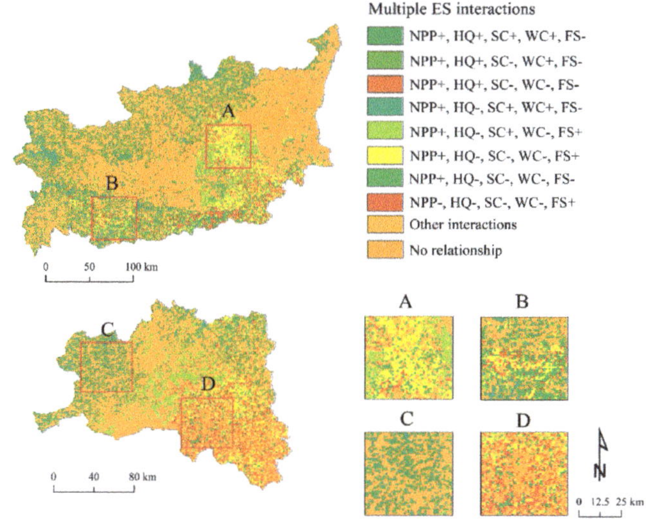

Figure 6. The spatial patterns of multiple interactions among ecosystem services (ESs) in two basins from 2000 to 2018. Because the ecosystem service assessments were based on the gird scales, the interactions among multiple ESs were complex. Some bigger area proportions and clear interactions were shown in the figure, and others are collectively named "other interactions"; '+' indicated an increase of service; '−' indicated a reduction. For example, "NPP+, HQ+, SC−, WC−, FS−" indicates that NPP (net primary production), HQ (habitat quality) increase simultaneously (suggesting synergies), SC (soil conservation), WC (water conservation), FS (food supply) decrease simultaneously, and the two services NPP, HQ both exhibit trade-off relationships with other three services WC, SC, FS. Moreover, We selected four evident trade-offs location (**A–D**) where the areas were 1km2, and further amplified and analyzed these interactions.

In the Hanzhong Basin, the relationship between multiple ESs was fairly simple than Guanzhong Basin but showed evident spatial differences. The trade-offs among FS and other four services (NPP, HQ, SC, WC, which simultaneously increased, suggesting synergies) was aggregated in the northwest of the basin accounting for 6.88% of all land use types (Figure 6C). NPP and HQ showed a simultaneous continuous increase, and SC, WC, and FS showed a simultaneous continuous decrease. It indicated that NPP and HQ had a synergistic relationship, and both presented trade-offs with the other three services (SC, WC, and FS). This phenomenon accounted for 5.89% of land use types that were in the southeast of the basin. HQ and WC showed a simultaneous continuous decrease, and NPP, SC, and FS showed a simultaneous continuous increase from 2000 to 2018. This phenomenon meant that NPP, SC, and FS services had a synergistic relationship, and the two services HQ and WC both had trade-off relationships with the other three services. It was aggregated in the central of the basin, which accounted for 7.99% of the land use types. The trade-offs among NPP and FS (simultaneously increased, suggesting synergies) and the three services (HQ, SC, and WC simultaneously decreased) were spatially aggregated in the southeast region of the basin, which accounted for 9.06% and were located in the west of the Xixiang County and the edge of the Zhenba County. Furthermore, the FS gradually increased, and the other four services (NPP, HQ, SC, and WC) showed a simultaneous continuous decrease from 2000 to 2018. These results were aggregated in the west of Xixiang County and the southeast of Zhenba County (Figure 6D).

3.4.2. Trade-Off Relationships in Various Land Use Types

The interactions among multiple ESs were complex based on the gird scale across two basins, so we further analyzed the patterns of trade-offs among multiple ESs in various land use types (Figure 7). In the Guanzhong Basin, the trade-offs among FS and other services (NPP, HQ, SC, and WC) (simultaneous increase) were mostly concentrated in the forest land and crop land, and occupied 51.34% and 36.90%, respectively. The trade-offs among NPP, HQ, WC (simultaneous increase, suggesting synergies), and SC, FS (simultaneous decrease) accounted for 40.44% that were located in the grass land, and 34.83% were located in the forest land. Meanwhile, the trade-offs among NPP, SC, and WC (simultaneous increase, suggesting synergies), and HQ and FS (simultaneous decrease) were mostly in the forest land, which accounted for 72.19%. it was discovered that the trade-offs among NPP and other services HQ, SC, WC, FS (simultaneous decrease) were concentrated in the forest land and settlements.

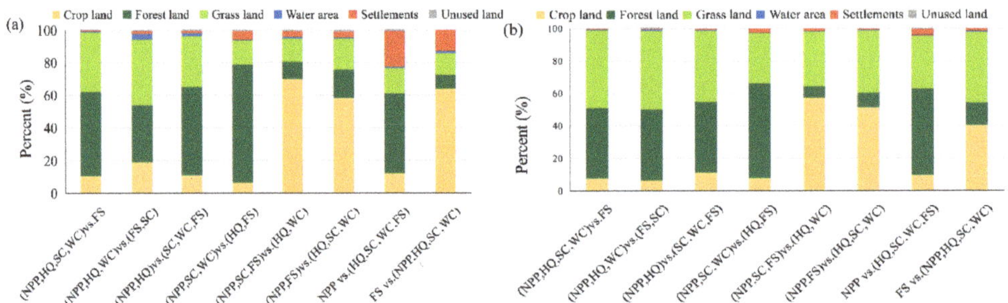

Figure 7. Trade-offs and synergies among multiple ecosystem services in different land use types.

In the Hanzhong Basin, the synergies occurred among the NPP, HQ, SC, and WC (simultaneous increase), and exhibited trade-offs with FS that were aggregated in the crop land and forest land, which accounted for 48.02% and 42.98%. At the same time, the trade-offs among NPP and HQ (simultaneous increase, suggesting synergies), and SC, WC, FS (simultaneous decrease) was also concentrated in crop land and forest land. While the trade-offs among NPP, SC, FS (simultaneous increase, suggesting synergies), and WC, HQ (simultaneous decrease) accounted for 57.10% and 31.29% in crop land and grass land, respectively. It was found in the forest land and crop land that accounted for 38.72% and 57.27% where the trade-offs occurred among NPP and FS (simultaneous increase, suggesting synergies), and HQ, SC, and WC (simultaneous decrease). Furthermore, the trade-offs among FS and other services NPP, HQ, SC, and WC (simultaneous decrease) were aggregated in grass land and crop land, which accounted for 43.89% and 40.55%. Overall, the trade-off relationships were mainly concentrated on crop land and forest land in the Guanzhong Basin, while, in the Hanzhong Basin, these were aggregated in forest land and grass land.

4. Discussion

4.1. Difference Analysis of the Guanzhong and Hanzhong Basin

Because of this unique geographical location and climate environment, the land use structure and vegetation coverage exit spatial differences in the Guanzhong Basin and Hanzhong Basin, which directly generate the spatial heterogeneity of ecosystem services and the spatial correlation relationships between ecosystem services in two basins. For example, in terms of the land use structure, crop land was dominated in the Guanzhong Basin that accounted for 45.17%, and provided the provisioning services even though it has gradually decreased in recent years, which was concentrated in the central area of the basin and surrounding urban land of the bigger city. In comparison, the ecological environment is better in the Hanzhong Basin. Vegetation coverage is relatively high and the land use types were dominated by grassland and forest land, which accounted for 41.08% and 30.11%, respectively, concentrated in the edge of the basin. Consequently, the provisioning service was higher in the Guanzhong Basin and the regulating service was little bigger in the Hanzhong Basin. Furthermore, accompanying with other natural environment factors such as rainfall or spatial patterns of vegetation and different soil types, it determined the distinct temporal and spatial distribution characteristics of each ecosystem service in two basins. Under the interaction of various ecosystem services, trade-offs, and synergies presented different spatial and temporal patterns in two basins. For instance, our results showed the correlation direction between paired ES, which was the same in two basins, but the strength and change rate of trade-offs and synergies was strong in the Guanzhong Basin than that in the Hanzhong Basin. As the rapid economic development and the expansion of construction land, ecosystem services have suffered a down trend in the Guanzhong Basin over the past years. Meanwhile, it increased the conflict of human demand for the ecosystem. At the same time, results showed the strong trade-offs were discovered in the central region of the Guanzhong Basin, particularly in the edge of Xi'an City as the capital of Shaanxi Province. The construction land was expended with the population growth. The demand of supply services gradually increased, while the value of regulating services continued to decline, so it became the higher decreasing areas of ecosystem services, which caused the strong trade-offs [42]. Furthermore, the regulating services were mostly provided by forest land, grassland, and water areas. The trade-off relationships became strong since these areas decreased. For example, in the southeast region of the Hanzhong Basin, the conversion of grassland to cropland lead to strong trade-offs between FS and SC. Therefore, the local government should formulate the corresponding target and ecological restoration approach, according to the regional ecological demand and social development, in order to realize the sustainable supply of ecosystem services [24,43].

4.2. Temporal and Spatial Changes of Trade-Offs and Synergies

Our results showed the synergies between NPP, habitat quality, soil conservation, and water conservation that existed in two basins. This result agrees with other published findings [18,33,41]. Nevertheless, when the trade-off or synergy is weak in a region, it could be easily changed from trade-offs to synergies or inverse under the influence of local policy and planning [20]. Since 2001, the government has implemented the Reforestation of Cultivated Land project in the Shaanxi Province. Some deep slope of crop land is converted into forest land or grassland. Consequently, services provide by crop land were reduced, whereas water conservation services by grassland were greatly improved. It was discovered that relationships between NPP and WC as well as SC and WC were changed into negative relationships in 2010 in two basins because, compared with 2005, the increased rate of WC was much higher than that of SC, while NPP showed a decreasing trend. Although its crop land has decreased, demand for food supply in the urban area was enhanced, and it could enhance the giving services in grassland and water areas. Consequently, the conflict relationship between FS and WC was reduced, even when presented with a synergic relationship in the Guanzhong Basin in 2010. Therefore, the research found the trade-offs and synergies between ESs have temporal variations under the influence of different land use structures, environment factors, and social development demand (e.g., urban expansion) [44].

In addition, the landscape structure determines the ability of the ecosystem to provide the service, which humans ultimately depend on [45]. Ecosystem service relationships are affected by land use conflict [46,47]. Furthermore, spatial correlation analysis illustrates the differences of spatial distribution of two ecosystem services for a given year, ignoring the prerequisites of dynamic changes of ecosystem services and true interaction when considering the trade-offs and synergies [25]. Therefore, based on the correlation coefficients of two time series on the gird scale, we could further explore the spatial distribution of trade-offs and synergies between paired ES. Among these trade-offs, those between regulating services and the food supply service have drawn more attention. For example, our results showed a negative relationship between the food supply and soil conservation based on the whole region. However, the spatial synergistic relationships between FS and SC were widely distributed in the two basins, especially in the central region of the Guanzhong Basin. This result may be interpreted as well managed and high yield farmland in the central region of the valley basin, which is beneficial for soil conservation. Both services showed an increasing trend simultaneously. On the other hand, it could be explained that the synergy disappeared, which showed a negative relationship between two services because of the data integration and land use conflict on the whole region scale. For instance, both forest land and grass land could generate the benefit of soil conservation and food supply where it may represent synergy, but their closeness to soil conservation and food supply was inconsistent. Moreover, the closeness may be changed over time [25]. Spatial scale and temporal change play the important roles in the relationship between paired ES. Meanwhile, Felipe suggested there was no single relevant scale to analyze the relationships among multiple ESs [48]. Consequently, more place-based studies with sensitivity analysis are needed for our further understanding of the spatiotemporal dynamic interactions among multiple ESs [49] in order to select the optimized spatial range for achieving the highest values of various ecosystem services.

4.3. Multiple ESs Interactions

There is no generalizable theoretical basis to ensure the balance of economic and ecology. Thus, acquiring knowledge of how multiple ESs interactions occur locally is more likely to achieve a win-win situation [50]. Contrasted with previous studies that focused on trade-offs and synergies between paired ESs [34,41], we explored the multiple ESs interactions based on the grid cells. Regarding the inherent complexity of integrated social and ecological systems, most of the ecosystem services interact with one another. A simple consideration of only ES might generate an unexpected and dramatic decline in other ESs [51]. Trade-offs often occur when provisioning service is increased as a consequence of the decrease use of other services. Nevertheless, oversupply of the services may result in the trade-offs with other services or an unsustainable eco-environment, which causes a conflict between human demand and ecological protection. Furthermore, each of the ecosystem services presented a distinct change based on the gird scales, which reflected the multiple interactions among ES. For instance, results show the trade-offs between NPP and FS (simultaneously increased, suggesting synergy) and HQ, SC, and WC (simultaneously decreased) were widespread in two basins when compared with other multiple interactions. This phenomenon indicated that, when it is transformed from grassland to forestland or from grassland to crop land, both NPP and food supply simultaneously increased and appeared as synergy. Based on the different land use change, identified the causes, and the locations of the multiple interactions among ES, which could help decision-makers develop targeted ecosystem management strategies [52].

4.4. Limitations and Future Study Directions

It is necessary to consider the influencing mechanism and future scenario predictions in a future study of ecosystem services [4]. For instance, future scenarios aim to provide a theoretical basis for the government's ecological planning and recommendations by the simulation of ecosystem services in different land structure allocations [11]. Furthermore, the current ecological protection policy directly increases NPP, SC, and WC and some regulating services lead to the decrease of food supply services and other provisioning services, which might be unable to meet human needs. Therefore, future research aims to develop an integrated framework for simulating ESs based on the combination of land use type, climate change, government policy, and topography factors. At the same time, it could analyze the region difference and multiple ESs spatial interactions, and then select the appropriate scenario.

In this study, the ecosystem services simulation was based on a raster dataset but lacked cultural services. Under the strategic background of "One Belt and One Road" and "Guanzhong-Tianshui economic zone," local social and economic developments have faced unprecedented opportunities. The social cultural value would play an important role in ecosystem functions, and there could also be a large uncertainty about the impact of human activities on the ecosystem process [38]. How to estimate the cultural services more reasonably, and how to simulate future ecosystem services under the influence of social-economic and eco-environment factors in particular, will be the main factors focused on in future research studies.

5. Conclusions

The Guanzhong Basin and Hanzhong Basin were selected as case study areas. Our research quantified the spatial relationship among multiple ESs (NPP, HQ, WC, SC, and FS) and explored the spatial distribution of multiple ESs interactions based on the gird scales. Results showed the direction of the correlation coefficient between paired ES was the same in two basins, but the extent of the correlation coefficient was stronger in the Guanzhong Basin than that in the Hanzhong Basin. Meanwhile, our results demonstrated the spatial trade-off relationships between paired ES were spatially aggregated in the central and the southwest of the Guanzhong Basin, and the southeast of the Hanzhong Basin. Furthermore, the multiple ES interactions were spatially heterogeneous on the gird scales across two basins. Moreover, land use change might cause the various trade-offs among multiple ES. For example, the conversion of crop land to forest land lead to NPP, HQ, SC, and WC to continuously increase and exhibited trade-offs with FS in the Guanzhong Basin. While in the Hanzhong Basin, the conversion of grassland to crop land lead to a continuous increase in NPP and FS and exhibited trade-off interactions with the three services SC, WC, and HQ. This information may help policymakers develop targeted and local ecological management measures. Furthermore, our funding could provide a theoretical basis for the sustainable development of society, economy, and ecology in Northwest China.

Author Contributions: Conceptualization, Y.S. and Z.R. Data curation, Y.D. Funding acquisition, J.L. and Z.R. Methodology, Y.S. Software, Y.S. and Y.D. Validation, J.L. and X.L. Formal analysis, Y.S. Investigation, Y.S. and X.L. Resources, Y.S. and Y.D. Writing-original draft preparation, Y.S. and J.L. Writing-review and editing, J.L., X.L., Z.R., Z.Z., and Y.D. Visualization, Y.S. Supervision, Z.R. Project administration, J.L., X.L., and Z.Z. All authors have read and agreed to the published version of the manuscript.

Funding: The National Natural Science Foundation of China (No. 41771198, 41771576 and 41801333), the Fundamental Research Funds for the Central Universities (No.GK 201901009), the NSFC - NRF Scientific Cooperation Program (Grant no. 41811540400), and the project supported by the Natural Science Basic Research Plan in Shaanxi Province of China (Program No. 2018JM4010), and China Postdoctoral Science Foundation (2019M650859 and 2019T120142) funded this research.

Acknowledgments: Acknowledgement for the meteorological data support from "National Earth System Science Data Center, National Science & Technology Infrastructure of China. (http://www.geodata.cn)." We thank the Resource and Environmental Data Cloud Platform for sharing the Land-Use/cover datasets, which could be downloaded from http://www.resdc.cn/. We are grateful to the anonymous reviewers and Jing Li (School of Geography and Tourism in Shaanxi Normal University) and Xianfeng Liu (School of Geography and Tourism in Shaanxi Normal University) for their valuable comments and suggestions that greatly improve the manuscript.

Conflicts of Interest: The authors declare no conflict of interest.

Appendix A

In this study, multiple interactions among ESs were calculated by spatial overlay analysis based on ArcGIS. The specific calculation steps are as follows.

Step 1 We set up a set of five digit codes and make each ES correspond to one digit, respectively.

Table A1. Different digits codes for each ecosystem service.

Digits Code	Ten Thousand Digits	Thousand Digits	Hundred Digits	Ten Digits	Single Digits
ES	NPP	HQ	SC	WC	FS

Step 2 We made the subtraction operations on each ES in 2000 and 2018 by the ArcGIS raster calculation, and reclassified the difference value 1, 2, and 3 to increased, reduced, and no change pixel, respectively.

2018 NPP			2010 NPP		
23	15	9	13	17	10
20	24	17	15	20	19
28	6	21	22	6	12
14	13	11	7	15	8

2018 NPP–2010NPP				NPP_Reclassified		
10	−2	−1		10,000	20,000	20,000
5	4	−2	*reclassify* ⇒	10,000	10,000	20,000
6	0	9		10,000	30,000	10,000
7	−2	3		10,000	20,000	10,000

Step 3 We performed spatial overlay operations for the reclassified layers, and recognized the interactions among multiple ecosystem services by interpreting the codes in the pixels of the output layers of overlay analysis.

NPP_Reclassified			HQ_Reclassified			SC_Reclassified		
10,000	20,000	20,000	02000	03000	02000	00200	00200	00200
10,000	10,000	20,000	01000	02000	02000	00200	00200	00100
10,000	30,000	10,000	02000	02000	02000	00100	00200	00200
10,000	20,000	10,000	02000	02000	01000	00200	00100	00100

WC_Reclassified			FS_Reclassified		
00010	00020	00020	00002	00001	00001
00020	00020	00020	00001	00002	00001
00020	00010	00020	00002	00001	00001
00010	00010	00010	00001	00001	00002

Spatial Overlay

ES_Layer_Overlaid		
12,212	23,221	22,221
11,221	12,222	22,121
12,122	32,211	12,221
12,211	22,111	11,112

The results of the table indicates the locations where multiple interactions among ecosystem services occurred based on the gird cell. For example: the code 12212" indicates that NPP, WC increased simultaneously (suggesting synergies), HQ, SC, FS decreased simultaneously, and the two services NPP, WC both exhibited trade-offs with the three services HQ, SC, and FS.

References

1. Costanza, R.; De Groot, R.; Farber, S.; Grasso, M.; Hanno, B.; Karin, L. The value of the world's ecosystem services and natural capital. *Nature* **1997**, *25*, 3–15.
2. Liu, J.; Li, J.; Qin, K.; Zhou, Z.; Yang, X.; Li, T. Changes in land-uses and ecosystem services under multi-scenarios simulation. *Sci. Total Environ.* **2017**, *586*, 522–526. [CrossRef] [PubMed]
3. Costanza, R.; De Groot, R.; Braat, L.; Kubiszewski, I.; Fioramonti, L.; Sutton, P.; Grasso, M. Twenty years of ecosystem services: How far have we come and how far do we still need to go? *Ecosyst. Serv.* **2017**, *28*, 1–16. [CrossRef]
4. Li, B.; Chen, N.; Wang, Y.; Wang, W. Spatio-temporal quantification of the trade-offs and synergies among ecosystem services based on grid-cells: A case study of Guanzhong Basin, NW China. *Ecol. Indic.* **2018**, *94*, 246–253. [CrossRef]
5. Li, S.; Zhang, C.; Liu, J.; Zhu, W.; Ma, C.; Wang, Y. The tradeoffs and synergies of ecosystem services: Research progress, development trend, and themes of geography. *Geogr. Res.* **2013**, *32*, 1379–1390.
6. Briner, S.; Huber, R.; Bebi, P.; Elkin, C.; Schmatz, D.R.; Grêt-Regamey, A. Trade-offs between ecosystem services in a mountain region. *Ecol. Soc.* **2013**, *18*, 35. [CrossRef]
7. Bennett, E.M.; Peterson, G.D.; Gordon, L.J. Understanding relationships among multiple ecosystem services. *Ecol. Let.* **2009**, *12*, 1394–1404. [CrossRef]
8. Turkelboom, F.; Thoonen, M.; Jacobs, S.; García-Llorente, M.; Martín-López, B.; Berry, P. Ecosystem services trade-offs and synergies (draft). Available online: www.openness-project.eu/library/reference-book (accessed on 16 January 2016).
9. Tilman, D.; Cassman, K.G.; Matson, P.A.; Naylor, R..; Polasky, S. Agricultural sustainability and intensive production practices. *Nature* **2002**, *418*, 671–677. [CrossRef]
10. Han, Z.; Song, W.; Deng, X.; Xu, X. Trade-Offs and Synergies in Ecosystem service within the Three-Rivers Headwater Region, China. *Water* **2017**, *9*, 588. [CrossRef]
11. Fu, Q.; Hou, Y.; Wang, B.; Bi, X.; Li, B.; Zhang, X. Scenario analysis of ecosystem service changes and interactions in a mountain-oasis-desert system: A case study in Altay Prefecture, China. *Sci. Rep. UK* **2018**, *8*, 12939. [CrossRef] [PubMed]
12. Torralba, M.; Fagerholm, N.; Hartel, T.; Moreno, G.; Plieninger, T. A social-ecological analysis of ecosystem services supply and trade-offs in European wood-pastures. *Sci. Adv.* **2018**, *4*, 2176. [CrossRef] [PubMed]
13. Ament, J.M.; Moore, C.A.; Herbst, M.; Cumming, G.S. Cultural ecosystem services in protected areas: Understanding bundles, trade-offs, and synergies. *Conserv. Lett.* **2017**, *10*, 440–450. [CrossRef]
14. Cord, A.F.; Bartkowski, B.; Beckmann, M.; Dittrich, A.; Hermans-Neumann, K.; Kaim, A.; Lienhoop, N.; Locher-Krause, K.; Priess, J.; Schröter-Schlaack, C.; et al. Towards systematic analyses of ecosystem service trade-offs and synergies: Main concepts, methods and the road ahead. *Ecosyst. Serv.* **2017**, *28*, 264–272. [CrossRef]
15. Bai, Y.; Zhuang, C.; Ouyang, Z.; Zheng, H.; Jiang, B. Spatial characteristics between biodiversity and ecosystem services in a human-dominated watershed. *Ecol. Complex.* **2011**, *8*, 177–183. [CrossRef]
16. Howe, C.; Suich, H.; Vira, B.; Mace, G.M. Creating win-wins from trade-offs? Ecosystem services for human well-being: A meta-analysis of ecosystem service trade-offs and synergies in the real world. *Glob. Environ. Chang.* **2014**, *28*, 263–275. [CrossRef]
17. Plieninger, T.; Torralba, M.; Hartel, T.; Fagerholm, N. Perceived ecosystem services synergies, trade-offs, and bundles in European high nature value farming landscapes. *Landscape Ecol.* **2019**, *34*, 1565–1581. [CrossRef]
18. Raudsepp-Hearne, C.; Peterson, G.D.; Bennett, E.M. Ecosystem service bundles for analyzing tradeoffs in diverse landscapes. *Proc. Natl. Acad. Sci. USA* **2010**, *107*, 5242–5247. [CrossRef]
19. Demestihas, C.; Plénet, D.; Génard, M.; Raynal, C.; Lescourret, F. A simulation study of synergies and tradeoffs between multiple ecosystem services in apple orchards. *J. Environ. Manag.* **2019**, *236*, 1–16. [CrossRef]
20. Wang, Y.; Li, X.; Zhang, Q.; Li, J.; Zhou, X. Projections of future land use changes: Multiple scenarios-based impacts analysis on ecosystem services for Wuhan city, China. *Ecol. Indic.* **2018**, *94*, 430–445. [CrossRef]

21. Grasso, M. Ecological–economic model for optimal mangrove trade-off between forestry and fishery production: Comparing a dynamic optimization and a simulation model. *Ecol. Model.* **1998**, *112*, 131–150. [CrossRef]

22. Seppelt, R.; Lautenbach, S.; Volk, M. Identifying trade-offs between ecosystem services, land use, and biodiversity: A plea for combining scenario analysis and optimization on different spatial scales Ralf Seppelt1, Sven Lautenbach2 and Martin Volk1. *Curr. Opin. Env. Sust.* **2013**, *5*, 458–463. [CrossRef]

23. Bohensky, E.L.; Reyers, B.; Jarrsveld, A.S.V. Conservation in practice: Future ecosystem services in a Southern African River Basin: A scenario planning approach to uncertainty. *Conserv. Biol.* **2006**, *20*, 1051–1061. [CrossRef] [PubMed]

24. Gong, J.; Liu, D.; Zhang, J.; Xie, Y.; Cao, E.; Li, H. Tradeoffs/synergies of multiple ecosystem services based on land use simulation in a mountain-basin area, western China. *Ecol. Indic.* **2019**, *99*, 283–293. [CrossRef]

25. Zheng, Z.; Fu, B.; Hu, H.; Sun, G. A method to identify the variable ecosystem services relationship across time: A case study on Yanhe Basin, China. *Landscape Ecol.* **2014**, *29*, 1689–1696. [CrossRef]

26. Zhang, Y.M.; Li, J.; Zeng, L.; Yang, X.; Liu, J.; Zhou, Z. Optimal protected area selection: Based on multiple attribute decision making method and ecosystem service research—Illustrated by Guanzhong-Tianshui Economic Region section of the Weihe River Basin. *Sci. Agric. Sin.* **2019**, *52*, 2114–2127.

27. Peng, J.; Hu, X.; Wang, X.; Meersmans, J.; Liu, Y.; Qiu, S. Simulating the impact of Grain-for-Green Programme on ecosystem services trade-offs in Northwestern Yunnan, China. *Ecosyst. Serv.* **2019**, *39*, 100998. [CrossRef]

28. Qin, K.; Li, J.; Liu, J.; Yan, L.; Huang, H. Setting conservation priorities based on ecosystem services - A case study of the Guanzhong-Tianshui Economic Region. *Sci. Total Environ.* **2019**, *650*, 3062–3074. [CrossRef]

29. Hadian, F.; Jafari, R.; Bashari, H.; Tartesh, M.; Clarke, K.D. Estimation of spatial and temporal changes in net primary production based on Carnegie Ames Stanford Approach (CASA) model in semi-arid rangelands of Semirom County, Iran. *J. Arid Land.* **2019**, *11*, 477–494. [CrossRef]

30. Potter, C.S.; Randerson, J.T.; Field, C.B.; Matson, P.A.; Vitousek, P.M.; Mooney, H.A.; Klooster, S.A. Terrestrial ecosystem production: A process model based on global satellite and surface data. *Global Biogeochem Cy.* **1993**, *7*, 811–841. [CrossRef]

31. Loreau, M.; Naeem, S.; Inchausti, P.; Bengtsson, J.; Grime, J.P.; Hector, A.; Hooper, D.U.; Huston, M.A.; Raffaelli, D.; Schmid, B.; et al. Biodiversity and ecosystem functioning: Current knowledge and future challenges. *Science* **2001**, *294*, 804–808. [CrossRef]

32. Hector, A.; Bagchi, R. Biodiversity and ecosystem multifunctionality. *Nature* **2007**, *448*, 188–190. [CrossRef] [PubMed]

33. Hou, Y.; Lü, Y.; Chen, W.; Fu, B. Temporal variation and spatial scale dependency of ecosystem service interactions: A case study on the central Loess Plateau of China. *Landscape Ecol.* **2017**, *32*, 1201–1217. [CrossRef]

34. Asadolahi, Z.; Salmanmahiny, A.; Sakieh, Y.; Mirkarimi, S.H.; Baral, H.; Azimi, M. Dynamic trade-off analysis of multiple ecosystem services under land use change scenarios: Towards putting ecosystem services into planning in Iran. *Ecol. Complex.* **2018**, *36*, 250–260. [CrossRef]

35. Bao, Y.; Liu, K.; Li, T. Effects of land use change on habitat based on InVEST model: Taking yellow river wetland nature reserve in Shaanxi province as an example. *Arid Zone Res.* **2015**, *32*, 622–629.

36. Biao, Z.; Wenhua, L.; Gaodi, X.; Yu, X. Water conservation of forest ecosystem in Beijing and its value. *Ecol. Econ.* **2010**, *69*, 1416–1426. [CrossRef]

37. Li, J.; Ren, Z. Spatiotemporal change of water conservation value of Loess Plateau ecosystem in northern Shaanxi Province. *Chin. J. Ecol.* **2008**, *27*, 240–244.

38. Shi, Z.H.; Cai, C.F.; Ding, S.W.; Wang, T.W.; Chow, T.L. Soil conservation planning at the small watershed level using RUSLE with GIS: A case study in the Three Gorge Area of China. *CATENA* **2004**, *55*, 33–48. [CrossRef]

39. Liu, B.Y.; Nearing, M.A.; Shi, P.J.; Jia, Z.W. Slope length effects on soil loss for steep slopes. *Soil Sci. Soc. Am. J.* **2000**, *64*, 1759. [CrossRef]

40. Yang, X.; Zhou, Z.; Li, J.; Fu, X.; Mu, X.; Li, T. Trade-offs between carbon sequestration, soil retention and water yield in the Guanzhong-Tianshui Economic Region of China. *J. Geogr. Sci.* **2016**, *26*, 1449–1462. [CrossRef]

41. Power, A.G. Ecosystem services and agriculture: Tradeoffs and synergies. *Philos. Trans. R. Soc. B Biol. Sci.* **2010**, *365*, 2959–2971. [CrossRef]

42. Sun, Y.; Ren, Z.; Zhao, S.; Zhang, J. Spatial and temporal changing analysis of synergy and trade-off between ecosystem services in valley basins of Shaanxi Province. *Acta Geogr. Sin.* **2017**, *72*, 521–532.

43. Zhu, Y.; Zhongke, F.; Lu, J.; Liu, J. Estimation of forest biomass in Beijing (China) using multisource remote sensing and forest inventory data. *Forest* **2020**, *11*, 163. [CrossRef]

44. Qiao, X.; Gu, Y.; Zou, C.; Xu, D.; Wang, L.; Ye, X.; Yang, Y.; Huang, X. Temporal variation and spatial scale dependency of the trade-offs and synergies among multiple ecosystem services in the Taihu Lake Basin of China. *Sci. Total Environ.* **2019**, *651*, 218–229. [CrossRef] [PubMed]

45. Fan, M.; Xiao, Y. Impacts of the grain for Green Program on the spatial pattern of land uses and ecosystem services in mountainous settlements in southwest China. *Glob. Ecol. Conserv.* **2020**, *21*, e806. [CrossRef]

46. Xu, S.; Liu, Y.; Wang, X.; Zhang, G. Scale effect on spatial patterns of ecosystem services and associations among them in semi-arid area: A case study in Ningxia Hui Autonomous Region, China. *Sci. Total Environ.* **2017**, *598*, 297–306. [CrossRef]

47. Li, S.; Bing, Z.; Jin, G. Spatially explicit mapping of soil conservation service in monetary units due to land use/cover change for the Three Gorges Reservoir Area, China. *Remote Sens. (Basel)* **2019**, *11*, 468. [CrossRef]

48. Felipe-Lucia, M.R.; Comín, F.A.; Bennett, E.M. Interactions among ecosystem services across land uses in a floodplain agroecosystem. *Ecol. Soc.* **2014**, *19*, 20. [CrossRef]

49. Yi, H.; Güneralp, B.; Kreuter, U.P.; Güneralp, İ.; Filippi, A.M. Spatial and temporal changes in biodiversity and ecosystem services in the San Antonio River Basin, Texas, from 1984 to 2010. *Sci. Total Environ.* **2018**, *619–620*, 1259–1271. [CrossRef]

50. Li, T.; Lü, Y.; Fu, B.; Hu, W.; Comber, A.J. Bundling ecosystem services for detecting their interactions driven by large-scale vegetation restoration: Enhanced services while depressed synergies. *Ecol. Indic.* **2019**, *99*, 332–342. [CrossRef]

51. Fu, Q.; Li, B.; Hou, Y.; Bi, X.; Zhang, X. Effects of land use and climate change on ecosystem services in Central Asia's arid regions: A case study in Altay Prefecture, China. *Sci. Total Environ.* **2017**, *607–608*, 633–646. [CrossRef]

52. Aurenhammer, P.K. Nudging in the Forests—the Role and Effectiveness of NEPIs in Government Forest Initiatives of Bavaria. *Forests* **2020**, *11*, 168. [CrossRef]

Article

Influence of Site-Specific Conditions on Estimation of Forest above Ground Biomass from Airborne Laser Scanning

Jan Novotný [1,*], Barbora Navrátilová [1], Růžena Janoutová [1] and Filip Oulehle [1,2] and Lucie Homolová [1]

1   Global Change Research Institute of the Czech Academy of Sciences, Bělidla 986/4a,
    603 00 Brno, Czech Republic
2   Czech Geological Survey, Klárov 131/3, 118 21 Prague, Czech Republic
*   Correspondence: novotny.j@czechglobe.cz

Received: 30 January 2020; Accepted: 23 February 2020; Published: 27 February 2020

Abstract: Forest aboveground biomass (AGB) is an important variable in assessing carbon stock or ecosystem functioning, as well as for forest management. Among methods of forest AGB estimation laser scanning attracts attention because it allows for detailed measurements of forest structure. Here we evaluated variables that influence forest AGB estimation from airborne laser scanning (ALS), specifically characteristics of ALS inputs and of a derived canopy height model (CHM), and role of allometric equations (local vs. global models) relating tree height, stem diameter (DBH), and crown radius. We used individual tree detection approach and analyzed forest inventory together with ALS data acquired for 11 stream catchments with dominant Norway spruce forest cover in the Czech Republic. Results showed that the ALS input point densities (4–18 pt/m^2) did not influence individual tree detection rates. Spatial resolution of the input CHM rasters had a greater impact, resulting in higher detection rates for CHMs with pixel size 0.5 m than 1.0 m for all tree height categories. In total 12 scenarios with different allometric equations for estimating stem DBH from ALS-derived tree height were used in empirical models for AGB estimation. Global DBH models tend to underestimate AGB in young stands and overestimate AGB in mature stands. Using different allometric equations can yield uncertainty in AGB estimates of between 16 and 84 tons per hectare, which in relative values corresponds to between 6% and 37% of the mean AGB per catchment. Therefore, allometric equations (mainly for DBH estimation) should be applied with care and we recommend, if possible, to establish one's own site-specific models. If that is not feasible, the global allometric models developed here, from a broad variety of spruce forest sites, can be potentially applicable for the Central European region.

Keywords: norway spruce; LiDAR; allometric equation; individual tree detection; tree height; diameter at breast height; GEOMON

---

1. Introduction

Forests provide multiple ecosystem services at various spatial scales and constitute an important sink of sequestered atmospheric carbon [1,2]. Carbon stocks are an important input in climate models, and responsible forest management is a way to mitigate the impact of global climate change [3]. Accurate and consistent estimation of AGB can help to reduce current uncertainties regarding carbon fluxes [4,5]. Several methods exist for calculating AGB, these range from destructive methods to satellite-based estimations. Estimation can be improved by combining multiple approaches [6,7].

Traditionally, forest AGB and carbon stocks have been assessed by measuring tree dimensions in permanent field plots and then using allometric equations (e.g., [8,9]). In recent years, we have seen a move toward remote sensing as the primary tool for monitoring forest AGB and carbon stocks (e.g., [10]). The rapid development of laser scanning systems is making them great tools for estimating tree heights and forest biomass [11,12]. Airborne laser scanning (ALS), in particular, has been established as a standard technology for high-precision three-dimensional topographic data acquisition [13]. It has a powerful penetrating ability and can obtain vertical structure information for forests, thereby improving accuracy in estimating forest height and structure [14,15]. ALS is mostly operated in form of small-footprint discrete-return or waveform systems and allows for estimating AGB using area-based or individual tree detection approaches [16–20]. Because remote sensing-based AGB estimates are nevertheless indirect, high quality and large quantities of traditional field inventory data and AGB destructive sampling are still needed.

AGB estimates for both approaches depend upon canopy characteristics, such as tree branching and foliage structure, because these elements intercept laser pulses in various canopy height layers. Consequently, the accuracy of ALS-based AGB models can vary by species and forest types [21]. Bouvier et al. [22] demonstrated, that plot size significantly impacted AGB model performance within pine forest in southwestern France. The accuracy of ALS-based models can potentially vary with the ALS data point density. For instance, Montagnoli et al. [23] investigated the utility of low-density ALS data (<2 pt/m^2) for estimating AGB within mixed broad-leaved forest in Italy. Wu et al. [24] reported an overall decreasing trend in error of ALS-derived AGB estimates as LiDAR point density increased from 0.5 to 8 pt/m^2 within forested landscape in east-central Arizona. Brovkina et al. [25] used high point density LiDAR data (50 pt/m^2) for AGB estimation within mixed spruce and beech forest in the Czech Republic.

The main objective of this study is to obtain deeper insight into variables that influence estimation of aboveground biomass from ALS data. Specifically, we will evaluate (i) the effect of various ALS point densities and spatial resolutions of input canopy height model rasters on individual tree detection, and (ii) the impact and transferability of site-specific relationships between tree height and stem diameter at breast height (DBH) on estimates of Norway spruce forest AGB in the Czech Republic. The first part of our study should provide suggestions which ALS data to choose for operational AGB estimation. The second part should fill a gap in assessment of uncertainty in AGB estimates caused by usage of allometric equation from different sources as this issue is rarely addressed in scientific literature.

2. Materials and Methods

The overall methodology of spruce forest AGB estimation from ALS data using the individual tree detection approach and testing the effects of various ALS inputs and allometric equations is depicted in Figure 1 and is described in further detail in the following sections.

2.1. Study Sites

Forest inventory and airborne laser scanning (ALS) data for this study were collected at 11 small forested stream catchments that are located across the Czech Republic (Figure 2). The catchments are part of the GEOMON network that was established primarily to study long-term effects of decreasing acid depositions on the recovery of forested ecosystems, and on soil and water chemistry. Field observations was begun in 1994 [26,27]. The catchments are predominantly covered by productive forests of Norway spruce (Picea abies L. Karst) and they represent a large variety of environmental conditions in the Czech Republic. Catchments' characteristics are summarized in Table 1, and further details are provided by Oulehle et al. [27].

In addition to the sites from the GEOMON network, we also used forest inventory data from another site to test the impact of local allometric equations on AGB estimation. That external site, known as Bílý Kříž, is located in a mountainous area near the Czech Republic and Slovakia borders within the Moravian-Silesian Beskydy Mountains (49°30'N, 18°32'E, 800 to 920 m a.s.l.). It is an experimental research site for long-term monitoring of CO_2 and energy fluxes of a Norway spruce forest monoculture (40 years old in 2018) and, as such, it is included into Integrated Carbon Observation

System (ICOS). Site characteristics are reported in Table 1. Hereinafter, this site is referred to as "BK" and allometric equations computed for this site are labeled as "outside".

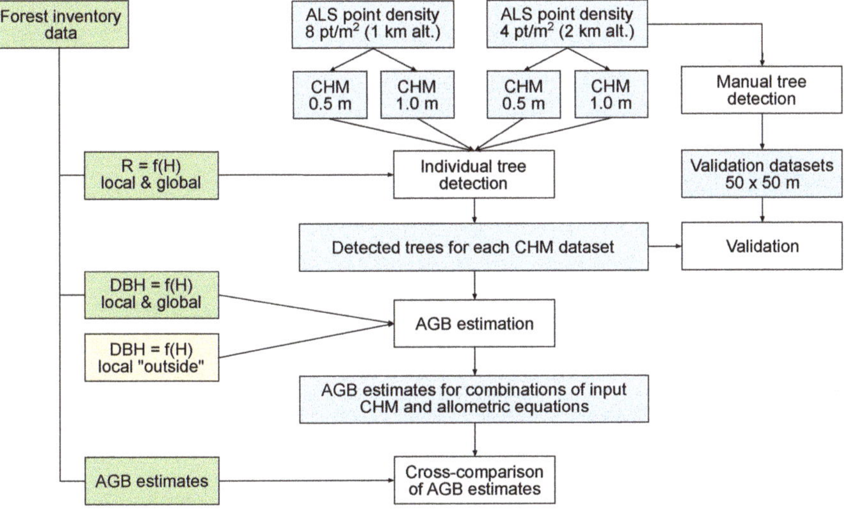

Figure 1. Methodological design of this study (DBH = diameter at breast height, R = crown radius, H = tree height, AGB = aboveground biomass, ALS = airborne laser scanning, CHM = canopy height model).

Figure 2. Location of small forested stream catchments within the GEOMON long-term monitoring network (circles with labels) and the Bílý Kříž site (cross with BK label) in the Czech Republic. The sites are characterized (and abbreviations explained) in Table 1.

Table 1. Characteristics of small forested stream GEOMON catchments located across the Czech Republic (mean characteristics computed for the period 1994–2016). Last row reports characteristics of an external Bílý Kříž site (computed for the period 1988–2018).

Catchment Name (Abbreviation)	Area (ha)	Elevation (m a.s.l.)	Mean temp. (°C)	Mean ann. precip. (mm)	Forest Floor pH	Mineral Soil pH	Bedrock Type	Forest Age (yr)
Anenský potok (ANE)	27	480–540	8.0	673	4.00	4.67	Paragneiss	40–60
Červík (CER)	185	640–960	6.0	1212	3.40	4.13	Sandstone, Claystone	40–60
Na Lizu (LIZ)	98	830–1025	5.5	892	3.50	4.38	Paragneiss	60–100
Loukov (LKV)	66	470–660	7.5	754	3.64	4.33	Granite	40–80
Lysina (LYS)	27	830–950	5.0	1005	3.37	4.33	Granite	40–60
Na Zeleném (NAZ)	60	740–800	6.0	800	3.86	5.07	Amphibolite	60–100
Pluhův bor (PLB)	22	690–800	6.0	861	3.85	5.82	Serpentinite	60–100
Polomka (POM)	69	510–640	7.0	697	3.77	4.37	Paragneiss, Orthogneiss	61–80
Salačova Lhota (SAL)	168	560–7745	7.0	675	3.46	4.30	Paragneiss	60–80
U dvou louček (UDL)	33	880–950	5.0	1502	3.75	4.43	Gneiss	20–40
Uhlířská (UHL)	187	780–870	5.5	1250	3.50	4.33	Granite, Granodiorite	20–40
Bílý Kříž (BK)	0.25	800–920	6.3	1230	n.a.	n.a.	Sandstone, Claystone	40

2.2. Field Data

Field data (i.e., tree species identification, stem diameter at breast height (DBH), tree height (H), crown radius (R), height to crown base and crown completeness) were measured at the GEOMON sites during the 2015 vegetation season and in the case of NAZ during 2016. These parameters were recorded for all trees with DBH greater than 5 cm within a circular plot of 500 m² (r = 12.6 m). In total, 81 forest inventory plots were sampled and the numbers of plots varied between 5 and 10 per catchment according to the area of each. Selection of the plots followed stratified random sampling, where strata were determined based on forest management maps, as well as local soil and orography conditions. The center of each plot was recorded using differential GPS with a horizontal accuracy of about 1 m. Mean values of the measured tree parameters per catchment are summarized in Table 2.

Field measurements for H and DBH of Norway spruce trees at the external site Bílý Kříž are recorded differently than at the GEOMON sites. This dataset contains H and DBH measurements that have been recorded in every vegetation season since 1997 for all trees with DBH > 2.0 cm growing within a fenced area (i.e., 290 trees measured in 2017, details in Table 2).

Field measurements of tree parameters were primarily used to construct empirical models (so-called allometric equations), which were either specific for each GEOMON catchment (local models) or for all catchments together (global models) or for the external site (outside models). The first set of allometric equations was established between R and H using a linear model (Equation (1)):

$$R = aH + b \tag{1}$$

Estimates of R were used to tune a tree detection algorithm (described in Section 2.4). The second set of allometric equations was established between DBH and H using an exponential model (Equation (2), further labeled as local1, global1, or outside1) and a power model (Equation (3), further labeled as local 2, global 2, or outside 2):

$$DBH = exp(a + bH) \tag{2}$$

$$DBH = aH^b \tag{3}$$

Table 2. Forest characteristics per stream catchment as measured in field surveys during 2015–2016. Mean values are presented together with standard deviation. Last row reports characteristics measured at the Bílý Kříž site in 2017.

Catchment	No. of Plots	No. of Sampled Trees	Mean Tree Height H(m)	Mean Stem DBH (cm)	Mean Crown Radius R(m)	Norway Spruce (%)	Beech (%)	Others (%)
ANE	5	108	30.1 ± 3.3	36.6 ± 8.8	3.0 ± 0.9	81.5	0.9	17.6 [1]
CER	9	359	16.2 ± 10.3	19.8 ± 14.3	2.3 ± 1.1	68.8	26.5	4.7
LIZ	10	431	21.6 ± 8.7	24.6 ± 12.1	2.0 ± 0.9	80.3	4.6	15.1 [2]
LKV	7	312	20.6 ± 9.2	23.7 ± 12.7	2.1 ± 1.2	81.7	4.2	14.1 [3]
LYS	5	358	11.8 ± 6.9	16.5 ± 11.5	1.8 ± 0.7	100.0	0.0	0.0
NAZ	7	216	19.1 ± 7.0	26.9 ± 11.1	1.7 ± 0.7	90.7	0.5	8.8
PLB	5	143	20.7 ± 5.7	27.8 ± 9.5	2.6 ± 0.9	93.7	5.6	0.7
POM	7	143	30.7 ± 6.5	39.4 ± 12.1	2.8 ± 1.2	94.4	3.5	2.1
SAL	10	345	25.2 ± 7.6	28.6 ± 12.9	2.4 ± 1.0	74.2	0.6	25.2 [4]
UDL	6	372	9.2 ± 2.8	15.9 ± 5.6	2.0 ± 0.5	100.0	0.0	0.0
UHL	10	399	10.7 ± 3.9	16.8 ± 6.8	1.7 ± 0.8	96.5	0.2	3.3
BK outside	1	290	16.9 ± 2.0	20.1 ± 4.1	1.9 ± 0.3	100.0	0.0	0.0

Other common tree species are [1] European larch (Larix decidua) 10% and Scots pine (Pinus sylvestris) 5%. [2] Silver birch (Betula pendula) 10% and European larch 4%. [3] Scots pine 9% and common alder (Alnus glutinosa) 3%. [4] Scots pine 17% and European larch 8%.

DBH estimates were used to compute a single tree biomass and subsequently forest AGB.

The allometric equations (Equations (1)–(3)) were established for two tree species: the dominant Norway spruce and less abundant European beech. The spruce models were applied to all other coniferous trees found in our study sites. The allometric equation for European beech was created only as a global model, as there was not enough data for site-specific local models. The beech global model was applied to all other deciduous trees found in our study sites.

Before the allometric equations were constructed, the input field data were harmonized such that a comparable number of samples was selected across all catchments and all tree height categories. In the first step, all dead and broken trees were removed from the forest inventory database. In the second step, we used stratified random sampling to eliminate the influence of tree height distribution. The remaining trees were assigned into height categories at 5 m intervals (<5 m, 5–10 m, 10–15 m, etc.). From each height category, n samples were randomly selected (n = 20 for spruce local allometric equations, n = 40 for spruce global equations, n = 100 for spruce local equation established for Bílý Kříž, and n = 20 for beech global equation). According to Sullivan et al. [28], stratified sampling should produce smaller prediction error for larger sample pools. In the third step, the models (Equations (1)–(3)) were fitted to input data and outlying samples were removed based on 95% confidence interval of these preliminary models. Final models were constructed then by fitting Equations (1)–(3) to the cleaned data. The quality of the models was assessed by root mean square error (RMSE) computed between the predicted (P) and observed (O) values:

$$RMSE = \sqrt{\frac{1}{n}\sum_{i=1}^{n}(P_i - O_i)^2}, \tag{4}$$

2.3. Airborne Laser Scanning Data

Airborne laser scanning data for 11 GEOMON sites (Table 2) were acquired during August 2017. ALS data were recorded using a Riegl LMS—Q780 scanner on board Flying Laboratory of Imaging Systems [29] that is operated by the Global Change Research Institute CAS (CzechGlobe). ALS data were recorded at two nominal pulse densities: 4 pulses per m^2 (flight altitude at 1000 m above the ground level, further labeled as 1000 AGL) and 2 pulses per m^2 (flight altitude at 2000 m above the ground level, further labeled as 2000 AGL). For larger catchments, the pulse densities were doubled due to 50% overlap between flight lines. Moreover, because the Riegl system allows for multiple echo recording, the real point cloud densities varied between 8 and 18 pt/m^2 for datasets acquired at 1000 AGL and between 4 and 8 pt/m^2 for datasets acquired at 2000 AGL.

Pre-processing of the raw ALS data encompassed full-waveform decomposition and georeferencing in RiProcess software (RIEGL GmbH) and exporting a point cloud in LAS format after a strip adjustment. In the next step, the point clouds were transformed into a canopy height model (CHM) raster. The CHM calculation included noise removal, ground filtering, and rasterization using the LAStools software package (Rapidlasso GmbH) and the approach reported by Khosravipour [30] to create a pit-free CHM. In total, four CHMs were prepared for each catchment (i.e., rasters with pixel sizes 0.5 m and 1.0 m for 1000 AGL and 2000 AGL data, respectively).

Because the assessment of ALS input data properties (point density and CHM pixel size) on tree detection and AGB estimates was part of this study, the ALS input datasets were compared at the level of point clouds and at the level of CHM first. We computed histograms for height distribution and descriptive statistics for the entire catchment area and per forest inventory plot.

2.4. Individual Tree Detection and Its Verification

Individual tree detection and estimation of tree H is a prerequisite step for AGB estimation. The individual tree detection was executed for four CHMs per catchment and furthermore expanded by two scenarios to evaluate the effect of local and global models estimating tree crown R. In total eight tree detection results were generated for each catchment. The tree detection algorithm is based on previous work of Novotný [31], and it was further extended and modified in this study. It is fully automated and implemented in Python. The algorithm requires several parameters to be specified by a user to optimize results of the tree detection. Parameter selection and description of basic steps in the algorithm are as follow:

1. Identification of local maxima in CHM.
2. Applying a forest mask to remove those local maxima that were not classified as trees. (We used a mask of coniferous, broadleaf, and non-forested areas that was prepared from airborne hyperspectral images acquired simultaneously with ALS. For the sake of brevity, this was not described in the methods section.)
3. Removal of those local maxima lower than a certain threshold H_{min}. (We used H_{min} = 5 m.)
4. Estimation of a crown R, which was computed as the weighted mean of R obtained from three methods:

    (a) allometric equation R = f(H) based on forest inventory data,
    (b) slope breaks—a distance from the local maxima (tree top) to the nearest place where the slope changes from decreasing to increasing,
    (c) semivariogram—a range value in a variogram model between height values and a distance from the local maxima.
    (We used weight factors w_a = 0.6, w_b = 0.2, w_c = 0.1.)

5. Removal of those local maxima closer to one another than an expected crown R of the higher tree.
6. Removal of false tree detection at the edge of forest stands.(Here, we applied a condition that maximally 5% of all pixels within an expected crown R can be outside the range of $0.1 \times H_{top}$ and $1.05 \times H_{top}$.)
7. Export to a point shapefile.

Accuracy of the individual tree detection could not be assessed by means of direct comparison with measured tree positions as these had not been recorded during the field surveys. Therefore, we selected six validation plots (50 × 50 m) from our ALS data and identified individual tree positions in the point cloud manually (hereafter termed reference trees). Each validation plot was located in a different catchment and these represented a variety of forest types in terms of mean tree height.

Two plots represented forests with mean tree height less than 15 m, two plots were selected for mean tree height between 15 and 25 m, and the last two were selected for mean tree height greater than 25 m. Trees from manual and automated detections were paired by searching the closest neighborhood of our reference trees (in case of CHM with pixel size 0.5 m the maximum distance of three pixels was searched; in case of CHM with pixel size 1.0 m the maximum distance was two pixels). For each validation plot we computed detection, omission, and commission rates.

2.5. Biomass Estimation from ALS

Total AGB (computed as sum of its components including stems, branches, leaves) was computed according to equations established by Wirth et al. [32] for Norway spruce and by Wutzler et al. [33] for European beech and other deciduous tree species. All equations for AGB computation were chosen in their simplest forms, using DBH as the only input variable (summarized in Table 3).

Table 3. Parameters of the allometric models used for tree biomass (and its components) for estimating spruce and beech biomass and its components. W is dry mass of the biomass component, DBH is diameter at breast height, and H is tree height.

	Spruce Simple Model (Wirth et al. [32])			Beech (Wutzler et al. [33])			
Components	**ln(W)**	α	β	**W**	α	β	γ
stem	$\alpha + \beta \ln(DBH)$	-2.50602	2.44277	$\alpha(DBH^2 H)^\beta$	0.0293	0.974	-
needles/leaves	$\alpha + \beta \ln(DBH)$	-3.19632	1.9162	$\alpha(DBH^\beta H)^\gamma$	0.0377	2.430	-0.913
living branches	$\alpha + \beta \ln(DBH)$	-3.96201	2.2552	$\alpha(DBH^\beta H)^\gamma$	0.1230	3.090	-1.170
dry branches	$\alpha + \beta \ln(DBH)$	-3.22406	1.6732	-	-	-	-

The DBH field measurements were used directly to compute AGB for each field-surveyed plot and subsequently scaled up to represent the entire catchment area (further labeled as "field-based AGB"). The results of individual tree detection with the estimates of tree H from ALS were used to retrieve DBH first and then compute ALS-based estimates of AGB. We used six models for DBH recalculation from ALS-based tree H (local 1 and 2, global 1 and 2, outside 1 and 2). In combination with two input point cloud densities, two CHM pixel sizes, and two allometric equations predicting R from H (local and global), we had 48 scenarios for assessing the effect of different parameters.

The ALS-based AGB estimates were cross-compared between the scenarios and compared with the field-based results. The actual variable to be compared is AGB in tons per hectare of forest. We calculated this value for field data as the sum of AGB for individual trees divided by plot size times the percentage of forest cover. The mean value computed from all plots was taken as an estimate of AGB per catchment. We calculated AGB per plot from ALS data as the sum of individual trees inside a circle with an area of 1000 m^2 around the plot center divided by the area times the percentage of forest cover. The decision to double the plot size (1000 for ALS vs. 500 for field) should minimize inaccuracies in geometric position (field GPS vs. aircraft IMU accuracy). We calculated AGB per catchment from ALS data as the sum of all individual trees divided by area of catchment times the percentage of forest cover. The forest cover was estimated from ALS point clouds and publicly available orthophoto maps using Google Earth to visually check differences in forest cover between 2015 and 2017.

Comparison between the ALS-based scenarios and with the field-based results was evaluated using linear fit, R^2, and RMSE. Additionally, we evaluated dispersion of points around the 1:1 line and computed 1:1 error as:

$$1:1\,\mathrm{err} = \frac{1}{n}\sum_{i=1}^{n}\frac{|\mathrm{AGB}_{\mathrm{field}}(i) - \mathrm{AGB}_{\mathrm{lidar}}(i)|}{\mathrm{AGB}_{\mathrm{field}}(i)}, \tag{5}$$

3. Results and Discussions

3.1. Allometric Equations

The 11 small forested catchments from the GEOMON network used in this study represented broad regional variability in Norway spruce dominated forest stands. Field data were used to establish allometric equations to predict tree crown R and stem DBH from tree H. All allometric equations are summarized in Table 4. Only a simple linear model was tested for the tree crown R. Two models, the exponential (Equation (2)) and the power (Equation (3)) models, were tested for predicting stem DBH from tree H. Figure 3 shows examples of allometric equations used in spruce DBH estimation for two local sites (LYS and BK) and the global model for spruce and beech trees (graphs with the local allometric equations for all other sites are presented in the Supplementary Material S1).

Table 4. Allometric equations for estimating tree crown radius (R) and stem diameter at breast height (DBH) from the measured tree height (H) for Norway spruce. Local equations are computed for each catchment and the global equation is computed for all catchments together. In the last row, we also included the local model that was computed outside the GEOMON catchments for the Bílý Kříž study site.

	R = aH + b			DBH = exp(aH + b)			DBH = aH^b		
	(Local/Global)			(Local1/Global1/Outside1)			(Local2/Global2/Outside2)		
Catchment	a	b	RMSE	a	b	RMSE	a	b	RMSE
ANE	0.124	−0.841	0.639	0.065	1.640	4.484	0.051	1.933	4.641
CER	0.067	0.947	0.322	0.050	2.038	3.848	0.688	1.182	4.431
LIZ	0.038	1.174	0.618	0.055	1.900	3.276	0.753	1.125	3.523
LKV	0.076	0.517	0.657	0.063	1.714	3.063	0.487	1.266	3.630
LYS	0.095	0.668	0.268	0.073	1.821	3.995	1.002	1.121	3.325
NAZ	0.063	0.443	0.471	0.044	2.401	4.654	1.051	1.096	3.828
PLB	0.111	0.259	0.576	0.060	1.971	4.398	0.891	1.120	4.983
POM	0.069	0.458	0.846	0.051	2.044	5.911	0.265	1.450	6.294
SAL	0.072	0.559	0.593	0.064	1.623	2.499	0.332	1.366	2.798
UDL	0.135	0.709	0.336	0.122	1.595	2.875	2.431	0.846	3.152
UHL	0.142	0.193	0.479	0.085	1.786	2.540	1.518	1.096	2.624
ALL global	0.056	1.052	0.553	0.050	2.086	4.651	1.036	1.059	4.736
BK outside	n.a.	n.a.	n.a.	0.970	1.417	1.591	1.232	0.991	1.095

The allometric equation for tree crown R estimation from tree H is important in our algorithm for individual tree detection from ALS data. However, as this relationship has scarcely any practical use in forest management, it is difficult to find reliable data or existing models in the scientific literature. Therefore, this linear relationship was determined solely from our available field data, with RMSE varying between 0.268 and 0.657 m for local models and equal to 0.553 m for the global model. Its use in tree detection is further discussed in Section 3.3, but, as the detection rates with the global model were generally higher for all tree height categories, we can consider the global model to be more robust and to slightly outperform the local models.

In forestry practice, the relationship between stem DBH and tree H is generally reported as H = f(DBH) and mathematically expressed using various models, including power, exponential, hyperbolic, and others [34]. For the purpose of AGB estimation from ALS, however, the reverse form of DBH = f(H) is required. Simple mathematical inversion is not always feasible, though, due to the asymptotic nature of the published models. Thanks to our own field data collected for a large variety of spruce trees, we could test two mathematical models: the exponential and power expression. RMSE of the exponential model varied between 2.499 and 4.654 m and RMSE of the power model varied between 2.624 and 4.983 m. Although for most of the catchments the exponential model resulted in lower RMSE than did the power model, for biomass estimation, the exponential model seemed to be more sensitive to tree H. For tall trees, therefore, it could yield unrealistic (too high) estimates of DBH and thus of AGB (further discussed in Section 3.4).

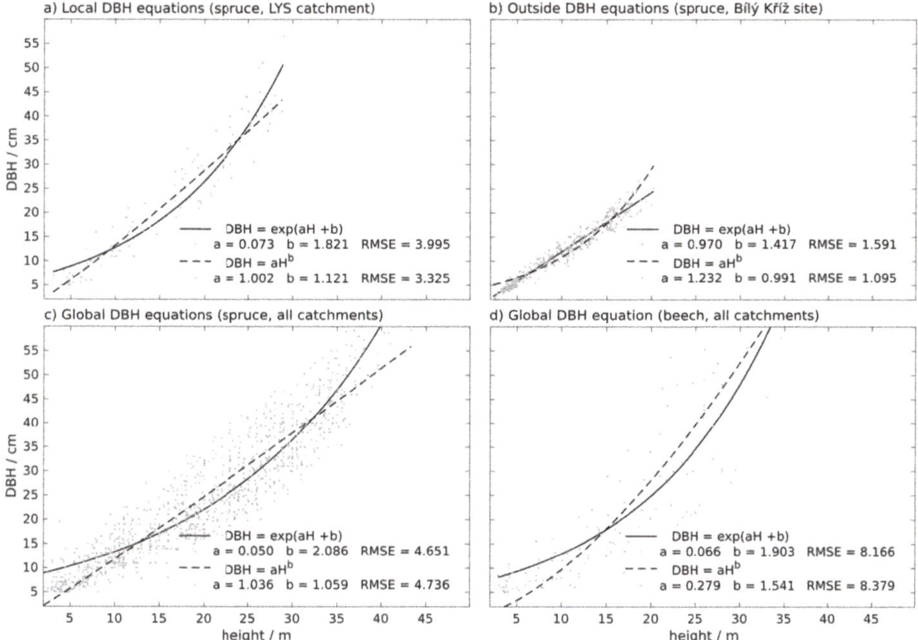

Figure 3. Allometric equations for prediction of stem diameter at breast height (DBH) from tree height (H). (**a**) local model for Norway spruce established for the LYS catchment, (**b**) local model for the external site Bílý Kříž, (**c**) global model for Norway spruce computed from all catchments, and (**d**) global model for European beech computed from all catchments. Graphs for local equations from other sites are presented in Supplementary Material S1.

3.2. Comparison of ALS Input Data

Before the AGB estimation from the ALS data, we evaluated differences between two ALS input datasets with different point densities (1000 AGL, 2000 AGL). We compared histograms for height distributions extracted from point clouds, as well as from CHM rasters, for the entire catchment area and for each forest inventory plot. Figure 4 shows an example from the CER catchment (for brevity's sake, the other sites are not presented). The histograms are not identical for point clouds and CHMs because the point cloud data contain all points (including multiple returns from the canopy profile), whereas the CHM rasters refer only to the top-of-canopy layer. Height distribution in CHM with 1 m resolution tends to be slightly shifted to the right compared to CHM with resolution 0.5 m. This is because the rasterization technique we used lays just one 1m pixel over a bunch of points instead of four 0.5 m pixels, which makes the tree tops (and crown borders) less sharp.

Comparison at the level of point clouds and CHM rasters showed very similar canopy height distributions across all the sites. Therefore, we could assume that the point densities of the ALS input data (1000 AGL vs. 2000 AGL) will likely not affect the results of individual tree detection and tree height estimation.

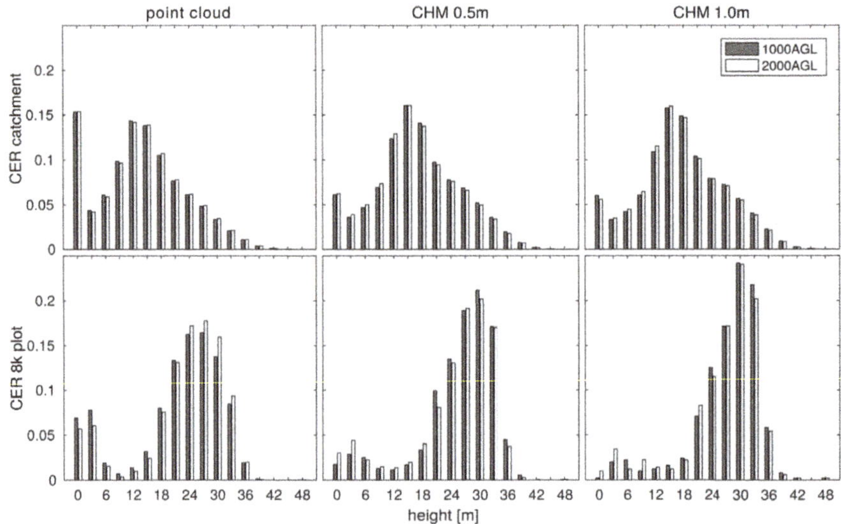

Figure 4. Histograms for height distributions of ALS input data acquired at point density of 8 pt/m^2 (1000 AGL) and 4 pt/m^2 (2000 AGL) at the level of the entire Červík (CER) catchment (**top graphs**) and at the level of a single forest inventory plot (CER 8k) of size 500 m^2 (**bottom graphs**). Comparison of height distribution is shown for point cloud data (**left**) and for the canopy height model rasters with pixel sizes of 0.5 m (**middle**) and 1.0 m (**right**).

3.3. Tree Detection

Results of tree detection were obtained for eight scenarios combining four CHMs (pixel size 0.5 and 1.0 m, point density from 1000 AGL and 2000 AGL) and two allometric equations for tree crown R estimation. Results from the automated tree detection were verified for six validation squares distributed across three tree height categories (more precisely, located within ANE, CER, LIZ, PLB, UDL, and UHL catchments). In each case, detection and commission rates were computed for each tree height category and for all validation plots together (Figure 5). We achieved detection rates from 0.75 to 0.90, which is in line with other studies. For instance, Luo et al. [35] reported overall accuracy of 0.87 for comparison between automatically segmented trees and visual examination in Tahoe National Forest (USA). Wang et al. [7] reported R^2 of 0.8–0.9 for comparison of plot-level field-based and ALS-based tree counts in Heihe River Basin (China).

Almost no differences in detection and commission rates were found between the results obtained from the ALS data of different point densities (1000 AGL, 2000 AGL). This had been expected after comparison of the ALS input data (Figure 4, Section 3.2). Similar conclusions had been reached by Bouvier et al. (2019), who reported almost no influence of ALS pulse densities (0.5, 1, 2, and 4 pulses per m^2) on metrics derived for AGB estimation in a pine forest.

Pixel size of the input CHM raster had a stronger effect on tree detection. Detection rates varied between 0.78 and 0.90 for the CHM with 0.5 m pixel size, and lower values between 0.55 and 0.79 were found for the CHM with 1.0 m pixel size. On the contrary, higher commission rates (0.17–0.39) were found for the CHM with 0.5 m pixel size and lower values for the CHM with 1.0 m pixel size (0.05–0.19). Higher detection rates were found for the global model when we quantified the effect of global vs. local allometric equation used to estimate tree crown R in the tree detection phase.

Detection rates also varied between the tree height categories. Detection rates in forests with tree height below 15 m were lowest, whereas those rates for forests with tree height above 25 m were highest. Similar results had been found by Kaartinen et al. [36], who analyzed accuracies of tree detection from ALS data with point densities of 2, 4 and 8 pt/m^2 in Southern Finland forests with prevailing Scots pine, Norway spruce, and silver birch. The best accuracy was observed for taller trees. For the local maxima method, the RMSE in tree location was estimated as 1.2 m, 0.7 m, and 0.5 m for tree height classes 10–15 m, 15–20 m, and >20 m, respectively.

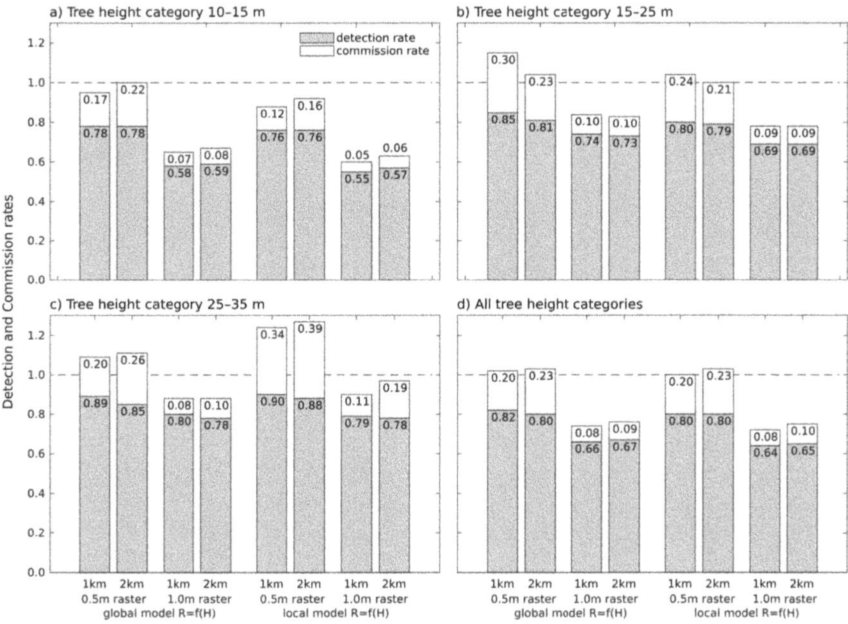

Figure 5. Verification of tree detection. Detection and commission rates calculated for three forest height categories (**a–c**) and all categories together (**d**) using manual identification of trees.

From the point of view of the final AGB product, the effect of tree detection omission and commission errors compensated one another to some extent. We saw this especially in the tree height category 15–25 m with the local model and in the tree height category 25–35 m with the global model. One more interaction worth noting is that underestimating of the tree count can be compensated by an overestimation of the DBH model, thus ending up with a false agreement between field- and ALS-based AGB totals. This happened in our case with CHMs having 1.0 m pixel size.

Based on the results of the tree detection we eliminated half of our scenarios. For AGB assessment, we retained only the results of tree detection from CHM with pixel size of 0.5 m from 1000 AGL.

3.4. Biomass Estimation

AGB at plot level and at entire catchment level was computed for 12 scenarios using as input the CHM raster with pixel size 0.5 m from 1000 AGL. The 12 scenarios considered two allometric equations for tree crown R estimation and six allometric equations for tree stem DBH estimation (the power and the exponential local, global, and outside models).

Intercomparison between field-based and ALS-based AGB estimates at plot level is summarized for all scenarios in Table 5. RMSE between field- and ALS-based estimates varied between 93 and 171 t/ha (if we omit two outstanding scenarios with exceptionally high RMSE > 326 t/ha). In general, we can conclude that all ALS-based scenarios overestimated AGB especially for plots with taller trees and higher biomass in comparison with the field-based estimates. ALS-based AGB estimation derived using local allometric equations for both crown radius and DBH was the scenario with best precision as expressed by steep slope (1.054) and low offset (8.561) in linear fit which was also the tightest

(R^2 = 0.667, RMSE = 93 t/ha). All the plots of all three height classes were evenly distributed around the 1:1 line, which is expressed by the lowest value of 1:1 error equal to 0.264 (see Figure 6, and we refer our readers to the Supplementary Material S2 to see the same figure for the other scenarios).

Two comparisons between ALS-based estimates are presented in Figure 7. Both graphs show AGB estimates using the local allometric equations for both crown R and stem DBH vs. global allometric equations, but the two plots show difference by equation type (exponential vs. power type). Most plots differed within the range of ±50%, while the majority of those in height class >25 m laid within the range ±20%.

Table 5. Numerical comparison of different scenarios for AGB estimation from ALS data. The first two columns describe the scenario itself by specifying which type of allometric equation was used in the tree detection phase for estimating tree crown radius (R = f(H)) and for stem diameter estimation (DBH = f(H)). The remaining variables evaluate the agreement between field- and ALS-based estimates of AGB (R^2 and RMSE were computed based on linear regression with equation $AGB_{ALS} = a \cdot AGB_{field} + b$; 1:1 error tells how far the points deviate from the 1:1 line).

Scenario Definition		Agreement between Field- and ALS-Based AGB Estimates				
R = f(H)	**DBH = f(H)**	**a**	**b**	**R^2**	**RMSE**	**1:1 Err**
global	local 1	1.361	−45.178	0.566	148	0.331
global	local 2	1.206	−10.008	0.584	127	0.318
global	global 1	1.513	−82.611	0.549	171	0.421
global	global 2	1.236	−26.731	0.606	124	0.352
global	outside 1	1.775	−180.363	0.314	326	0.569
global	outside 2	1.308	−20.776	0.594	134	0.396
local	local 1	1.185	−23.365	0.631	113	0.272
local	local 2	1.054	8.561	0.667	93	0.264
local	global 1	1.327	−55.018	0.580	140	0.365
local	global 2	1.111	−10.644	0.630	106	0.322
local	outside 1	1.565	−150.356	0.301	296	0.507
local	outside 2	1.175	−3.875	0.620	114	0.354

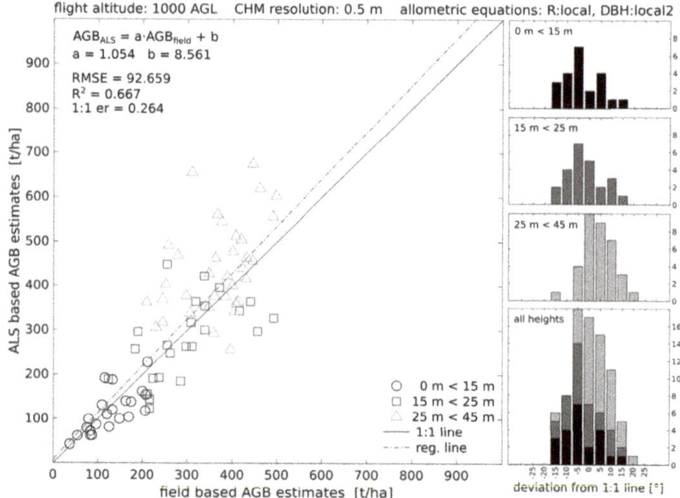

Figure 6. AGB comparison at plot level between estimates based on field data and ALS data (scenario based on the CHM raster with pixel size of 0.5 m and combination of two local models, one for tree crown R estimation and the other for the stem diameter at breast height estimation). The linear regression line (dash-dot) is presented together with the 1:1 line (solid) and symbols differentiate tree height categories. Histograms on the right-hand side show deviation from the 1:1 line for different tree height categories.

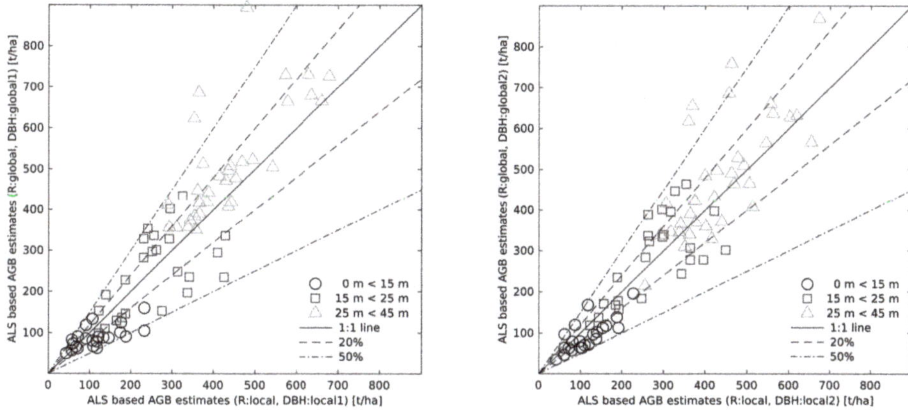

Figure 7. Comparison of AGB estimates at plot level based solely on ALS data. The effect of local vs. global model for DBH estimation is shown: exponential eq. (**left**) and power eq. (**right**). Solid line highlights 1:1 line and the dashed lines indicate 20% and 50% deviation from 1:1 line.

Total AGB estimates for the entire catchment areas under 12 scenarios using ALS data are summarized in Table 6. Comparison between the field- and ALS-based AGB estimates is presented in Figure 8 for two contrasting scenarios, one using the local allometric equations, the other one using only the global equations. Figures for the other scenarios are presented in the Supplementary Material S3.

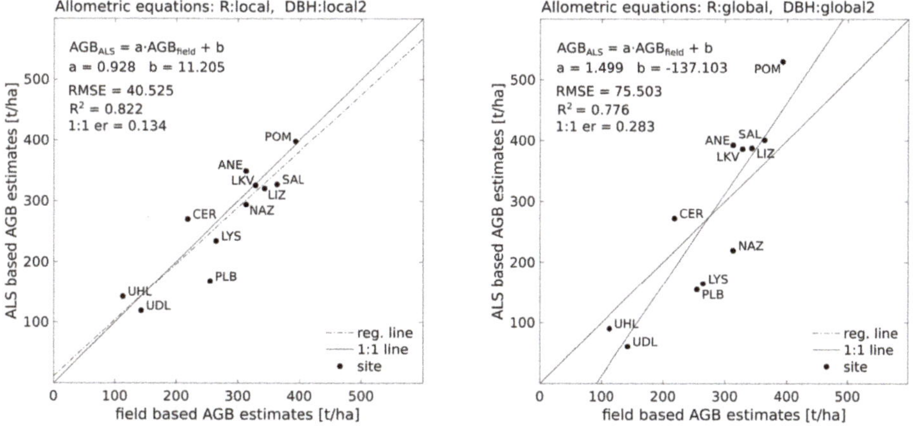

Figure 8. Comparison of total AGB for entire catchment areas based on estimates obtained from field data and from ALS data using local allometric equations (**left**) and global allometric equations (**right**). Linear regression line (dashed) is presented together with a 1:1 line (solid).

Forest AGB (as estimated from the R:local, DBH:local 2 scenario) varied between 119 and 398 t/ha according to forest stand age. For example, the lowest AGB was observed at the UDL and UHL sites with mean forest age between 20 and 40 years. In general, global DBH models tend to underestimate AGB in young stands and overestimate AGB in mature stands. This will likely remain true despite the fact that there might be some differences between field- and ALS-based estimates because of the two year gap between the data acquistion. We assume that yearly increment in AGB is not very high (less than 10% for younger stands and less than 5% for mature stands). An exception was seen in the PLB catchment, where AGB was underestimated every time. We know that there was a logging activity between 2015 and 2017 at that site. Furthermore, specific bedrock chemistry (alkaline serpentinite, see Table 1) suppress tree growth due to nutrient imbalance [37] leading to shorter trees with the same DBH compared to other sites.

We achieved linear fit between field-based and ALS-based AGB estimate at catchment level varying from 0.822 to 0.673 in terms of R^2 and from 40 to 90 t/ha in terms of RMSE. ALS-based AGB estimation derived using local allometric equations for both crown R and stem DBH had the greatest precision as expressed by steep slope (a = 0.928) and low offset (b = 11.205) in linear fit which was also the tightest (R^2 = 0.822, RMSE = 40 t/ha). Our results are in agreement with those from other studies, such as that of Mutwiri et al. [17], who reported R^2 of 0.7–0.8 in comparing field- and ALS-based AGB estimates within montane forest in Kenya. Similar levels of accuracy were reported by Shao et al. [20], with R^2 of 0.6 for AGB estimates in Yellowwood State Forest (USA), and Wang et al. [7] reporting R^2 of 0.6–0.8 for estimates in Heihe River Basin (China). Ferraz et al. [18] analyzed differences between field- and ALS-based AGB estimates for different forest types and reported highly variable results with R^2 ranging between 0.37 and 0.99.

Based on the results presented in Table 6 it is clear that the exponential local allometric equation (type 1) for DBH estimation that was established outside the GEOMON catchments produces unrealistic estimates of AGB, especially for sites with mature and tall spruce trees. This could be expected, because the dataset at the Bílý Kříž site contains data only for spruce trees younger than 40 years. Therefore, the allometric equation for DBH estimation, especially the exponential model, is scarcely transferable to older trees. If we remove the results for the "outside 1" model from Table 6, we are able to assess uncertainty or variability in AGB estimation from the remaining scenarios. The uncertainty (measured as standard deviation) in AGB estimates varies from 16 t/ha (observed at catchment CER) up to 84 t/ha (observed at catchment POM). In relative values, it goes from 6% to 37% of the field-based AGB estimation per catchment. Similarly, Bellan et al. [38] reported underestimation of 10% in total AGB when transferring field-based allometric equations AGB = f(DBH) from one spruce stand in the Czech Republic to another.

Table 6. Total aboveground biomass estimates per catchment (t/ha) obtained directly from the field data (in the first row) and from airborne laser scanning data for 12 scenarios, differing in their combinations of local or global allometric equations for (i) estimation of tree crown radius R in tree detection and (ii) stem diameter estimation in biomass computation.

Scenario Definition		Total Aboveground Biomass Per Catchment (t/ha)										
R = f(H)	DBH = f(H)	ANE	CER	LIZ	LKV	LYS	NAZ	PLB	POM	SAL	UDL	UHL
field	field	313	218	343	328	265	313	255	393	363	143	113
global	local 1	371	239	330	346	241	275	160	563	355	195	158
global	local 2	359	273	319	332	240	302	170	547	326	127	150
global	global 1	359	268	355	341	134	192	121	572	376	67	80
global	global 2	393	272	387	386	165	220	155	530	401	61	90
global	outside 1	2470	2680	2165	2081	359	1040	348	6082	2717	57	108
global	outside 2	344	242	346	344	152	197	142	461	352	61	87
local	local 1	362	236	331	335	234	268	158	403	354	176	147
local	local 2	349	270	321	326	235	294	168	398	327	119	142
local	global 1	352	264	356	334	131	186	120	410	377	64	77
local	global 2	387	270	389	382	161	214	153	393	403	56	86
local	outside 1	2373	2564	2171	1958	346	973	341	4051	2694	52	100
local	outside 2	339	240	347	341	149	193	140	343	354	57	83

4. Conclusions

This study explored variables influencing estimation of AGB for Norway spruce dominated forests from ALS data. Specifically, we looked at the influence of ALS data point densities and spatial resolution of corresponding CHM for individual tree detection and role of allometric equations for tree crown radius and stem DBH estimation in forest AGB estimation. We conclude that variable point densities of ALS data (4–18 pt/m^2) did not influence the results for individual tree detection. Therefore, we recommend using ALS data with point density around 5 pt/m^2 for operational estimation of spruce forest AGB in the Central European region. Spatial resolution of the input CHM rasters had a greater impact on tree detection, higher detection rates were achieved for CHMs with pixel size 0.5 m than 1.0 m for all tree height categories.

AGB estimation from ALS data is largely dependent on the allometric equations that estimate basic tree parameters, such as crown radius and stem diameter, from an easily retrievable tree height. In common forestry practice, these allometric equations often do not exist (such as for crown radius) or they are reported in inverse form (such as estimating tree height from stem diameter). Thanks to the large database of forest inventory parameters that was compiled for small forested catchments within the GEOMON monitoring network in the Czech Republic (about 3000 trees), we could establish our own allometric equations for Norway spruce and European beech. Because the GEOMON sites cover a broad variety of environmental conditions and spruce-dominated forests, we can consider the global allometric models to be representative for the Czech Republic, and potentially for the Central European region.

We computed AGB at the catchment level for 12 scenarios with varying allometric equations and the results showed that uncertainty varied from 16 up to 84 t/ha. In relative values, it goes from 6% to 37% of the field-based AGB estimation per catchment. Therefore, allometric equations mainly for DBH estimation should be applied with care. We can recommend establishing one's own site-specific models, but this requires investments into field data collection. Direct comparison between ALS-based AGB estimates using local and global models showed variation as much as 20% (both positive and negative) for the highest tree height class and as much as 50% for the lowest tree height class.

The main advantages of the ALS-based approach for estimating forest AGB are full spatial coverage and an ability to detect real spatial distribution of trees whereas field-based estimation relies on a limited number of sampling plots and requires extrapolation to the stand level.

Supplementary Materials: The following are available online at http://www.mdpi.com/1999-4907/11/3/268/s1. Complete set of figures (Figure 3, Figure 6, Figure 8) is provided for different scenarios and study sites. [S1] Local allometric equations; [S2] Comparison of AGB estimates at plot level; [S3] Comparison of AGB estimates at stream catchment level.

Author Contributions: Conceptualization, J.N. and L.H.; methodology, J.N.; software, J.N.; formal analysis, B.N. and R.J.; resources, F.O.; data curation, B.N.; writing–original draft preparation, J.N. and B.N.; writing–review and editing, F.O. and L.H.; visualization, B.N. and R.J.; project administration, L.H. All authors have read and agreed to the published version of the manuscript.

Funding: This work was supported by the Czech Science Foundation, grant number 17-05743S and the Ministry of Education, Youth and Sports of the Czech Republic within the National Sustainability Program I (NPU I), grant number LO1415.

Acknowledgments: We acknowledge the site manager P. Krám (LYS, PLB, NAZ) for help with field work preparation. We would like to thank field researchers R. Hédl, L. Hédlová, M. Chudomelová, J. Šebesta, I. Paukertová, I. Müllerová, A. Kučerová, and J. Šipoš, who collected the forest inventory data at the GEOMON sites and our colleague J. Krejza for sharing forest inventory data from the Bílý Kříž site.

Abbreviations

The following abbreviations are used in this manuscript:

AGB	Aboveground biomass
1000AGL	Flight altitude at 1000 m above the ground level
ALS	Airborne laser scanning
ANE	Anenský potok site
BK	Bílý Kříž site
CAS	Czech Academy of Sciences
CER	Červík site
CHM	Canopy height model
CHM	Canopy height model
CzechGlobe	Global Change Research Institute CAS
DBH	Diameter at breast height
GEOMON	Network of study sites (see Section 2.1)
H	Tree height
ICOS	Integrated Carbon Observation System
LIZ	Na Lizu site
LKV	Loukov site
LYS	Lysina site
NAZ	Na Zeleném site
PLB	Pluhův bor site
POM	Polomka site
R	Crown radius
RMSE	Root mean square error
SAL	Salačova Lhota site
UDL	U dvou louček site
UHL	Uhlířská site

References

1. Pan, Y.; Birdsey, R.A.; Fang, J.; Houghton, R.; Kauppi, P.E.; Kurz, W.A.; Phillips, O.L.; Shvidenko, A.; Lewis, S.L.; Canadell, J.G.; et al. A Large and Persistent Carbon Sink in the World's Forests. *Science* **2011**, *333*, 988–993. [CrossRef]
2. Goodale, C.L.; Apps, M.J.; Birdsey, R.A.; Field, C.; Heath, L.; Houghton, R.A.; Jenklngsu, N. Forest Carbon Sinks in the Northern Hemisphere. *Ecog. Appl.* **2002**, *12*, 897–899. [CrossRef]
3. Agrawal, A.; Nepstad, D.; Chhatre, A. Reducing Emissions from Deforestation and Forest Degradation. *Ann. Rev. Environ. Resour.* **2011**, *36*, 373–396. [CrossRef]
4. Baccini, A.; Goetz, S.J.; Walker, W.S.; Laporte, N.T.; Sun, M.; Sulla-Menashe, D.; Hackler, J.; Beck, P.S.A.; Dubayah, R.; Friedl, M.A.; et al. Estimated carbon dioxide emissions from tropical deforestation improved by carbon-density maps. *Nat. Clim. Chang.* **2012**, *2*, 182–185. [CrossRef]
5. Harris, N.L.; Brown, S.; Hagen, S.C.; Saatchi, S.S.; Petrova, S.; Salas, W.; Hansen, M.C.; Potapov, P.V.; Lotsch, A. Baseline Map of Carbon Emissions from Deforestation in Tropical Regions. *Science* **2012**, *336*, 1573–1576. [CrossRef]
6. Figueiredo, E.O.; d'Oliveira, M.V.N.; Braz, E.M.; de Almeida Papa, D.; Fearnside, P.M. LIDAR-based estimation of bole biomass for precision management of an Amazonian forest: Comparisons of ground-based and remotely sensed estimates. *Remote Sens. Environ.* **2016**, *187*, 281–293. [CrossRef]
7. Wang, Z.; Liu, L.; Peng, D.; Liu, X.; Zhang, S.; Wang, Y. Estimating woody aboveground biomass in an area of agroforestry using airborne light detection and ranging and compact airborne spectrographic imager hyperspectral data: Individual tree analysis incorporating tree species information. *J. Appl. Remote Sens.* **2016**, *10*, 036007. [CrossRef]
8. Anderson-Teixeira, K.J.; Davies, S.J.; Bennett, A.C.; Gonzalez-Akre, E.B.; Muller-Landau, H.C.; Joseph Wright, S.; Abu Salim, K.; Almeyda Zambrano, A.M.; Alonso, A.; Baltzer, J.L.; et al. CTFS-ForestGEO: A worldwide network monitoring forests in an era of global change. *Glob. Chang. Biol.* **2015**, *21*, 528–549. [CrossRef] [PubMed]

9. Malek, S.; Miglietta, F.; Gobakken, T.; Næsset, E.; Gianelle, D.; Dalponte, M. Prediction of stem diameter and biomass at individual tree crown level with advanced machine learning techniques. *iForest Biogeosci. Forestry* **2019**, *12*, 323–329. [CrossRef]

10. Avitabile, V.; Herold, M.; Heuvelink, G.B.M.; Lewis, S.L.; Phillips, O.L.; Asner, G.P.; Armston, J.; Ashton, P.S.; Banin, L.; Bayol, N.; et al. An integrated pan-tropical biomass map using multiple reference datasets. *Glob. Chang. Biol.* **2016**, *22*, 1406–1420. [CrossRef] [PubMed]

11. Lefsky, M.A.; Cohen, W.B.; Harding, D.J.; Parker, G.G.; Acker, S.A.; Gower, S.T. Lidar remote sensing of above-ground biomass in three biomes. *Glob. Ecol. Biogeogr.* **2002**, *11*, 393–399. [CrossRef]

12. Asner, G.P.; Mascaro, J.; Muller-Landau, H.C.; Vieilledent, G.; Vaudry, R.; Rasamoelina, M.; Hall, J.S.; van Breugel, M. A universal airborne LiDAR approach for tropical forest carbon mapping. *Oecologia* **2012**, *168*, 1147–1160. [CrossRef] [PubMed]

13. Jochem, A.; Höfle, B.; Rutzinger, M. Extraction of Vertical Walls from Mobile Laser Scanning Data for Solar Potential Assessment. *Remote Sens.* **2011**, *3*, 650–667. [CrossRef]

14. Gwenzi, D.; Lefsky, M.A. Modeling canopy height in a savanna ecosystem using spaceborne lidar waveforms. *Remote Sens. Environ.* **2014**, *154*, 338–344. [CrossRef]

15. Hao, H.; Li, W.; Zhao, X.; Chang, Q.; Zhao, P. Estimating the Aboveground Carbon Density of Coniferous Forests by Combining Airborne LiDAR and Allometry Models at Plot Level. *Front. Plant Sci.* **2019**, *10*, 917. [CrossRef] [PubMed]

16. Garcia, M.; Saatchi, S.; Ferraz, A.; Silva, C.A.; Ustin, S.; Koltunov, A.; Balzter, H. Impact of data model and point density on aboveground forest biomass estimation from airborne LiDAR. *Carbon Balance Manag.* **2017**, *12*, 4. [CrossRef]

17. Mutwiri, F.K.; Odera, P.A.; Kinyanjui, M.J. Estimation of Tree Height and Forest Biomass Using Airborne LiDAR Data: A Case Study of Londiani Forest Block in the Mau Complex, Kenya. *Open J. Forestry* **2017**, *7*, 255–269. [CrossRef]

18. Ferraz, A.; Saatchi, S.; Mallet, C.; Jacquemoud, S.; Gonçalves, G.; Silva, C.; Soares, P.; Tomé, M.; Pereira, L. Airborne Lidar Estimation of Aboveground Forest Biomass in the Absence of Field Inventory. *Remote Sens.* **2016**, *8*, 653. [CrossRef]

19. Hansen, E.; Ene, L.; Mauya, E.; Patočka, Z.; Mikita, T.; Gobakken, T.; Næsset, E. Comparing Empirical and Semi-Empirical Approaches to Forest Biomass Modelling in Different Biomes Using Airborne Laser Scanner Data. *Forests* **2017**, *8*, 170. [CrossRef]

20. Shao, G.; Shao, G.; Gallion, J.; Saunders, M.R.; Frankenberger, J.R.; Fei, S. Improving Lidar-based aboveground biomass estimation of temperate hardwood forests with varying site productivity. *Remote Sens. Environ.* **2018**, *204*, 872–882. [CrossRef]

21. Deo, R.; Russell, M.; Domke, G.; Andersen, H.E.; Cohen, W.; Woodall, C. Evaluating Site-Specific and Generic Spatial Models of Aboveground Forest Biomass Based on Landsat Time-Series and LiDAR Strip Samples in the Eastern USA. *Remote Sens.* **2017**, *9*, 598. [CrossRef]

22. Bouvier, M.; Durrieu, S.; Fournier, R.A.; Saint-Geours, N.; Guyon, D.; Grau, E.; de Boissieu, F. Influence of Sampling Design Parameters on Biomass Predictions Derived from Airborne LiDAR Data. *Canad. J. Remote Sens.* **2019**, *45*, 650–672. [CrossRef]

23. Montagnoli, A.; Fusco, S.; Terzaghi, M.; Kirschbaum, A.; Pflugmacher, D.; Cohen, W.B.; Scippa, G.S.; Chiatante, D. Estimating forest aboveground biomass by low density lidar data in mixed broad-leaved forests in the Italian Pre-Alps. *Forest Ecosyst.* **2015**, *2*, 10. [CrossRef]

24. Wu, Z.; Dye, D.; Stoker, J.; Vogel, J.; Velasco, M.; Middleton, B. Evaluating Lidar Point Densities for Effective Estimation of Aboveground Biomass. *Int. J. Adv. Remote Sens. GIS* **2016**, *5*, 1483–1499. [CrossRef]

25. Brovkina, O.; Novotny, J.; Cienciala, E.; Zemek, F.; Russ, R. Mapping forest aboveground biomass using airborne hyperspectral and LiDAR data in the mountainous conditions of Central Europe. *Ecol. Eng.* **2017**, *100*, 219–230. [CrossRef]

26. Fottová, D. Regional evaluation of mass element fluxes: Geomon network of small catchments. *Environ. Monitor. Assess.* **1995**, *34*, 215–221. [CrossRef]

27. Oulehle, F.; Chuman, T.; Hruška, J.; Krám, P.; McDowell, W.H.; Myška, O.; Navrátil, T.; Tesař, M. Recovery from acidification alters concentrations and fluxes of solutes from Czech catchments. *Biogeochemistry* **2017**, *132*, 251–272. [CrossRef]

28. Sullivan, M.J.P.; Lewis, S.L.; Hubau, W.; Qie, L.; Baker, T.R.; Banin, L.F.; Chave, J.; Cuni-Sanchez, A.; Feldpausch, T.R.; Lopez-Gonzalez, G.; et al. Field methods for sampling tree height for tropical forest biomass estimation. *Methods Ecol. Evol.* **2018**, *9*, 1179–1189. [CrossRef]

29. Hanuš, J.; Fabiánek, T.; Fajmon, L. Potential of airborne imaging spectroscopy at CzechGlobe. *ISPRS Int. Arch. Photogram. Remote Sens. Spat. Inf. Sci.* **2016**, *XLI-B1*, 15–17. [CrossRef]

30. Khosravipour, A.; Skidmore, A.K.; Isenburg, M.; Wang, T.; Hussin, Y.A. Generating Pit-free Canopy Height Models from Airborne Lidar. *Photogramm. Eng. Remote Sens.* **2014**, *80*, 863–872. [CrossRef]

31. Novotný, J. MAtematické Metody Segmentace Obrazu Pro DáLkový PrůZkum Země (Mathematical Methods of Image Segmentation for Remote Sensing Applications). Doctoral's Thesis, Faculty of Mechanical Engineering, Brno University of Technology, Brno, Czech Republic, 2015.

32. Wirth, C.; Schumacher, J.; Schulze, E.D. Generic biomass functions for Norway spruce in Central Europe—A meta-analysis approach toward prediction and uncertainty estimation. *Tree Physiol.* **2004**, *24*, 121–139. [CrossRef] [PubMed]

33. Wutzler, T.; Köstner, B.; Bernhofer, C. Spatially explicit assessment of carbon stocks of a managed forest area in eastern Germany. *Eur. J. Forest Res.* **2007**, *126*, 371–383. [CrossRef]

34. Adamec, Z.; Drápela, K. Comparison of parametric and nonparametric methods for modeling height-diameter relationships. *iForest Biogeosc. Forestry* **2017**, *10*, 1–8. [CrossRef]

35. Luo, L.; Zhai, Q.; Su, Y.; Ma, Q.; Kelly, M.; Guo, Q. Simple method for direct crown base height estimation of individual conifer trees using airborne LiDAR data. *Opt. Express* **2018**, *26*, A562. [CrossRef] [PubMed]

36. Kaartinen, H.; Hyyppä, J.; Yu, X.; Vastaranta, M.; Hyyppä, H.; Kukko, A.; Holopainen, M.; Heipke, C.; Hirschmugl, M.; Morsdorf, F.; et al. An International Comparison of Individual Tree Detection and Extraction Using Airborne Laser Scanning. *Remote Sens.* **2012**, *4*, 950–974. [CrossRef]

37. Krám, P.; Oulehle, F.; Štědrá, V.; Hruška, J.; Shanley, J.B.; Minocha, R.; Traister, E. Geoecology of a Forest Watershed Underlain by Serpentine in Central Europe. *Northeast. Nat.* **2009**, *16*, 309–328. [CrossRef]

38. Bellan, M.; Světlík, J.; Krejza, J.; Rosík, J.; Marková, I. Chyba odhadu nadzemní biomasy v případě použití lokálních alometrických rovnic na příkladu dvou mladých smrkových porostů (Error of above-ground biomass estimation by using site specific allometric equations on the example of the example of two young spruce stands). *Zprávy LesnickéHo VýZkumu* **2018**, *63*, 173–183.

 *forests*

Article

Multi-Sensor Prediction of Stand Volume by a Hybrid Model of Support Vector Machine for Regression Kriging

Lin Chen [1,2], Chunying Ren [1,*], Bai Zhang [1] and Zongming Wang [1,3]

[1]   Key Laboratory of Wetland Ecology and Environment, Northeast Institute of Geography and Agroecology, Chinese Academy of Sciences, Changchun 130102, China; chenlin@iga.ac.cn (L.C.); zhangbai@iga.ac.cn (B.Z.); zongmingwang@iga.ac.cn (Z.W.)
[2]   University of Chinese Academy of Sciences, Beijing 100049, China
[3]   National Earth System Science Data Center, Beijing 100101, China
*   Correspondence: renchy@iga.ac.cn; Tel.: +86-431-8554-2297

Received: 23 January 2020; Accepted: 5 March 2020; Published: 6 March 2020

Abstract: Quantifying stand volume through open-access satellite remote sensing data supports proper management of forest stand. Because of limitations on single sensor and support vector machine for regression (SVR) as well as benefits from hybrid models, this study innovatively builds a hybrid model as support vector machine for regression kriging (SVRK) to map stand volume of the Changbai Mountains mixed forests covering 171,450 ha area based on a small training dataset (n = 928). This SVRK model integrated SVR and its residuals interpolated by ordinary kriging. To determine the importance of multi-sensor predictors from ALOS and Sentinel series, the increase in root mean square error (RMSE) of SVR was calculated by removing the variable after the standardization. The SVRK model achieved accuracy with mean error, RMSE and correlation coefficient in −2.67%, 25.30% and 0.76, respectively, based on an independent dataset (n = 464). The SVRK improved the accuracy of 9% than SVR based on RMSE values. Topographic indices from L band InSAR, backscatters of L band SAR, and texture features of VV channel from C band SAR, as well as vegetation indices of the optical sensor were contributive to explain spatial variations of stand volume. This study concluded that SVRK was a promising approach for mapping stand volume in the heterogeneous temperate forests with limited samples.

Keywords: ALOS-2 L band SAR; Sentinel-1 C band SAR; Sentinel-2 MSI; ALOS DSM; stand volume; support vector machine for regression; ordinary kriging

---

1. Introduction

Forest stand volume, as an ecosystem service, forms the basis for decision-making at diverse levels [1]. Spatial explicit information on forest stand volume is critical for indirect estimation of aboveground biomass for quantifying carbon sequestration and carbon dioxide exchange [2]. Field-based inventories of forest stand volume, the conventional approach, is costly and spatially limited [3]. Progress has been made in mapping forest volume by remote sensing modeling based on multisource satellite and inventory data for spatially continuous and temporally uniform predictions [4,5]. Those remote sensing algorithms were divided into two categories, i.e., physical and empirical models, and the latter included statistical regressions, machine learning techniques, and hybrid approaches [6–8]. Physically based models depend on numerous geometry and biochemistry factors, which may not be readily available [9,10]. Statistical regressions model stand volume by estimating equation parameters related to remote sensing variables [11,12]. These regressions have advantages on modeling explicit relationships and applications at large scales [13,14]. Machine learning algorithms have no assumption on input variable distribution, type and number, which achieve robust and accurate predictions on complex relationships [15,16]. Among the various machine learning techniques, support vector machine is acclaimed for its capacity of dealing with small training datasets in remote sensing-based classification [17,18]. After the re-design to predict quantitative outputs and solve regression problems, this algorithm came to be the support vector machine for regression (SVR) and acquired wide successes in stand volume modeling [19,20]. Hybrid approaches involve either the statistical regression or machine learning model between the target variable and remote sensing predictors, interpolating residuals of predictions by kriging, and combining them [21–23]. Those two-step approaches both consider the spatial heterogeneity conveyed by remote sensing predictors and autocorrelation of neighboring observed data [24,25]. Those approaches, especially machine learning combined ordinary kriging of residuals such as artificial neural network kriging (ANNK) and random forest kriging (RFK), have yielded accurately spatial predictions [26,27]. However, support vector machine for regression kriging (SVRK) modeling for mapping forest volume has rarely been tested and reported.

Stand volume modeling with open-access satellite data has been comparable, repeatable, and has long-term monitoring [28–30]. With the global coverage, Sentinel-1 C band synthetic aperture radar (SAR) and Sentinel-2 multispectral instrument (MSI) images provide capabilities for stand volume modeling using both active and passive remote sensing techniques [31,32]. The Advanced Land Observing Satellite (ALOS/ALOS-2) Phased Array type L band SAR (PALSAR/PALSAR-2) from L band SAR have penetrability, which contain comprehensive information on the orientation and structure of tree canopy and stems within the pixel [33,34]. It makes the ALOS/ALOS-2 images with global observations particularly useful for stand volume mapping [35,36]. The ALOS digital surface model (DSM) from L band interferometric SAR (InSAR) with accurate values of elevation and can provide useful topographic indices to estimate stand volume [37–39]. Reported studies have explored the potential of multi-sensor data using the SVR in volume mapping [31,40,41]. However, how volume predictions would be affected by using SAR and MSI predictors based on the SVR deserves further exploration.

The Changbai Mountains Mixed forests, as the richest eco-region in temperate forests of northeastern China, play a key role in carbon cycles and ecosystem services both at regional and global scales [42–45]. Hence, in this study, we innovatively developed a SVRK model based on limited samples and open-access satellite predictors, and adopted it to map stand volume of the Changbai Mountains Mixed forests, a vital eco-region of temperate ecosystems. The specific objectives were to: (1) determine and compare the relationships of forest volume with multi-sensor variables from ALOS-2, Sentinel-1, Sentinel-2 and ALOS DSM; (2) map stand volume by the SVRK modeling; and (3) analysis spatial variations of stand volume and provide managerial suggestions for forest farms in the study area.

2. Materials and Methods

2.1. Study Area and Field-Measured Stand Volume

The study area covers 171,450 ha and 12 forest farms belonging to Forestry Bureau of Dunhua County (Figure 1). The site is located within the western mountainous area of Yanbian Korean Autonomous Prefecture of Jilin Province, northeast China. The climate is four-season, monsoon-influenced and humid continental, with an annual average temperature and precipitation of 3.28 °C and 632 mm, respectively [25]. Characterized by the dense cover of the Changbai Mountains mixed forests, the major forest types include deciduous broadleaved forest and mixed broadleaf-conifer forest with natural vegetation [43]. Dominant tree species include *Tilia amurensis* (Rupr.), *Juglans mandshurica* (Maxim.), *Fraxinus mandschurica* (Rupr.), *Mongolian oak* (Quercus spp.) and *Betula platyphylla* (Suk.). Typical soils are dark-brown earths, meadow, bog, chernozem, and peat soil.

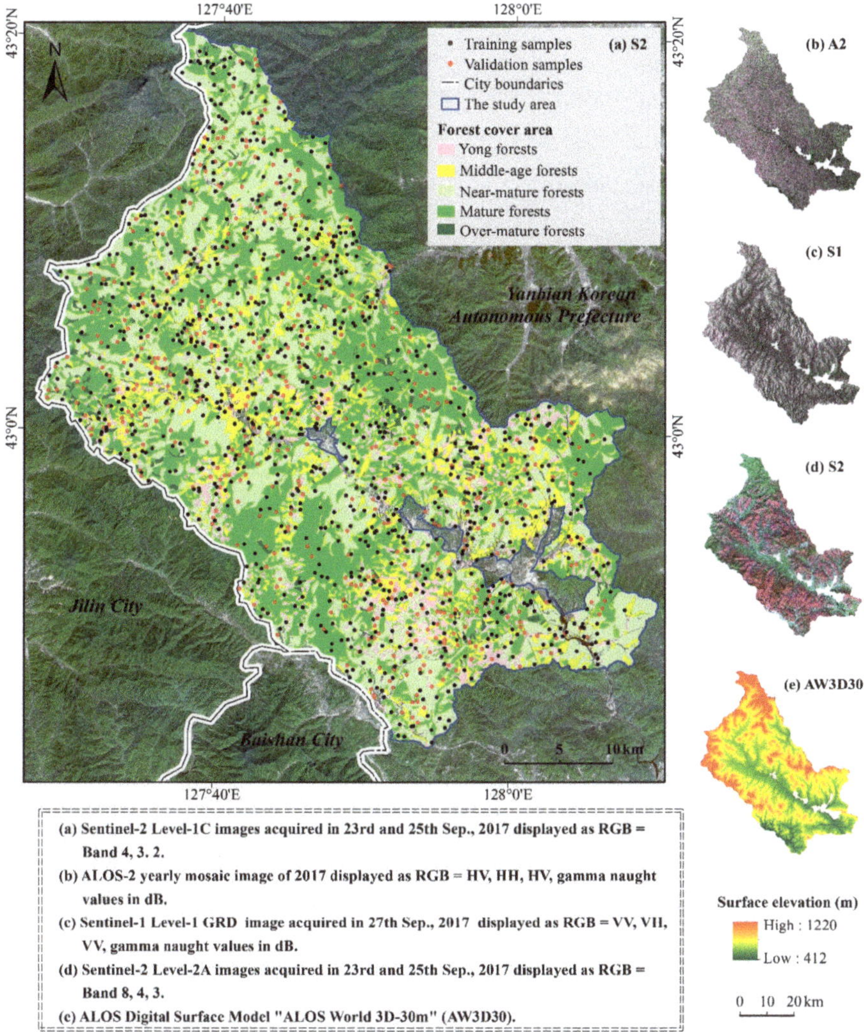

(a) Sentinel-2 Level-1C images acquired in 23rd and 25th Sep., 2017 displayed as RGB = Band 4, 3. 2.
(b) ALOS-2 yearly mosaic image of 2017 displayed as RGB = HV, HH, HV, gamma naught values in dB.
(c) Sentinel-1 Level-1 GRD image acquired in 27th Sep., 2017 displayed as RGB = VV, VH, VV, gamma naught values in dB.
(d) Sentinel-2 Level-2A images acquired in 23rd and 25th Sep., 2017 displayed as RGB = Band 8, 4, 3.
(e) ALOS Digital Surface Model "ALOS World 3D-30m" (AW3D30).

Figure 1. The outline of the study area, sampling sites, and employed open-access satellite remote sensing data derived from Advanced Land Observing Satellite (ALOS), ALOS-2, Sentinel-1, and Sentinel-2 series. Date include (**a**) Sentinel-2 Level-1C, (**b**) ALOS-2 yearly mosaic, (**c**) Sentinel-1 Level-1 GRD, (**d**) Sentinel-2 Level-2A, and (**e**) ALOS Digital Surface Model products.

The field campaign was carried out in September 2017. Stratified sampling design was used by masking non-forest areas and randomly generating the distribution of sampling plots in forest areas, while the plots that were impossible to access were replaced by the nearest sites. Following the national guidelines for forest resource survey [46], eight teams took part in collecting measured data under the same protocol. A total of 1392 squared 30 m by 30 m samples were established (Figure 1a). At each sample site, tree species, diameter at breast height (DBH, the diameter at 1.3 m from the ground), and tree height were measured and recorded. Age classes from young to over-mature were acquired from the forest manager's archives at the local forestry bureau for further analysis (Figure 1a). Stand volume was estimated by DBH and tree height according to the National Standard of China: Tree volume tables (LY/T 1353–1999) [47]. The field-measured stand volume was from 1 to 499.8 m³/ha, and was divided into six levels with the same frequency, with the median and standard deviation (SD) value of 146.3 and 56.2 m³/ha, respectively (Figure 2a). The values of measured volume were mainly below 200 m³/ha with 85.57 % (Figure 2b). The 1,392 sampling sites were randomly divided into training (n = 928) and validation (n = 464) sets (Figure 1a) for training and assessing the models.

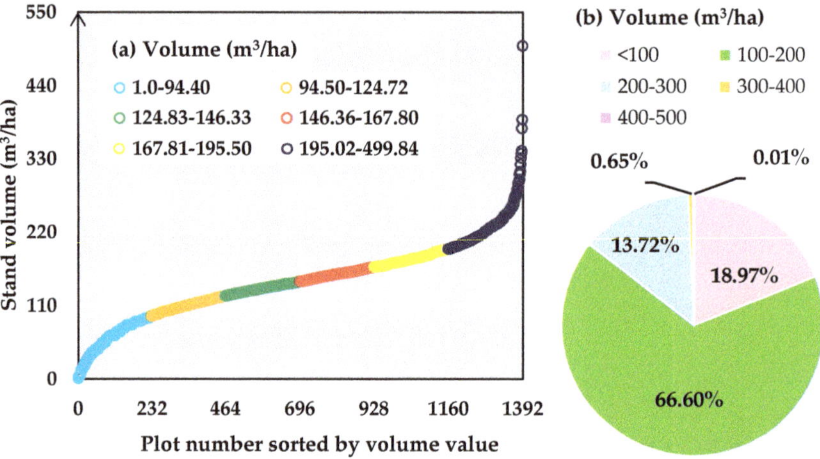

Figure 2. The values of measured stand volume. (**a**) Field sample profiles of volume in the study site from Plot 1 to 1,392; (**b**) Components of volume.

2.2. Satellite Data Pre-Processing and Derived Variables

The adopted multi-sensor satellite data are listed in Table 1. The 25-m ALOS-2 L band SAR yearly mosaic images of 2017 were downloaded from the ALOS Research and Application Project of EORC, the Japan Aerospace Exploration Agency to acquire the normalized backscatter coefficients (gamma naught values) (Table 2), which was sensitive to stand volume [36,48,49]. Images were converted to gamma naught values in decibel unit (dB) from 16-bit digital number (DN) (Figure 1b) using the following Equation (1) [50]:

$$\gamma^0 = 10 \log_{10}\left(DN^2\right) - 83 \tag{1}$$

where γ^0 is gamma naught backscatter coefficient of horizontal transmit-horizontal channel (HH) or horizontal transmit-vertical channel (HV); *DN* is the polarization data in HH or HV.

Table 1. The adopted ALOS-2, Sentinel-1, Sentinel-2, and digital surface model (DSM) data.

Sensors	Elements	Time	Spatial Resolution (m)
ALOS-2	N043E127/N043E128/ N044E127/N044E128	2017	25
Sentinel-1	D633_FCEE of Sentinel-1B	20170927	10
Sentinel-2	T52TCP/T52TCN of Sentinel-2A, T52TDN of Sentinel-2B	20170923 20170925	10
ALOS	N042E127/N042E128/ N043E127/N043E128	Derived from PALSAR data during 2006 to 2011	30

Table 2. Remote sensing indices from the ALOS and Sentinel series data for volume mapping.

Source Image	Relevant Variables		Description
ALOS-2 L band SAR	Polarization	HV	Normalized backscatter coefficient of horizontal transmit-vertical channel in dB
		HH	Normalized backscatter coefficient of horizontal transmit-horizontal channel in dB
Sentinel-1 C band SAR	Polarization	VV	Normalized backscatter coefficient of vertical transmit-vertical channel in dB
		VH	Normalized backscatter coefficient of vertical transmit-horizontal channel in dB
	Texture	VV/VH_CON	Contrast
		VV/VH_DIS	Dissimilarity
		VV/VH_HOM	Homogeneity
		VV/VH_ASM	Angular second moment
		VV/VH_ENE	Energy
		VV/VH_MAX	Maximum probability
		VV/VH_ENT	Entropy
		VV/VH_MEA	Gray-level co-occurrence matrix (GLCM) mean
		VV/VH_VAR	GLCM variance
		VV/VH_COR	GLCM correlation
Sentinel-2 MSI	Multispectral bands	B2	Blue, 490 nm
		B3	Green, 560 nm
		B4	Red, 665 nm
		B5	Red edge, 705 nm
		B6	Red edge, 749 nm
		B7	Red edge, 783 nm
		B8	Near infrared, 842 nm
		B8a	Near infrared, 865 nm
		B11	Short-wave infrared, 1610 nm
		B12	Short-wave infrared, 2190 nm
	Vegetation indices	RVI	Ratio vegetation index, B8/B4
		DVI	Difference vegetation index, B8–B4
		PVI	Perpendicular vegetation index, $\sin(45°)\times B8 - \cos(45°)\times B4$
		NDVI	Normalized difference vegetation index, $(B8 - B4)/(B8 + B4)$
		SAVI	Soil adjusted vegetation index, $1.5 \times (B8 - B4)/(B8 + B4 + 0.5)$
		NDVI5	Normalized difference vegetation index with bands 4 and 5, $(B5 - B4)/(B5 + B4)$
		NLI5	Non-linear vegetation index with bands 4 and 5, $(B5^2 - B4)/(B5^2 + B4)$
		NDVI6	Normalized difference vegetation index with bands 4 and 6, $(B6 - B4)/(B6 + B4)$
		NDVI7	Normalized difference vegetation index with bands 4 and 7, $(B7 - B4)/(B7 + B4)$
		NDVI8a	Normalized difference vegetation index with bands 4 and 8a, $(B8a - B4)/(B8a + B4)$
		MSI	Moisture stress index, B8/B11
		EVI5	Enhanced vegetation index with bands 4, 5 and 2, $2.5 * (B5 - B4) / (B5 + 6 * B4 - 7.5 * B2 + 1)$
		S2REP	Sentinel-2 red-edge position index, $705 + 35 \times [(B4 + B7)/2 - B5] \times (B6 - B5)$
	Transform indices	TCW	Tasseled cap wetness, $0.1509 * B2 + 0.1973 * B3 + 0.3279 * B4 + 0.3406 * B8 + 0.7112 * B11 + 0.4572 * B12$
		TCB	Tasseled cap brightness, $0.3037 * B2 + 0.2793 * B3 + 0.4743 * B4 + 0.5585 * B8 + 0.5082 * B11 + 0.1863 * B12$
		TCG	Tasseled cap greenness, $-0.2848 * B2 - 0.2435 * B3 - 0.5436 * B4 + 0.7243 * B8 + 0.0840 * B11 - 0.1800 * B12$
ALOS DSM	Topographic indicators	H	Elevation
		S	Slope
		A	Aspect
		M	Surface roughness
		SPI	Stream power index, $Ln[Ac \times \tan\beta \times 100]$

Sentinel series images were downloaded from the Copernicus Sentinel Scientific Data Hub. The data included one Sentinel-1 C-band SAR and three Sentinel-2 MSI images. The SAR image was at a high-resolution (HR) Level-1 ground range detected (GRD) processing level with a pixel size of 10 m [51]. Promising results demonstrated that the normalized backscatter coefficients and texture features from Sentinel-1 images could improve forest parameter estimation [32,52]. Sentinel-1 Toolbox in SNAP software (version 6.0, European Space Agency, Paris, France) was used to acquire Sentinel-1 variables with a map projection (Figure 1c) by image calibration, speckle reduction using the Refined Lee Filter, terrain correction by the Range-Doppler, and grey level co-occurrence matrix analysis with 3 × 3-pixel window [52–54]. The Sentinel-2 Level 1C data were top-of-atmosphere reflectance, which were processed by orthorectification and registration [55]. The MSI data had 13 spectral bands, including four in 10 m (bands 2–4, 8), six in 20 m (band 5–7, 8a, 11–12), and three in 60 m (band 1, 9–10) spatial resolutions, respectively [55]. The 10-m Sentinel-2 Level 2A data (Figure 1d) were atmospherically corrected from the Level 1C data by the radiative transfer model-based SEN2COR atmospheric correction processor (version 2.5.5, European Space Agency, Paris, France), and were resampled by Sentinel-2 Toolbox in SNAP. Spectral indices were strongly related to reflectance, and were useful in volume mapping, especially some with red edge bands (band 5, 6, 7, and 8A) [56,57]. Totally, 26 variables from Sentinel-2 were selected and extracted based on previous findings (Table 2) [58,59].

The ALOS Global Digital Surface Model (AW3D30) used in this study was a global dataset generated from L band SAR images collected using the ALOS from 2006 to 2011 (Figure 1e). The data were download from the Japan Aerospace Exploration Agency to extract topographic indices from previous researches by Spatial Analyst of ArcGIS software (version 10.0, ESRI, RedLands, CA, USA) [60,61]. All remote sensing variables were re-projected into UTM Zone 52 WGS84, and then resampled to the 30 m pixel size by ArcGIS.

2.3. Support Vector Machine for Regression Kriging (SVRK) and Modeling Evaluation

The pairwise Pearson's product-moment correlation analysis was operated to determine predictor variables from multi-sensor indices. It consisted of two steps: the selection of variables which were significantly related to field-measured volume ($p < 0.05$) as candidates; the disposal of candidates that were collinear ($r \geq 0.8$), except the one that had the largest correlation coefficient with volume [62]. Those analyses were performed in SPSS software (version 21.0, IBM, Armonk, NY, USA).

The SVRK model built in this study is the extension of SVR, which integrated SVR prediction and estimation of the residuals by ordinary kriging using Equation (2). SVRK considers spatial parametric non-stationarity with the effects of multi-sensor predictors derived from the benefits of SVR. It also added the spatial dependence of the residuals interpolated through ordinary kriging to the estimated trend, as part of the spatial autocorrelation:

$$V_{SVRK} = V_{SVR} + R_{OK}. \tag{2}$$

where: V_{SVRK}, V_{SVR} are predication of stand volume based on SVRK and SVR, respectively; R_{OK} is the estimated residuals of volume from the SVR prediction.

The implementation of SVRK includes two steps, as shown in Figure 3. SVR is firstly used to model the relationship between stand volume and multi-sensor predictors, as a non-linear machine learning method. It uses kernel functions to project the training data onto a new hyperspace where complex non-linear patterns can be simply illustrated (Figure 3a) [63,64]. The optimal hyperspace, constructed by SVR, fits training data and predicts with minimal empirical risk [65]. The SMO (sequential minimal optimization) algorithm is used to solve the quadratic programming optimization problem step-by-step. It updates the SVR function, as shown in Equation 3, to reflect the new values until the Lagrange multipliers converged [66]:

$$f(x) = \sum_{k=1}^{n} (\alpha_k - \alpha_k^*) K(x_k, x_j) + b \tag{3}$$

where x is a vector of the input predictors, $f(x)$. is an optimal function developed by SVR, b is a constant threshold, $K(x_k, x_j)$ is the radial basis function (RBF) kernel with the best bandwidth parameter σ, and α_k are α_k^* the weights (Lagrange multipliers) with the constraints given in Equation 4.

$$\begin{cases} \sum_{k=1}^{n}\left(\alpha_k + \alpha_k^*\right) = 0 \\ 0 \le \alpha_k, \ \alpha_k^* \le C \end{cases} \tag{4}$$

where C is the regularization parameter for balancing between the training error and model complexity.

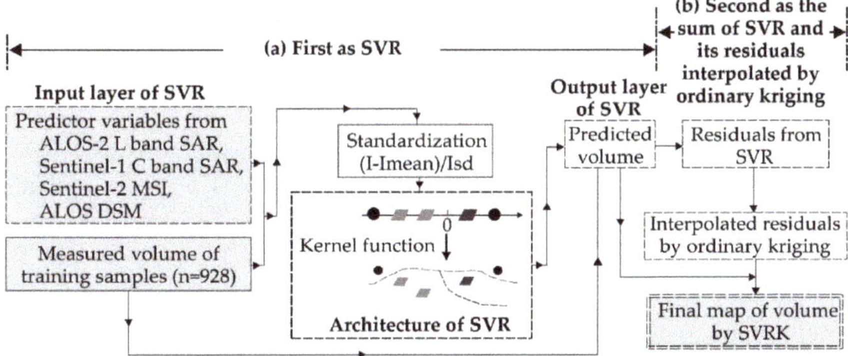

Figure 3. Illustration of support vector machine for regression kriging (SVRK). It includes (**a**) the first step as the support vector machine for regression (SVR) algorithm and (**b**) the second step as the sum of SVR and residuals interpolated by ordinary kriging. I, Imean, and Isd are raw, mean and standard deviation values, respectively.

In this study, SVR was conducted in WEKA software (version 3.8, The University of Waikato, Hamilton, New Zealand). Parameters of SVR, C, and σ, were selected by the smallest root mean square error (RMSE) based on field-measured volume in training dataset (Figure 1a). In order to determine the importance of multi-sensor predictors on volume mapping, the training data was standardized (Figure 3a), and then the increases in RMSEs were calculated as the predictors were excluded one by one from the SVR model.

At the second step, the residuals resulting from SVR are estimated using the ordinary kriging approach (R_{OK}) (Figure 3b). Ordinary kriging, a widely used geostatistical technique, generates an optimal unbiased estimation by the semivariogram [67]. The semivariogram can be modeled by spherical, exponential, and Gaussian functions with three parameters—nugget, range and sill [68]. The nugget is an observation error, and sill is the magnitude of spatial autocorrelation [69]. Thus, the stronger spatial autocorrelation is denoted by the larger value of sill relative to nugget [69]. The range parameter shows on which distance the spatial autocorrelation does not influence any more [69]. The Kolmogorov-Smirnov test (K-S) was used to examine the distribution of residuals based on the stationarity assumption of ordinary kriging. The interpolation of residuals by ordinary kriging was conducted in ArcGIS with the smallest RMSE. Finally, volume prediction by SVRK (V_{SVRK}) were acquired as the sum of V_{SVR} and R_{OK}.

The validation set (Figure 1a) was used to test the performance of volume mapping by SVRK based on the mean error (ME), RMSE, and correlation coefficient between the measured and predicted parameters (r) [70]. In order to better estimate accuracy, the mean measured value of stand volume (146.1 m^3/ha in Figure 2a) was applied to divide the ME and RMSE. The relative improvement (RI) based on RMSE of SVRK over SVR was used as another index for accuracy evaluation [25].

3. Results

3.1. Relationship between Field-Measured Volume and Remote Sensing Variables

In total, 31 variables were significantly related to stand volume ($p < 0.05$) (Table 3), including seven from SAR, 20 from MSI, and four from DSM. The backscatters from different wavelengths all had strongly positive correlations with volume, while variables from ALOS-2 had the closer relationship than that from Sentinel-1. The backscatter from HH was more sensitive to volume than that from HH. Among 10 kinds of texture features from Sentinel-1, only the GLCM mean and variation were actively related to volume. In other words, the increasing texture regularity and variety of VV and VH backscatters indicated the growth of stand volume. It was shown that backscatter texture from VV was more relevant to volume than that from VH.

Table 3. Related variables and predictors derived from multi-sensor satellite data for stand volume mapping. * denotes significance with a *p*-value of the t-test being below 0.05; ** denotes strong significance with a *p*-value below 0.01.

Source Image	Related Variables	*r*	Collinear With	Predictors
ALOS-2	HV	0.138 **	/	Yes
	HH	0.181 **	/	Yes
Sentinel-1	VV	0.075 **	/	Yes
	VV_MEA	0.090 **	VV/VH_ VAR, VH_MEA	Yes
	VV_VAR	0.087 **	VV/VH_MEA, VH_VAR	No
	VH_MEA	0.057 *	VV/VH_ VAR, VV_MEA	No
	VH_VAR	0.061 *	VV/VH_MEA, VV_VAR	No
Sentinel-2	B2	−0.192 **	/	Yes
	B3	−0.111 **	B5, TCW	Yes
	B4	−0.162 **	/	Yes
	B5	−0.079 **	B3, B11, TCW	No
	B11	−0.111 **	B5, B12, TCW	No
	B12	−0.145 **	B11	Yes
	RVI	0.145 **	NDVI, NDVI5, NDVI6, NDVI7, NDVI8a	No
	DVI	0.087 **	PVI, SAVI, TCG	No
	PVI	0.087 **	DVI, SAVI, TCG	No
	NDVI	0.175 **	RVI, NDVI5, NDVI6, NDVI7, NDVI8a	Yes
	SAVI	0.110 **	DVI, PVI, TCG	Yes
	NDVI5	0.151 **	RVI, NDVI, NDVI6, NDVI7, NDVI8a	No
	NLI5	0.065 *	/	Yes
	NDVI6	0.165 **	RVI, NDVI5, NDVI, NDVI7, NDVI8a	No
	NDVI7	0.166 **	RVI, NDVI5, NDVI, NDVI6, NDVI8a	No
	NDVI8a	0.167 **	RVI, NDVI5, NDVI, NDVI6, NDVI7	No
	MSI	0.105 **	/	Yes
	S2REP	0.063 *	/	Yes
	TCW	−0.074 **	B3, B5, B11	No
	TCG	0.087 **	DVI, PVI, SAVI	No
ALOSDSM	H	0.252 **	/	Yes
	S	0.154 **	M	Yes
	A	0.091 **	/	Yes
	M	0.117 **	S	No

As for Sentinel-2 variables, the reflectance of B2–B5, B11, and B12 as well as TCW were negatively related to volume, while the other 13 variables represented the positive correlation. All Sentinel-2 volume-related variables displayed the strong correlation ($p < 0.01$), excluding NLI5 ($p < 0.05$). The vegetation indices that were calculated by characteristic red-edge bands of Sentinel-2 closely connected with volume. Variables from Sentinel-2 had similar performances with that from ALOS-2, which showed the greater sensitivity to volume than Sentinel-1 indices.

All four topographic indicators from ALOS DSM showed the strongly positive influence on the increase of volume. It was indicated that variables from DSM was distinguished in the correlation analysis with volume than that from MSI and SAR. Above all, elevation, ALOS-2 backscatters, the texture features of VV channel of Sentinel-1, and the vegetation indices from Sentinel-2 were comparatively vital for stand volume prediction.

3.2. Modeling Forest Volume by SVRK

3.2.1. Support Vector Machine for Regression (SVR) Modeling for Volume Mapping

To degrade the redundancy, 15 variables that had r values of the correlation analysis among predictor candidates above 0.8 were disposed [62]. The predictors involved in modeling were the following 16 list in Table 3. After standardization of training data, the optimal SVR model was built by C and σ setting as 1000 and 0.01, respectively, with the minimum RMSE being 40.58 m^3/ha. Based on the magnitude of increase in RMSEs (Figure 4), the SVR model showed topographic indicators as the most important predictor for explaining the spatial variations of stand volume, followed by ALOS-2 backscatters, Sentinel-2 indices and texture features of VV channel from Sentinel-1. The VV backscatter from Sentinel-1 was marginal in volume prediction by SVR.

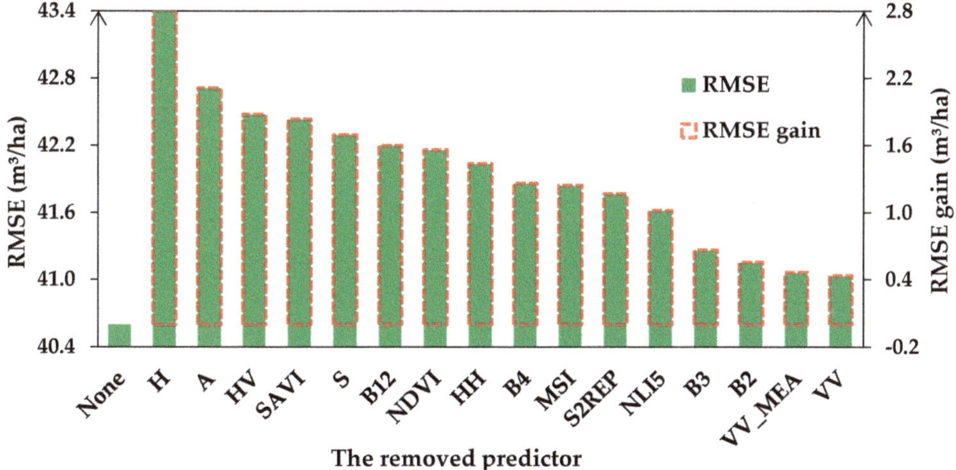

Figure 4. Variable importance shown by increase in the root mean square errors (RMSEs) of SVR models after excluding a predictor.

By the optimal SVR model, the predicted values of stand volume in the study area ranged from 5.37 to 523.84 m^3/ha, with the mean and SD of 150.26 and 26.04 m^3/ha, respectively (Figure 5a). Predicted values were divided into six levels by intervals of field-measured volume values in Figure 2a. The map depicted that the high-altitude region (Figure 1e) was the large forest volume area, with values ranging from 195.51 to 523.94 m^3/ha. Zones with small values of volume (5.37 to 94.40 m^3/ha) were located close to the non-forest area. Among six levels of stand volume, the smallest and largest occupy the minority of the study area. It was revealed that the SVR model overestimated the small values, and underestimated the large volume, compared to field-measured data.

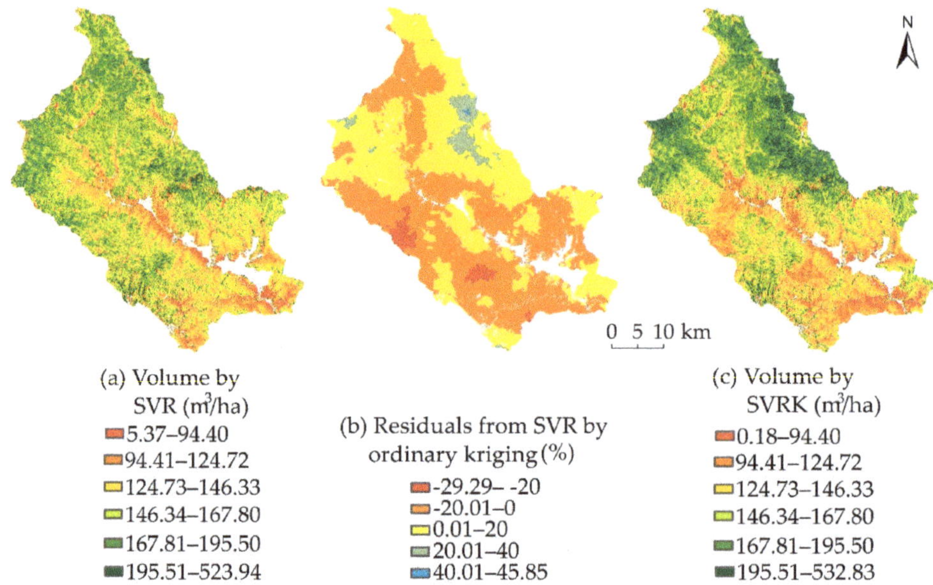

Figure 5. Stand volume mapping by SVR (**a**), and its residuals interpolated by ordinary kriging (%, divided by the mean measured value of stand volume) (**b**) as well as final prediction by SVRK (**c**).

3.2.2. Integration of SVR Prediction and its Residuals by Ordinary Kriging

The residuals were calculated by the field-measured volume and SVR predicted values based on training data. The result of K-S showed that volume residuals from the SVR model possessed a normal distribution ($p < 0.05$), which could be used to calculate experimental semivariograms for ordinary kriging interpolation (Figure 6). Nugget values of spherical, Gaussian, and exponential models were 1478.7, 1509.1, and 1500.5, respectively. Range values were 16.97, 11.44, and 3.16 km, respectively. Sill values were 1689.04, 1673.94, and 1550.52, respectively. The strongest spatial autocorrelation is shown in the spherical model with the largest values of sill relative to nugget. While, the exponential model of ordinary kriging in Figure 6c was chosen to interpolate residuals from SVR with smallest RMSE 39.18 m³/ha. Based on Equation (2), the SVRK model was built.

By the optimal ordinary kriging model (Figure 6c), the distribution of volume residuals from the SVR model were obtained (Figure 5b). The interpolated values of volume residuals ranged from –29.29 to 45.85% (–42.79 to 66.99 m3/ha), with the mean and SD of 0.74 and 14.56 m3/ha, respectively. It was demonstrated that the overestimation of small volume values was located in the western and southern parts of the study area with residuals ranging from –29.29% to –20%. While the SVR model underestimated large volume values in the northern part of the study area, and residuals were from 40.01% to 45.85%.

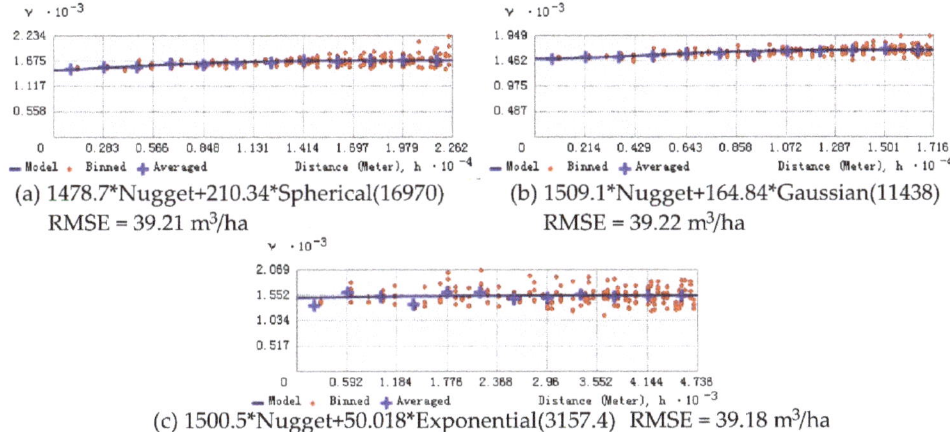

(a) 1478.7*Nugget+210.34*Spherical(16970)
RMSE = 39.21 m^3/ha

(b) 1509.1*Nugget+164.84*Gaussian(11438)
RMSE = 39.22 m^3/ha

(c) 1500.5*Nugget+50.018*Exponential(3157.4) RMSE = 39.18 m^3/ha

Figure 6. Experimental variograms and fitted models of residuals from SVR by (**a**) spherical, (**b**) Gaussian, and (**c**) exponential models.

3.3. Models Assessment and Volume Mapping

Table 4 presented the accuracy of the SVR and SVRK models for estimating volume of the validation set (n = 464). The comparison of SVR and SVRK models demonstrated that additional prediction of residuals by ordinary kriging as the spatial autocorrelation, was more accurate than only considering influences of predictor variables from multi-sensor satellite data (Figure 7). It was indicated by ME values that both two models overestimated stand volume. SVRK remarkably improved accuracy of volume prediction over SVR by 9% (3.77 m^3/ha) based on RMSE values.

Table 4. Accuracy assessment of stand volume modeling based on independent validation data.

Model	ME		RMSE		*r*	RI
	m^3/ha	%	m^3/ha	%		
SVR	−4.49	−3.07	40.73	27.88	0.70	/
SVRK	−3.9	−2.67	36.96	25.30	0.76	0.09

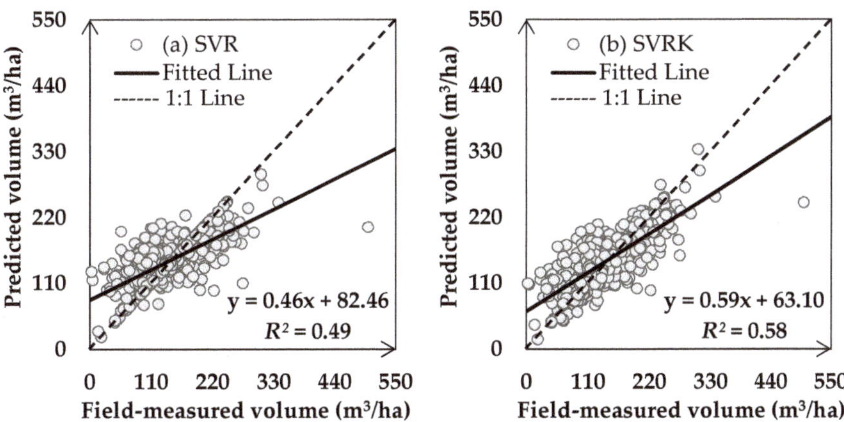

Figure 7. Scatter plots of predicted versus observed volume from validation data based on SVR (**a**) and SVRK (**b**) models.

The distribution of forest volume based on the SVRK model was acquired by combing the Figure 5a,b as the Figure 5c. By the optimal SVRK model, the predicted values of stand volume in the study area ranged from 0.18 to 532.83 m^3/ha, with the mean and SD of 150.99 and 30.83 m^3/ha, respectively (Figure 5c). Based on SVRK mapping, the northern part of the study area with high altitude had the largest volume values ranging from 195.51 to 532.83 m^3/ha. In the south with low altitude and nearby the non-forest area, the smallest volume values ranged from 0.18 to 94.40 m^3/ha. The map showed different distribution of stand volume with the SVR result, while values remained similar (Figure 5a,c). The six levels of stand volume in the SVRK map covered relatively equal areas than that in the SVR result, especially the largest ($\geq$ 195.91 m^3/ha). It was illustrated that the error, which was caused by the SVR model with the overestimation of small values and underestimation of large volume, was reduced. Forest volume of the SVRK map showed the greater spatial variation than that of the SVR. Namely, combining interpolation values of residuals, the spatial distribution of forest volume was much closer to the measured data (SD = 56.2 m^3/ha).

4. Discussion

4.1. Multi-Sensor Satellite Predictors of Forest Volume Mapping

The role of multi-sensor variables on volume mapping was revealed by correlation coefficients (Table 3) and importance (Figure 4). SAR was able to penetrate forest canopy to a certain depth, and related to roughness and water content of vegetation [71], so that its variables were valuable for volume prediction. The elevation as a proxy of InSAR height, was dominant in volume prediction of this study. It was support from previous findings that InSAR height and its slope parameter were directly proportional to volume [72,73]. It was found that HV was more contributive to stand volume prediction than HH and VV channel; yet, VH backscatter was not significantly related. It was owing to the stronger sensitivity of HV backscatter to the forest growth stage than the HH and VV polarizations [74,75]. All backscatters showed positive relationships with volume. This was likely due to an increase in the volume of scattering with the growth of trees [76].

The penetrability was weaker with shorter wavelength. It resulted in the weaker capability of C band SAR for volume prediction than that of L band SAR according to our findings. It also revealed saturation problems of backscatters. The measured forest volume values in the study area were partially above 200 Mg/ha (Figure 2a), which were larger than the common saturation value of C band and smaller than that of L band SAR backscatters [48,77]. The results indicated that texture features of SAR data were much more helpful than original backscatters to forest volume prediction, which was consistent with existing researches [52,78]. However, textural indices from Sentinel-1 was marginal in this study for volume mapping compared to the previous finding [79]. It was resulted from the decrease in the heterogeneity by texture analysis and large variations of stand volume in the study area.

As optical sensor data, MSI variables, i.e., reflectance and spectral indices, were powerful for the retrieval of horizontal forest structures such as vegetation types, canopy cover and DBH [80,81]. Results revealed that the short-wave infrared (SWIR) band was the highly ranked variable for predicting stand volume. It was explained by the closer relationship between SWIR spectral band and vegetation properties, i.e., canopy biomass and water content, compared to other electromagnetic spectrum regions [82]. In line with existing studies, reflectance of optical bands and spectral indices were quite helpful in volume prediction [83,84], whereas, the role of red-edge bands and their vegetation indices in this study was minor than that in previous researches [58,85]. This may be caused by the diversity of tree species in the study area with different responses to various red-edge bands, the average relationships of which were weaker. In a word, topographic indices from L band InSAR, backscatters of L band SAR, texture features of VV channel from C band SAR and vegetation indices of MSI were recommended for stand volume mapping based on open-access satellite data in the heterogeneous temperate forests.

4.2. SVR versus SVRK

It was a pioneering study that built the SVRK hybrid model and utilized it to map stand volume. The optimal RBF-kernel SVR model trained in this study as the first step achieved higher accuracy than multiple linear regressions and SVR models with various kernels based on similar multi-sensor satellite data [31,32], while the SVR model in this study was less accurate than that built by ALOS optical and SAR variables [20]. It was attributed to coarser spatial resolution of L band SAR data and the complex composition of tree species in the study area. Moreover, the density of training dataset of this study (928 samples/171450 ha) was quite smaller than that of the reported research (77 samples/83.71 ha).

Results demonstrated that SVRK improved the mapping accuracy by incorporating interpolation values of residuals to SVR models (Figure 5 and Table 4). The value of the accuracy improvement of SVRK was smaller than that of RFK and ANNK for soil carbon prediction as reported [86–88]. It was resulted from the weaker autocorrelation of residuals from SVR compared with that of soil attributes, as well as the smaller sampling density. It was denoted that stand volume was influenced more by multi-sensor variables, and values of nearby sites affected less. The autocorrelation of volume residuals from SVR was weaker than that of biomass errors from RF, while the accuracy improvement of SVRK (RI = 0.09) for volume mapping was much more than that of RFK (RI = 0.07) for biomass prediction [25]. This is due to the higher spatial heterogeneity and the smaller training dataset of this study. The study concluded that SVRK was a promising approach for mapping stand volume with a small training dataset in heterogeneous temperate forests.

4.3. Spatial Variations of Stand Volume and Forest Management

The spatial distribution of forest volume derived by SVRK with more equal area of each level than the result of the commonly used SVR model was much closer to the measured data (Figures 2 and 5a,c). Whereas, the stand volume map (SD = 30.83 m^3/ha) displayed smaller spatial variations than measured data (SD = 56.2 m^3/ha). The large variations of measured values of stand volume was resulted from the positively skew distribution with the majority below 300 m^3/ha (Figure 2b). The maximum measured volume of 499.84 m^3/ha belonged to a mature *Pinus koraiensis* (Sieb. et Zucc.) dominant natural forest site in the northern part of the study area. The smaller variations of SVRK-derived volume mainly resulted from the coarse mapping resolution as 30 m in this highly heterogeneous forest landscape. The smallest and largest levels of stand volume (< 94.90 and > 195.51 m^3/ha) still occupied smaller areas than other four levels. It was illustrated that 30-m multi-sensor data from mosaic L band SAR and InSAR, C band SAR and MSI displayed a saturation problem in detecting small and large values of stand volume.

The volume values of different forest ages were summarized as Table 5 from the SVRK prediction by multi-sensor satellite data. Based on spatial and age variations of stand volume, certain measures can be taken for the sustainable forest management. In young forests, minimum, maximum and mean were all the smallest among five classes, while the variation was largest. The larger values of stand volume in young forests was mainly attributed to the high stand density. With tree growth, more space and resource competition occur among individual trees. Thus, young forests with volume above 195.51 m^3/ha should be the critical focus areas, which need thinning (Figure 5c). However, the young forests with volume below 94.90 m^3/ha should be enclosed for cultivation. The volume variation of middle-age forests was second-largest. Middle-age forests with volume above 195.51 m^3/ha also require thinning, while the increment felling should be conducted in smaller volume areas. Near-mature forests obtained the smallest maximum of stand volume. Management measures in these forests can include the artificial promotion of natural regeneration and beforehand regeneration. Forest manager can selectively cut weak, pest-infested, and diseased trees in mature and over-mature forests.

Table 5. Stand volume of forests with different ages in the study area.

Age	Minimum	Maximum	Mean	Standard Deviation	Coefficient of Variation (%)
Yong	0.18	358.78	134.79	29.91	22.19
Middle-age	0.21	520.56	138.30	29.79	21.54
Near-mature	1.18	463.83	152.51	28.47	18.67
Mature	1.71	532.83	156.38	30.09	19.24
Over-mature	7.91	521.99	160.48	30.64	19.09

5. Conclusions

Machine learning modeling with remote sensing data combined sample plot data has become a well adopted method to generate spatially explicit estimates of forest parameters. Among that, SVR has achieved wide success in application and has been praised for its ability to deal with small training datasets. A major shortcoming of machine learning is that it ignores the spatial autocorrelation of neighboring observed data. The main objective of the study was to build a hybrid model, i.e., SVRK, which integrated SVR and its residuals by ordinary kriging, based on a small training dataset. Then SVRK was used to map stand volume, the most common forest parameter needed for sustainable forest management at all scales. This study also determined the potential of open-access satellite predictors from multi-frequency SAR data in predicting volume in the heterogeneous temperate forests. As the first exploration of the SVRK modeling, this study provides an informative foundation for decision makers and other researchers on stand volume mapping with limited samples in northeastern China.

Based on the results of this study, the following was concluded:

(1) SVRK can accurately predict stand volume of the heterogeneous Changbai Mountains Mixed forests with RMSE of 25.3% based on the low sampling density of 928 samples/171,450 ha, which improved accuracy of 9% than SVR.

(2) Topographic indices from ALOS DSM as L band InSAR, backscatters of ALOS-2 as L band SAR, and texture features of VV channel from Sentinel-1 as C band SAR, as well as vegetation indices of Sentinel-2 MSI as the optical sensor were vital for explaining the observed variability of stand volume.

(3) The northern part of the study area with high altitude had the largest volume values ranging from 195.51 to 532.83 m^3/ha. In the south with low altitude and near a non-forest area, the smallest volume values ranged from 0.18 to 94.40 m^3/ha.

(4) Yong forests should be paid attention to and certain measures can be taken for sustainable forest management. Indeed, young forests with large volume need thinning, while that with small values should be enclosed for cultivation.

Author Contributions: L.C., C.R., and B.Z. designed this research. L.C. and C.R. conducted field sampling. L.C. performed the experiments, conducted the analysis, and drafted the manuscript. L.C., C.R., B.Z., and Z.W. revised and finalized the manuscript. All authors have read and agreed to the published version of the manuscript.

Funding: This study was supported by the National Key Research and Development Project of China (No. 2016YFC0500300), the Jilin Scientific and Technological Development Program (No. 20170301001NY), the funding from Youth Innovation Promotion Association of Chinese Academy of Sciences (No. 2017277, 2012178), and National Earth System Science Data Center of China.

Acknowledgments: The authors are grateful to the support from colleagues and local forestry bureau who participated in the field surveys and data collection. We thank the National Earth System Science Data Center (http://www.geodata.cn) for providing geographic information data. This study is supported by the National Key Research and Development Project of China (No. 2016YFC0500300), the Jilin Scientific and Technological Development Program (No. 20170301001NY), funding from Youth Innovation Promotion Association of Chinese Academy of Sciences (No. 2017277, 2012178), and National Earth System Science Data Center of China.

Conflicts of Interest: The authors declare no conflict of interest.

References

1. Dube, T.; Sibanda, M.; Shoko, C.; Mutanga, O. Stand-volume estimation from multi-source data for coppiced and high forest *Eucalyptus Spp.* silvicultural systems in KwaZulu-Natal, South Africa. *ISPRS J. Photogramm. Remote Sens.* **2017**, *132*, 162–169. [CrossRef]
2. Chirici, G.; Mura, M.; McInerney, D.; Py, N.; Tomppo, E.O.; Waser, L.T.; Travaglini, D.; McRoberts, R.E. A meta-analysis and review of the literature on the k-nearest neighbors technique for forestry applications that use remotely sensed data. *Remote Sens. Environ.* **2016**, *176*, 282–294. [CrossRef]
3. Boisvenue, C.; Smiley, B.P.; White, J.C.; Kurz, W.A.; Wulder, M.A. Integration of Landsat time series and field plots for forest productivity estimates in decision support models. *For. Ecol. Manag.* **2016**, *376*, 284–297. [CrossRef]
4. Lehmann, E.A.; Caccetta, P.; Lowell, K.; Mitchell, A.; Zhou, Z.S.; Held, A.; Milne, T.; Tapley, I. SAR and optical remote sensing: Assessment of complementarity and interoperability in the context of a large-scale operational forest monitoring system. *Remote Sens. Environ.* **2015**, *156*, 335–348. [CrossRef]
5. Main, R.; Mathieu, R.; Kleynhans, W.; Wessels, K.; Naidoo, L. Hyper-temporal C Band SAR for baseline woody structural assessments in deciduous savannas. *Remote Sens.* **2016**, *8*, 661. [CrossRef]
6. Chen, G.; Hay, G.J.; St-Onge, B. A GEOBIA framework to estimate forest parameters from lidar transects, Quickbird imagery and machine learning: A case study in Quebec, Canada. *Int. J. Appl. Earth Obs. Geoinf.* **2012**, *15*, 28–37. [CrossRef]
7. Viana, H.; Aranha, J.; Lopes, O.; Cohen, W.B. Estimation of crown biomass of Pinus pinaster stands and shrubland above-ground biomass using forest inventory data, remotely sensed imagery and spatial prediction models. *Ecol. Model.* **2012**, *226*, 22–35. [CrossRef]
8. Santoro, M.; Wegmüller, U.; Askne, J. Forest stem volume estimation using C-band interferometric SAR coherence data of the ERS-1 mission 3-days repeat-interval phase. *Remote Sens. Environ.* **2018**, *216*, 684–696. [CrossRef]
9. Roy, A.; Royer, A.; Wigneron, J.-P.; Langlois, A.; Bergeron, J.; Cliche, P. A simple parameterization for a boreal forest radiative transfer model at microwave frequencies. *Remote Sens. Environ.* **2012**, *124*, 371–383. [CrossRef]
10. Sharma, R.C.; Kajiwara, K.; Honda, Y. Estimation of forest canopy structural parameters using kernel-driven bi-directional reflectance model based multi-angular vegetation indices. *ISPRS J. Photogramm. Remote Sens.* **2013**, *78*, 50–57. [CrossRef]
11. Alrababah, M.A.; Alhamad, M.N.; Bataineh, A.L.; Bataineh, M.N.; Suwaileh, A.F. Estimating east Mediterranean forest parameters using Landsat ETM. *Int. J. Remote Sens.* **2011**, *32*, 1561–1574. [CrossRef]
12. Nilsson, M.; Nordkvist, K.; Jonzén, J.; Lindgren, N.; Axensten, P.; Wallerman, J.; Egberth, M.; Larsson, S.; Nilsson, L.; Eriksson, J.; et al. A nationwide forest attribute map of Sweden predicted using airborne laser scanning data and field data from the National Forest Inventory. *Remote Sens. Environ.* **2017**, *194*, 447–454. [CrossRef]
13. Leboeuf, A.; Fournier, R.A.; Luther, J.E.; Beaudoin, A.; Guindon, L. Forest attribute estimation of northeastern Canadian forests using QuickBird imagery and a shadow fraction method. *For. Ecol. Manag.* **2012**, *266*, 66–74. [CrossRef]
14. Abdullahi, S.; Kugler, F.; Pretzsch, H. Prediction of stem volume in complex temperate forest stands using TanDEM-X SAR data. *Remote Sens. Environ.* **2016**, *174*, 197–211. [CrossRef]
15. Tamm, T.; Remm, K. Estimating the parameters of forest inventory using machine learning and the reduction of remote sensing features. *Int. J. Appl. Earth Obs. Geoinf.* **2009**, *11*, 290–297. [CrossRef]
16. Wang, M.J.; Sun, R.; Xiao, Z.Q. Estimation of forest canopy height and aboveground biomass from spaceborne LiDAR and Landsat imageries in Maryland. *Remote Sens.* **2018**, *10*, 344. [CrossRef]
17. Pal, M.; Mather, P.M. Support vector machines for classification in remote sensing. *Int. J. Remote Sens.* **2005**, *26*, 1007–1011. [CrossRef]
18. Mountrakis, G.; Im, G.; Ogole, C. Support vector machines in remote sensing: A review. *ISPRS J. Photogramm. Remote Sens.* **2011**, *66*, 247–259. [CrossRef]
19. dos Reis, A.A.; Carvalho, M.C.; de Mello, J.M.; Gomide, L.R.; Ferraz Filho, A.C.; Acerbi, F.W. Spatial prediction of basal area and volume in Eucalyptus stands using Landsat TM data: An assessment of prediction methods. *N. Z. J. For. Sci.* **2018**, *48*, 1. [CrossRef]

20. de Souza, G.S.A.; Soares, V.P.; Leite, H.G.; Gleriani, J.M.; do Amaral, C.H.; Ferraz, A.S.; de Freitas Silveira, M.V.; dos Santos, J.F.C.; Silveira Velloso, S.G.; Domingues, G.F.; et al. Multi-sensor prediction of Eucalyptus stand volume: A support vector approach. *ISPRS J. Photogramm. Remote Sens.* **2019**, *156*, 135–146. [CrossRef]

21. Kumar, S.; Lai, R.; Liu, D.S. A geographically weighted regression kriging approach for mapping soil organic carbon stock. *Geoderma* **2012**, *189–190*, 627–634. [CrossRef]

22. Viscarra Rossel, R.A.; Webster, R.; Kidd, D. Mapping gamma radiation and its uncertainty from weathering products in a Tasmanian landscape with a proximal sensor and random forest kriging. *Earth Surf. Proc. Land.* **2013**, *39*, 735–748. [CrossRef]

23. Tarasov, D.A.; Buevich, A.G.; Sergeev, A.P.; Shichkin, A.V. High variation topsoil pollution forecasting in the Russian Subarctic: Using artificial neural networks combined with residual kriging. *Appl. Geochem.* **2018**, *88*, 188–197. [CrossRef]

24. Meng, Q.M.; Cieszewski, C.; Madden, M. Large area forest inventory using Landsat ETM+: A geostatistical approach. *ISPRS J. Photogramm. Remote Sens.* **2009**, *64*, 27–36. [CrossRef]

25. Chen, L.; Wang, Y.Q.; Ren, C.Y.; Zhang, B.; Wang, Z.M. Assessment of multi-wavelength SAR and multispectral instrument data for forest aboveground biomass mapping using random forest kriging. *For. Ecol. Manag.* **2019**, *447*, 12–25. [CrossRef]

26. Fayad, I.; Baghdadi, N.; Bailly, J.S.; Barbier, N.; Gond, V.; Hérault, B.; Hajj, M.E.; Fabre, F.; Perrin, J. Regional scale rain-forest height mapping using regression-kriging of spaceborne and airborne LiDAR data: Application on French Guiana. *Remote Sens.* **2016**, *8*, 240. [CrossRef]

27. Li, Q.Q.; Zhang, X.; Wang, C.Q.; Li, B.; Gao, X.S.; Yuan, D.G.; Luo, Y.L. Spatial prediction of soil nutrient in a hilly area using artificial neural network model combined with kriging. *Arch. Agron. Soil Sci.* **2016**, *62*, 1541–1553. [CrossRef]

28. Fazakas, Z.; Nilsson, M.; Olsson, H. Regional forest biomass and wood volume estimation using satellite data and ancillary data. *Agric. For. Meteorol.* **1999**, *98–99*, 417–425. [CrossRef]

29. Wilhelm, S.; Hüttich, C.; Korets, M.; Schmullius, C. Large area mapping of boreal growing stock volume on an annual and multi-temporal level using PALSAR L-band backscatter mosaics. *Forests* **2014**, *5*, 1990–2015. [CrossRef]

30. Chrysafis, I.; Mallinis, G.; Tsakiri, M.; Patias, P. Evaluation of single-date and multi-seasonal spatial and spectral information of Sentinel-2 imagery to assess growing stock volume of a Mediterranean forest. *Int. J. Appl. Earth Obs. Geoinf.* **2019**, *77*, 1–14. [CrossRef]

31. Ataee, M.S.; Maghsoudi, Y.; Latifi, H.; Fadaie, F. Improving estimation accuracy of growing stock by multi-frequency SAR and multi-spectral data over Iran's heterogeneously-structured broadleaf Hyrcanian forests. *Forests* **2019**, *10*, 641. [CrossRef]

32. Mauya, E.W.; Koskinen, J.; Tegel, K.; Hämäläinen, J.; Kauranne, T.; Käyhkö, N. Modelling and predicting the growing stock volume in small-scale plantation forests of Tanzania using multi-sensor image synergy. *Forests* **2019**, *10*, 279. [CrossRef]

33. Peregon, A.; Yamagata, Y. The use of ALOS/PALSAR backscatter to estimate above-ground forest biomass: A case study in Western Siberia. *Remote Sens. Environ.* **2013**, *137*, 139–146. [CrossRef]

34. Reiche, J.; Lucas, R.; Mitchell, A.L.; Verbesselt, J.; Hoekman, D.H.; Haarpaintner, J.; Kellndorfer, J.M.; Rosenqvist, A.; Lehmann, E.A.; Woodcock, C.E.; et al. Combining satellite data for better tropical forest monitoring. *Nat. Clim. Chang.* **2016**, *6*, 120–122. [CrossRef]

35. Chowdhury, T.A.; Thiel, C.; Schmullius, C. Growing stock volume estimation from L-band ALOS PALSAR polarimetric coherence in Siberian forest. *Remote Sens. Environ.* **2014**, *155*, 129–144. [CrossRef]

36. Santoro, M.; Cartus, O.; Fransson, J.E.S.; Wegmüller, U. Complementarity of X-, C-, and L-band SAR backscatter observations to retrieve forest stem volume in boreal forest. *Remote Sens.* **2019**, *11*, 1563. [CrossRef]

37. Stage, A.R.; Salas, C. Interactions of elevation, aspect, and slope in models of forest species composition and productivity. *For. Sci.* **2007**, *53*, 486–492.

38. Rahlf, J.; Breidenbach, J.; Solberg, S.; Næsset, E.; Astrup, R. Comparison of four types of 3D data for timber volume estimation. *Remote Sens. Environ.* **2014**, *155*, 325–333. [CrossRef]

39. Tadono, T.; Takaku, J.; Tsutsui, K.; Oda, F.; Nagai, H. Status of "ALOS World 3D (AW3D)" global DSM generation. In Proceedings of the 2015 IEEE International Geoscience and Remote Sensing Symposium (IGARSS), Milan, Italy, 26–31 July 2015.

40. Xu, C.; Manley, B.; Morgenroth, J. Evaluation of modelling approaches in predicting forest volume and stand age for small-scale plantation forests in New Zealand with RapidEye and LiDAR. *Int. J. Appl. Earth Obs. Geoinf.* **2018**, *73*, 386–396. [CrossRef]

41. Schumacher, J.; Rattay, M.; Kirchhöfer, M.; Adler, P.; Kändler, G. Combination of multi-temporal sentinel 2 images and aerial image based canopy height models for timber volume modelling. *Forests* **2019**, *10*, 746. [CrossRef]

42. Olson, D.M.; Dinerstein, E.; Wikramanayake, E.D.; Burgess, N.D.; Powell, G.V.N.; Underwood, E.C.; D'Amico, J.A.; Itoua, I.; Strand, H.E.; Morrison, J.C.; et al. Terrestrial ecoregions of the world: A new map of life on Earth. *Bioscience* **2001**, *51*, 933–938. [CrossRef]

43. Wang, Y.Q.; Wu, Z.F.; Yuan, X.; Zhang, H.Y.; Zhang, J.Q.; Xu, J.W.; Lu, Z.; Zhou, Y.Y.; Feng, J. Resources and ecological security of the Changbai Mountain region in Northeast Asia. In *Remote Sensing of Protected Lands*; Wang, Y.Q., Ed.; CRC Press: Boca Raton, FL, USA, 2011.

44. Cai, H.Y.; Di, X.Y.; Chang, S.X.; Wang, C.K.; Shi, B.K.; Geng, P.F.; Jin, G.Z. Carbon storage, net primary production, and net ecosystem production in four major temperate forest types in northeastern China. *Can. J. For. Res.* **2016**, *45*, 143–151. [CrossRef]

45. Ma, J.; Xiao, X.M.; Qin, Y.W.; Chen, B.Q.; Hu, Y.M.; Li, X.P.; Zhao, B. Estimating aboveground biomass of broadleaf, needleleaf, and mixed forests in Northeastern China through analysis of 25-m ALOS/PALSAR mosaic data. *For. Ecol. Manag.* **2017**, *389*, 199–210. [CrossRef]

46. MOF (Ministry of Forestry). *Standards for Forestry Resource Survey*; China Forestry Publisher: Beijing, China, 1982.

47. Forestry Administration of China. *Tree Volume Tables (National Standard # LY/T 1353-1999)*; Forestry Administration of China: Beijing, China, 1999.

48. Santi, E.; Paloscia, S.; Pettinato, S.; Chirici, G.; Mura, M.; Maselli, F. Application of neural networks for the retrieval of forest woody volume from SAR multifrequency data at L and C bands. *Eur. J. Remote Sens.* **2015**, *48*, 673–687. [CrossRef]

49. Urbazaev, M.; Cremer, F.; Migliavacca, M.; Reichstein, M.; Schmullius, C.; Thiel, C. Potential of multi-temporal ALOS-2 PALSAR-2 ScanSAR data for vegetation height estimation in tropical forests of Mexico. *Remote Sens.* **2018**, *10*, 1227. [CrossRef]

50. Shimada, M.; Isoguchi, O.; Tadono, T.; Isono, K. PALSAR radiometric and geometric calibration. *IEEE Trans. Geosci. Remote Sens.* **2009**, *47*, 3915–3932. [CrossRef]

51. Sentinel-1_Team. *Sentinel-1 User Handbook*; European Space Agency: Paris, France, 2013.

52. Dos Reis, A.A.; Franklin, S.E.; de Mello, J.M.; Junior, F.W.A. Volume estimation in a Eucalyptus plantation using multi-source remote sensing and digital terrain data: A case study in Minas Gerais State, Brazil. *Int. J. Remote Sens.* **2019**, *4*, 2683–2702. [CrossRef]

53. Veci, L. *Sentinel-1 Toolbox: SAR Basics Tutorial*; ARRAY Systems Computing, Inc.: Toronto, ON, Canada; European Space Agency: Paris, France, 2015.

54. Laurin, G.V.; Balling, J.; Corona, P.; Mattioli, W.; Papale, D.; Puletti, N.; Rizzo, M.; Truckenbrodt, J.; Urban, M. Above-ground biomass prediction by Sentinel-1 multitemporal data in central Italy with integration of ALOS2 and Sentinel-2 data. *J. Appl. Remote Sens.* **2018**, *12*, 016008. [CrossRef]

55. Sentinel-2_Team. *Sentinel-2 User Handbook*; European Space Agency: Paris, France, 2015.

56. Puliti, S.; Saarela, S.; Gobakken, T.; Ståhl, G.; Naesset, E. Combining UAV and Sentinel-2 auxiliary data for forest growing stock volume estimation through hierarchical model-based inference. *Remote Sens. Environ.* **2018**, *204*, 485–497. [CrossRef]

57. Astola, H.; Häme, T.; Sirro, L.; Molinier, M.; Kilpi, J. Comparison of Sentinel-2 and Landsat 8 imagery for forest variable prediction in boreal region. *Remote Sens. Environ.* **2019**, *223*, 257–273. [CrossRef]

58. Chrysafis, I.; Mallinis, G.; Siachalou, S.; Patias, P. Assessing the relationships between growing stock volume and Sentinel-2 imagery in a Mediterranean forest ecosystem. *Remote Sens. Lett.* **2017**, *8*, 508–517. [CrossRef]

59. Wittke, S.; Yu, X.W.; Karjalainen, M.; Hyyppä, J.; Puttonen, E. Comparison of two-dimensional multitemporal Sentinel-2 data with three dimensional remote sensing data sources for forest inventory parameter estimation over a boreal forest. *Int. J. Appl. Earth Obs. Geoinf.* **2019**, *76*, 167–178. [CrossRef]

60. Wijaya, A.; Kusnadi, S.; Gloaguen, R.; Heilmeier, H. Improved strategy for estimating stem volume and forest biomass using moderate resolution remote sensing data and GIS. *J. For. Res.* **2010**, *21*, 1–12. [CrossRef]

61. Chen, L.; Wang, Y.Q.; Ren, C.Y.; Zhang, B.; Wang, Z.M. Optimal combination of predictors and algorithms for forest above-ground biomass mapping from Sentinel and SRTM data. *Remote Sens.* **2019**, *11*, 414. [CrossRef]

62. Xu, L.; Shi, Y.J.; Fang, H.Y.; Zhou, G.M.; Xu, X.J.; Zhou, Y.F.; Tao, J.X.; Ji, B.Y.; Xu, J.; Li, C.; et al. Vegetation carbon stocks driven by canopy density and forest age in subtropical forest ecosystems. *Sci. Total Environ.* **2018**, *631–632*, 619–626. [CrossRef]

63. Gunn, S.R. *Support Vector Machines for Classification and Regression*; Technical Report; University of Southampton: Southampton, UK, 1998.

64. Williams, G. *Data Mining with Rattle and R: The Art of Excavating Data for Knowledge Discovery, use R*; Springer Science+Business Media, LLC: New York, NY, USA, 2011.

65. Were, K.; Bui, D.T.; Dick, Ø.B.; Singh, B.R. A comparative assessment of support vector regression, artificial neural networks, and random forests for predicting and mapping soil organic carbon stocks across an Afromontane landscape. *Ecol. Indic.* **2015**, *52*, 394–403. [CrossRef]

66. Platt, J. *Fast Training of Support Vector Machines Using Sequential Minimal Optimization*; MIT Press: Cambridge, MA, USA, 1999.

67. Isaaks, E.H.; Srivastava, R.M. *An Introduction to Applied Geostatistics*; Oxford University Press: Oxford, England, 1989.

68. Ou, Y.; Rousseau, A.N.; Wang, L.X.; Yan, B.X. Spatio-temporal patterns of soil organic carbon and pH in relation to environmental factors-A case study of the Black Soil Region of Northeastern China. *Agric. Ecosyst. Environ.* **2017**, *245*, 22–31. [CrossRef]

69. Tang, G.A.; Yang, X. *ArcGIS Experimental Course for Spatial Analysis*, 2nd ed.; Science Press: Beijing, China, 2013.

70. Chen, L.; Ren, C.Y.; Zhang, B.; Wang, Z.M.; Wang, Y.Q. Mapping spatial variations of structure and function parameters for forest condition assessment of the Changbai Mountain National Nature Reserve. *Remote Sens.* **2019**, *11*, 3004. [CrossRef]

71. Lu, D.S.; Chen, Q.; Wang, G.X.; Liu, L.J.; Li, G.Y.; Moran, E. A survey of remote sensing-based aboveground biomass estimation methods in forest ecosystems. *Int. J. Digit. Earth* **2016**, *9*, 63–105. [CrossRef]

72. Gama, F.F.; dos Santos, J.R.; Mura, J.C. Eucalyptus biomass and volume estimation using interferometric and polarimetric SAR data. *Remote Sens.* **2010**, *2*, 939–956. [CrossRef]

73. Solberg, S.; Astrup, R.; Breidenbach, J.; Nilsen, B.; Weydahl, D. Monitoring spruce volume and biomass with InSAR data from TanDEM-X. *Remote Sens. Environ.* **2013**, *139*, 60–67. [CrossRef]

74. Aslan, A.; Rahman, A.F.; Warren, M.W.; Robeson, S.W. Mapping spatial distribution and biomass of coastal wetland vegetation in Indonesian Papua by combining active and passive remotely sensed data. *Remote Sens. Environ.* **2016**, *183*, 65–81. [CrossRef]

75. Sinha, S.; Santra, A.; Sharma, L.; Jeganathan, C.; Nathawat, M.S.; Das, A.K.; Mohan, S. Multi-polarized Radarsat-2 satellite sensor in assessing forest vigor from above ground biomass. *J. For. Res.* **2018**, *29*, 1139–1145. [CrossRef]

76. Pham, T.D.; Yoshion, K. Aboveground biomass estimation of mangrove species using ALOS-2 PALSAR imagery in Hai Phong City, Vietnam. *J. Appl. Remote Sens.* **2017**, *11*, 026010. [CrossRef]

77. Fransson, J.E.S.; Ispraelsson, H. Estimation of stem volume in boreal forests using ERS-1 C- an JERS-1 L-band SAR data. *Int. J. Remote Sens.* **1999**, *20*, 123–137. [CrossRef]

78. Thiel, C.; Schmullius, C. The potential of ALOS PALSAR backscatter and InSAR coherence for forest growing stock volume estimation in Central Siberia. *Remote Sens. Environ.* **2016**, *173*, 258–273. [CrossRef]

79. Morin, D.; Planells, M.; Guyon, D.; Villard, L.; Mermoz, S.; Bouvet, A.; Thevenon, H.; Dejoux, J.-F.; Toan, T.L.; Dedieu, G. Estimation and mapping of forest structure parameters from open access satellite images: Development of a generic method with a study case on coniferous plantation. *Remote Sens.* **2019**, *11*, 1275. [CrossRef]

80. Peña, M.A.; Brenning, A.; Sagredo, A. Constructing satellite-derived hyperspectral indices sensitive to canopy structure variables of a Cordilleran Cypress (Austrocedrus chilensis) forest. *ISPRS J. Photogramm. Remote Sens.* **2012**, *74*, 1–10. [CrossRef]

81. Lausch, A.; Erasmi, S.; King, D.J.; Magdon, P.; Heurich, M. Understanding forest health with remote sensing-part II—A review of approaches and data models. *Remote Sens.* **2017**, *9*, 129. [CrossRef]

82. Dube, T.; Mutanga, O.; Abdel-Rahman, E.A.; Ismail, R.; Slotow, R. Predicting *Eucalyptus spp.* Stand volume in Zululand, South Africa: An analysis using a stochastic gradient boosting regression ensemble with multi-source data sets. *Int. J. Remote Sens.* **2015**, *36*, 3751–3772. [CrossRef]

83. Latifi, H.; Nothdurft, A.; Koch, B. Non-parametric prediction and mapping of standing timber volume and biomass in a temperate forest: Application of multiple optical/LiDAR-derived predictors. *Forestry* **2010**, *83*, 395–407. [CrossRef]

84. Mura, M.; Bottalico, F.; Giannetti, F.; Bertani, R.; Giannini, R.; Mancini, M.; Orlandini, S.; Travaglini, D.; Chiricia, G. Exploiting the capabilities of the Sentinel-2 multi spectral instrument for predicting growing stock volume in forest ecosystems. *Int. J. Appl. Earth Obs. Geoinf.* **2018**, *66*, 126–134. [CrossRef]

85. Korhonen, L.; Packalen, H.P.; Rautiainen, M. Comparison of Sentinel-2 and Landsat 8 in the estimation of boreal forest canopy cover and leaf area index. *Remote Sens. Environ.* **2017**, *195*, 259–274. [CrossRef]

86. Dai, F.Q.; Zhou, Q.G.; Lv, Z.Q.; Wang, X.M.; Liu, G.C. Spatial prediction of soil organic matter content integrating artificialneural network and ordinary kriging in Tibetan Platea. *Ecol. Indic.* **2014**, *45*, 184–194. [CrossRef]

87. Guo, P.T.; Li, M.F.; Luo, W.; Tang, Q.F.; Liu, Z.W.; Lin, Z.M. Digital mapping of soil organic matter for rubber plantation at regional scale: An application of random forest plus residuals kriging approach. *Geoderma* **2015**, *237–238*, 49–59. [CrossRef]

88. Tziachris, P.; Aschonitis, V.; Chatzistathis, T.; Papadopoulou, M. Assessment of spatial hybrid methods for predicting soil organic matter using DEM derivatives and soil parameters. *Catena* **2019**, *174*, 206–216. [CrossRef]

Article

Applying LiDAR to Quantify the Plant Area Index Along a Successional Gradient in a Tropical Forest of Thailand

Siriruk Pimmasarn [1,*], Nitin Kumar Tripathi [1], Sarawut Ninsawat [1] and Nophea Sasaki [2]

1   Department of Information & Communication Technologies, School of Engineering and Technology (SET), Asian Institute of Technology, Pathum Thani 12120, Thailand; nitinkt@ait.asia (N.K.T.); sarawutn@ait.ac.th (S.N.)
2   Department of Development and Sustainability, School of Environment, Resources and Development (SERD), Asian Institute of Technology, Pathum Thani 12120, Thailand; nopheas@ait.ac.th
*   Correspondence: Siriruk.pim@gmail.com; Tel.: +66-88-252-3523

Received: 31 March 2020; Accepted: 28 April 2020; Published: 6 May 2020

Abstract: Long-term monitoring of vegetation is critical for understanding the dynamics of forest ecosystems, especially in Southeast Asia's tropical forests, which play a significant role in the global carbon cycle and have continually been converted into various stages of secondary forests. In Thailand, long-term monitoring of forest dynamics during the successional process is limited to plot scales assuming from the distinct structure of successional stages. Our study highlights the potential of coupling airborne light detection and ranging (LiDAR) technology and stand age data derived from Landsat time-series to track back forest succession, and infer patterns in the plant area index (PAI) recovery. Here, using LIDAR data, we estimated the PAI of the 510 sample plots of a seasonal evergreen forest dispersed over the study area in Khao Yai National Park, Thailand, capturing a successional gradient of tropical secondary forests. The sample plots age was derived from the available Landsat time-series dataset (1972–2017). We developed a PAI recovery model during the first 42 years of the succession process. We investigated the relationship between the model residuals and PAI values with topographic factors, such as elevation, slope, and topographic wetness index. The results show that the PAI increased non-linearly (pseudo-R^2 of 0.56) during the first 42 years of forest succession, and all three topographic factors have less influence on PAI variability. These results provide valuable information of the spatio-temporal PAI patterns during the successional process and help understand the dynamics of tropical secondary forests in Khao Yai National Park, Thailand. Such information is essential for forest management and local, regional, and global PAI synthesis. Moreover, our results provide significant information for ground-based spatial sampling strategies to enable more accurate PAI measurements.

Keywords: forest succession; LiDAR; leaf area index; plant area index

---

1. Introduction

Among the world's tropical forests, specifically Southeast Asia's tropical forests are considered to contribute significantly to carbon storage and climate mitigation, but they also belong to the major deforestation areas [1,2]. After deforestation, these forests feature patches of deforested areas and various stages of recovered forest through succession [3,4], where each stage has different forest structural attributes, species diversity, species composition, and capability in the forest ecosystem [3,5].

Currently, collecting spatiotemporal data, data on patterns, and data on the rate of forest recovery still represent a challenge, especially in tropical forests, while it is required for a better understanding of the functionality and dynamics of forest ecosystems. For decades, many studies have attempted to explore the recovery of forests after disturbance [3,6,7]. Among the forest attributes, the leaf area index (LAI), defined as the one-sided leaf area per unit ground surface area [8,9], is one of the most significant biophysical variables [10] that is required for quantitative analysis of physical and biological processes related to vegetation dynamics and the modelling of ecosystem processes at local, regional, and global scales [11,12]. Therefore, long-term LAI monitoring following stand regrowth would further enhance our understanding of the dynamic changes in tropical forests and the climate impacts on tropical forest ecosystems.

Data on directly measured LAI in the field, especially in tropical forests, are very rare [13] due to the labor intensive work that is limited in time and space [14,15]. With the advancement of technology, indirect methods are now widely used, mainly through optical sensors such as the LI-COR LAI-2000 Plant Canopy Analyzer (which produced by LI-COR Inc., Lincoln, NE, USA) and digital hemispherical photography (DHP) [15,16]. However, the use of these methods is still limited in inaccessible sites, due to their spatial scale, and they often involve large uncertainties, especially related to the dense, tall, and heterogeneous canopies in tropical forests [14,15]. At regional and global scales, remotely sensed data meet the requirement for spatial and temporal estimates of LAI over large landscape areas. Most of the estimated LAIs via passive remote sensing data rely on the empirical relationships between field-based measured LAIs and spectral information [9,17,18]. Although passive satellite-based approaches are highly suitable for a wide range of observations, they suffer from several factors, such as weather conditions, the signal saturation of spectral reflectance in dense tropical forests, and an inability to capture the vertical LAI [10,19].

The light detection and ranging (LiDAR) technology collects the three-dimensional (3D) forest structure and helps to overcome many shortcomings of passive optical sensors. LiDAR has, hence, attracted much attention because of the advantage of monitoring forest structure and secondary succession of many different types [20–22]. Currently, LiDAR data is being widely and successfully used to derive LAIs [23–25]. In previous efforts, some studies estimated LAI from LiDAR data based on the empirical relationship between the ground-based measured LAI and LiDAR-derived metrics, such as mean height, maximum height, percentile height, and canopy density metrics [23–25]. Including the leaf, stems, twigs, and fine branches, the plant area index (PAI) is usually measured through the use of optical sensors and the value of PAI is comparable to the LAI value because leaf area is generally much larger than branch area, and the majority of branches are shaded by leaves [26]. With the advent of such voxel-based approaches, the term PAI instead of LAI was used. Due to a lack of recent literature on PAI, LAI will also be used for comparison where possible throughout the paper. Recently, voxel-based approaches have been developed to improve the estimation of PAI in tropical forests. With this approach, LiDAR data are reconstructed in the form of a 3D voxelized space of the forest structure and converted into PAIs by applying the Beer-Lambert law [27]. These approaches now become a promising method to obtain the PAI with high accuracy because they are not influenced by the orientation of the sun, or the spatial distribution, size, or shape of canopy components. Such methods can, thus, overcome the problem of the leaf clumping effect and signal saturation.

Although LiDAR data can be used effectively to retrieve the vertical PAI with high accuracy in tropical forests, they are limited in areas intended for long-term monitoring, as this technology is relatively costly. To investigate long-term PAI dynamics in successional gradients, data with a high temporal and spatial resolution is required. There are many studies that investigated the combination of high temporal resolution satellite images (e.g., Landsat) and high spatial resolution LiDAR data for estimating the dynamics of aboveground forest [28,29]. Recently, this approach has also been successfully applied to the assessment of forest aboveground biomass and its resilience in Thailand [30]. Until recently, however, there have not been any studies examining the long-term vertical PAI in successional gradients of tropical forests. Therefore, combining Landsat time-series datasets with high spatial resolution LiDAR data may be a promising way to achieve highly accurate PAI values for the past years along secondary forest successions.

The main aim of this study was to estimate the PAI and model the PAI recovery using LiDAR data combined with a Landsat time-series dataset in the secondary forest site in Khao Yai National Park, Thailand. Here, the landscape contains a mosaic of secondary forest with different ages surrounded by old-growth forest. The effects of topographic factors, including elevation, slope, and topographic wetness index (TWI) on the model residuals were also investigated. These factors are well known to influence forest growth and LAI, especially TWIs, which are widely used as topography-based indicator of soil-moisture. This is the first attempt to study the PAI and its recovery modeling in Thailand and Southeast Asian using LiDAR data in successional gradients.

2. Materials and Methods

2.1. Study Area

The study area is located in Khao Yai National Park in central Thailand (Figure 1). It covers an area of approximately 64 km^2, with an elevation ranging from 700 to 800 m above mean sea level [7,31]. At this altitude, the forest is seasonal evergreen with a dry season from November to April. The average annual precipitation is ca. 2100 mm, and the annual temperature ranges between 19 to 28 °C [7]. This area has been strongly affected by anthropogenic activities since the end of the 19th century to the establishment of the park in 1962, mostly through low-intensity agriculture. Since 1962, some areas have been maintained through the practice of open fires (for space opening to accelerate the growth of natural grasses) by the park managers. Therefore, the study area is a landscape mosaic consisting of forest patches of different ages, even if old-growth or relatively mature forest dominates the landscape [32].

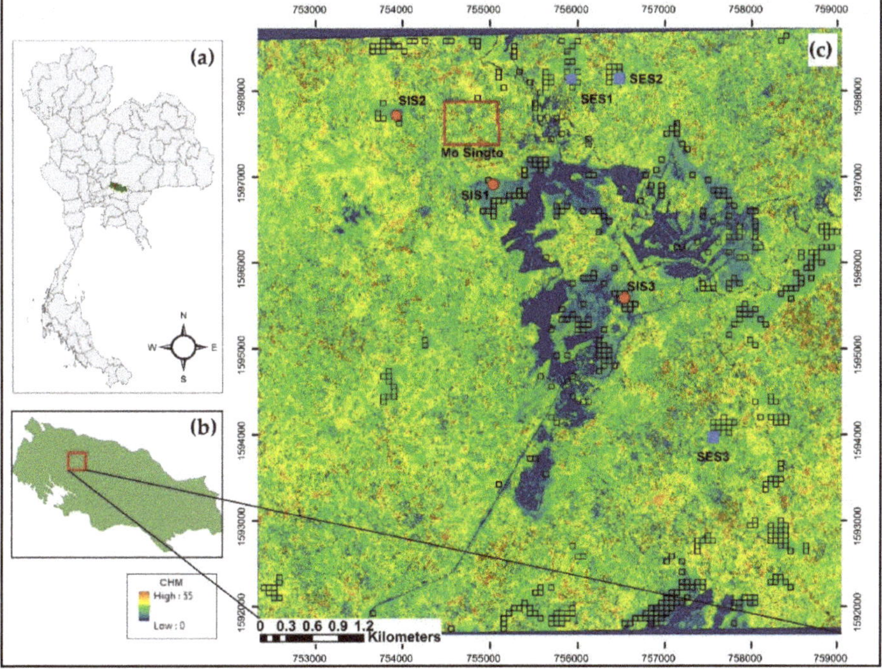

Figure 1. Location of the study site in Thailand (**a**) and in the Khao Yai area (**b**). The map (**c**) represents the canopy height in the study site with three stand initiation stage plots (SIS; red dots), three stem exclusion stage plots (SES; blue square), Mo Singto or permanent old growth stage plot (OGS; red rectangle), and 444 Landsat time-series plots (LTS; black squares).

2.2. Field Datasets

Two forest inventory datasets with different successional stages were used in this study to estimate PAI change patterns with forest age. The first dataset included six inventory of 60 m × 80 m (0.48 ha) size, established from March to May 2013 [32]. These plots were located across the study area in different successional stages, including three plots in the stand initiation stage (SIS) and three plots in the stand exclusion stage (SES). The second dataset comprised a 30-ha Mo Singto plot, which is defined as old growth stage (OGS). The Mo Singto plot is a permanent plot established in 1996 in the Mo Singto area located in a large area of old-growth forest in the central part of Khao Yai National Park. This permanent plot is part of the global network of large forest plots of the Center for Tropical Forest Science (CTFS) of the Smithsonian Tropical Research Institute (STRI). The plot age was estimated by Chanthorn et al. [32], based on Landsat imagery and interviewing the old rangers, as 15–20 years for the SIS stage, 35–40 years for the SES stage, and more than ~200 years for the OGS stage. To homogenize its scale, the Mo Singto plot was subdivided into subplots of 0.5 ha, resulting in 60 plots with 50 m × 100 m.

2.3. Forest Age Datasets

The age of secondary forest pixels in the study area was obtained from Jha et al. [30], whose study was based on LiDAR and a Landsat time-series dataset (LTS). Jha et al. [30] employed a random forest algorithm to classify the Landsat time-series (1972 to 2017) dataset using training pixels derived from the mean height of the Canopy height Model (CHM) from LiDAR (2017) at a 60-m resolution. In total, 34 images from Landsat 1-3 MSS (1972–1983), Landsat 4-5 TM (1984–2011), and Landsat 8 OLI and TIRS (2013–2017) (http://glovis.usgs.gov) were classified into forest and non-forest areas with over 90% accuracy. Using 34 classified Landsat time-series images and applying the quality filters, Jha et al. [30] further selected 550 secondary forest pixels with various recovery ages, in which pixels experienced a shift from non-forest to forest areas. Methodological details on the classification and selection of secondary pixels with respect to age are given in Jha et al. [30].

Based on Jha et al. [30], and after excluding pixels that were not located in our footprint of 1 m LiDAR-derived CHM, we obtained 444 secondary forest pixels (LTS pixels) for use in this study.

2.4. LiDAR Data Acquisition

Airborne discrete return LiDAR (ALS) data was collected by Asian Aerospace Services Limited of Bangkok on 10 April 2017 over an area of 64 km^2 (Figure 1c). LiDAR data was acquired using a RIEGL LMS Q680i full-waveform laser scanner (RIEGL Laser Measurement Systems GmbH, Horn, Austria), installed into a Diamond Aircraft "Airborne Sensors" DA-42 fixed-wing plane (Asian Aerospace Services, Bangkok, Thailand). The flight altitude was approximately 500–600 m above the ground level with a 60° field of view. The pulse repetition frequency was 400 kHz. The average point density was 22 pts m^2. Post processing of LiDAR data, including point cloud classification into ground, non-ground, and noise removal were done by AAS engineering (Aerospace Services Limited) using TerraScan, Terrasolid Version 14 (Terrasolid Ltd., Helsinki, Finland).

2.5. Overall Methodology

Firstly, all the sample plot coordinates were used to extract the point clouds for estimating the PAI. Secondly, all sample plot point clouds were used to estimate the PAI using the Amapvox software [27] and R programming software (which created by Ihaka, R. and Gentleman, R., University of Auckland, New Zealand). Then, the PAI was used to generate a recovery model by plotting against stand age. Simultaneously, the point clouds were also used to generate the digital surface model (DSM), digital terrain model (DTM), and canopy height model (CHM). The topographic factors were extracted from the DTM using the RSAGA package in R (which developed by Alexander Brenning, Jena, Germany). Finally, the topographic factors were used to assess their impact on the variability of the PAI recovery model, while PAI data were used to assess the correlation with the CHM. The overall methodology is illustrated in Figure 2; details of the methodology are explained in the next sections.

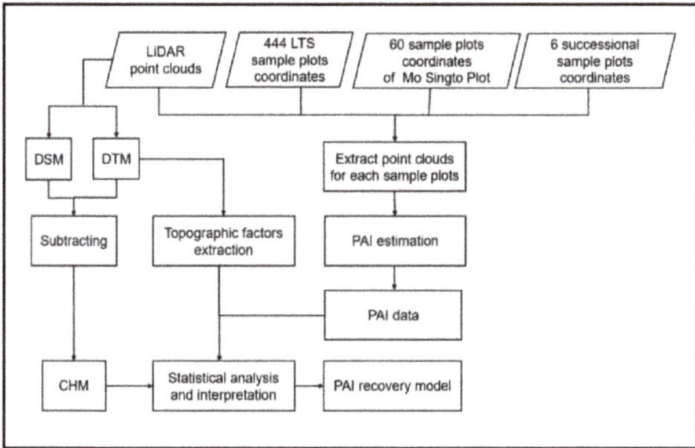

Figure 2. Overall methodology.

2.5.1. LiDAR Data and Topographic Factors Extraction

The point clouds classified as ground were interpolated to generate a DTM at 1 m resolution. The point clouds classified as non-ground points were used to generated a DSM with the same spatial resolution. A 1 m resolution CHM was then computed by subtracting the DTM from the DSM. Finally, we used the CHM to extract the canopy height at the plot level for assessing the correlation between forest canopy height and PAI data.

For creating topographic factors, the elevation was derived directly from the DTM, while both slope and TWI were computed from the DTM as the primary attribute using the RSAGA package in R. The slope was defined at each point in the DTM as a function of a gradient in the X and Y direction [33]:

$$\text{Slope} = \arctan \sqrt{(\text{fx})^2 + (\text{fy})^2} \tag{1}$$

TWI was calculated by combining the slope angle and specific catchment area (SCA), where SCA is considered as the factor that describes the tendency of the area to receive water [34]. The equation for TWI is computed as follows:

$$\text{TWI} = ln\left(\frac{\text{SCA}}{\tan\phi}\right) \tag{2}$$

where ln is the natural algorithm, SCA is the specific catchment area, and ϕ is the slope angle.

2.5.2. Estimation of LiDAR-Based Plant Area Index

We estimated the PAI of 510 sample plots, including 444 60×60 m pixels for which a forest starting recovery date was available, six 80×60 m successional sample plots, and 60 old-growth plots with an area of 50×100 m. For each plot, we first extracted all ALS points whose projected coordinates fell within the pixels using R Second, each plot containing point clouds was contructed in a form of a 3D voxelized space (m^3). Then, a local vegetation transmittance of each voxel was computed as the ratio of the sum of existing energy and the sum of entering energy normalized by the mean optical path length in each voxel, which was implemented in AMAPvox (version 1.3.5) [27], available at http://amap-dev.cirad.fr/projects/amapvox/files?sort=filename%2Csize. The equation for estimating transmittance P of an individual voxel is:

$$P = \left[\frac{\sum_i^n PFOut_i \times S_i \times l_i}{\sum_i^n PFEnt_i \times S_i \times l_i} \right]^{\frac{1}{\frac{1}{n}\sum_i^n l_i}} \tag{3}$$

where P is the transmittance, $PFEnt_i$ is the incoming fraction of pulse *I*, $PFOut_i$ is the exiting fraction of pulse i, l_i is the length of pulse i optical path, and S_i is the cross section of a pulse at the voxel center.

To control the uncertainty in plant area density (PAD) estimation, which is related to low sampling intensity, the transmittance of each voxel was then refined using a hierarchical linear mixed model in R. Individual voxels, which were nested in a $5 \times 5 \times 5$ m neighborhood, were treated as "random" effects, while the neighborhood was treated as a "fixed" effect. Subsequently, the PAD was computed for each voxel from the transmittance values by applying Beer-Lambert's turbid medium approximation, assuming isotropic transmittance as shown in Equation (4) [27]:

$$P(\theta) = e^{-G(\theta).PAD.l} \tag{4}$$

where $P(\theta)$ is the gap probability for inclination θ (view angle/shooting angle), $G(\theta)$ is the ratio of foliage area projected in direction θ to actual, and l is the optical path length.

Finally, the PAI values were obtained from the vertical integration of the PAD profiles (Figure 3).

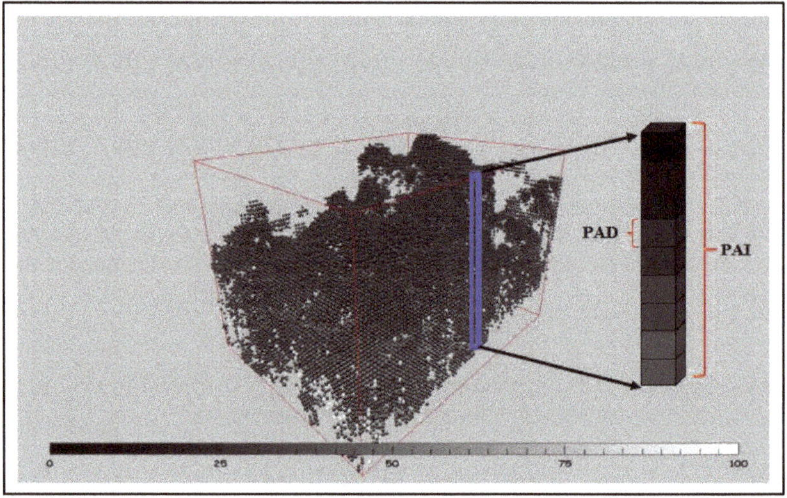

Figure 3. Example of PAD voxelization of LiDAR data point clouds applying AMAPVOX software.

2.5.3. Statistical Analyses

Nonlinear regression analyses were used to explore the change in PAI along the forest chronosequence. The PAI recovery model was, thus, of the form:

$$PAI = a + b(T^c) + \varepsilon \tag{5}$$

where PAI is the Plant Area Index, a, b, c are model parameters to be inferred, T is the forest age, and ε represents the model residuals.

For the statistical test, pseudo-R^2 was used to assess the fitted non-linear model. To evaluate the effects of topographic factors (elevation, slope, and TWI) on the variation of the PAI (in terms of PAI residuals and PAI values) over 42 years and the correlation between PAI and CHM, correlation through stepwise multiple regression analysis were performed. PAI residual values of the PAI recovery model were individually regressed through ordinary least square regressions. All of the statistical analyses were performed using the statistical program R version 3.6.2.

3. Results

3.1. Forest Successional Structure and Spatial PAI Variability

An example of the three successional stand characteristics with the LiDAR-derived CHM and the frequency of canopy height per plot is shown in Figure 4. The LiDAR-derived CHM characteristics are clearly differed in each of the successional sample plots. The crown canopy size and tree height increased from the initiation to the old-growth stage following regeneration, while the canopy was dense in the exclusion stage only but sparse in the initiation and old-growth stages due to disturbed areas and gap formation. In addition, the canopy indicates the characteristics of forest succession by the vertical stratification along with its height. Both initiation and old-growth stages had a heterogeneous upper canopy surface due to canopy gaps and variations in height, while the exclusion stage showed a smooth upper canopy surface, with a lack of understory.

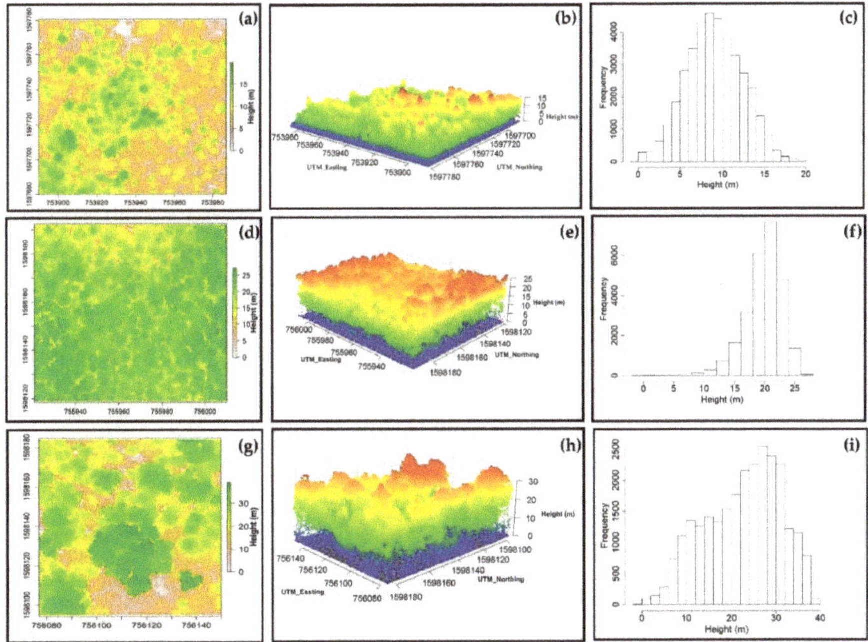

Figure 4. Example of the characteristics of the three forest successional stages. Each row represents the Canopy height Model (CHM), three-dimensional (3D) point clouds, and canopy height distribution of each stage, respectively. (**a**–**c**) represent the characteristics of SIS. (**d**–**f**) represent the characteristics of SES. (**g**–**i**) represent the characteristics of OGS.

Table 1 shows the characteristics of the successional sample plots, including mean PAI, mean canopy height, and stand age. We found that the PAI values varied with forest successional stage, canopy height, and age. The PAI values were lowest in the initiation stage plots and higher in the stem exclusion and old-growth stage plots, respectively. Moreover, the tree canopy height indicated that the stand height increased following the stand development stage. This result indicates that PAI values increase when the tree height increases. For all sample plots (secondary successional plots, Mo Singto sample plots, and LTS plots), the values of the secondary successional plots and those of OGS plots seem to fit into the trend of the LTS sample plots (R^2 0.75, $p < 0.001$; Figure 5). The mean PAI was 7.07 m^2 m^{-2} and ranged from 2.69 to 11.02 m^2 m^{-2} (s.d. =1.85 m^2 m^{-2}), while the mean canopy height was 16.57 m and ranged from 5.34 to 28.41 m (s.d. = 5.56 m).

Table 1. Means and standard deviations of the PAI, the canopy height, and the age.

	MeanPAI (m^2 m^{-2})	MeanCHM (m)	Age (Years)	Number of Sample Plots
Stand initiation stage plots (SIS)	4.19 ± 0.3	9.25 ± 0.3	8–20	3
Stem exclusion stage plots (SES)	6.63 ± 0.3	20.22 ± 1.8	35–41	3
Old-growth stage plots (OGS)	8.92 ± 0.65	21.87 ± 2.71	>200	60

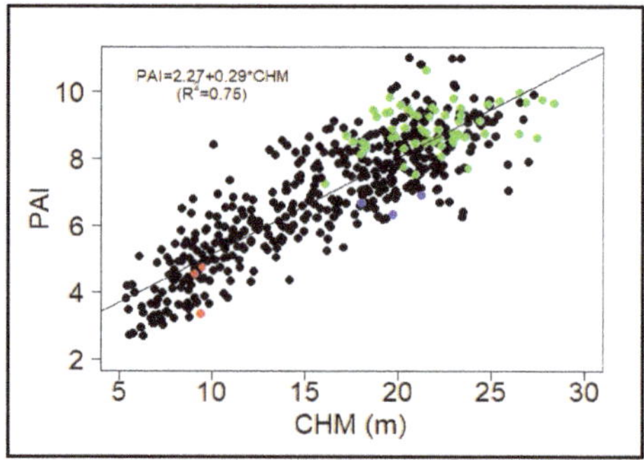

Figure 5. Relationship between PAI and canopy height model for all datasets. The regression model is illustrated by the black line. Red dots represent the PAI in SIS from the successional sample plots, blue dots represent the PAI in SES from successional sample plots, green dots represent PAI in OGS from Mo Singto sample plots, and black dots represent the PAI from LTS sample plots.

3.2. PAI Recovery Analysis

The LiDAR-derived PAI recovery along the successional gradient is illustrated in Figure 6. The LiDAR-derived PAI of the plots ranged from 1.36 to 11.02 m^2 m^{-2}, with a mean of 6.81 m^2 m^{-2}. We found that the PAI accumulation increased non-linearly through time during the 42 years. The relationship between PAI and age was best modeled with a non-linear power model and an exponent higher than 1. However, as shown in Figure 6, the exponent value indicated a close to linear relationship with a pseudo-R^2 of 0.56, indicating an increase in the PAI with recovery time during the 42 first years of the succession. Note that the model predicted a non-null PAI in year zero because we defined forests as areas with a mean top canopy height greater than 5 m, following the Food and Agriculture Organization of the United Nations (FAO) [35] definition of forests. After 20 years, the PAI was predicted to be 6.30 m^2 m^{-2}, and 7.58 m^2 m^{-2} after 40 years. For the neighboring old-growth forest in which the forest age was more than 200 years, the median of the estimated PAI was 8.87 m^2 m^{-2} and ranged from 8.05 to 10.63 m^2 m^{-2} (Figure 6).

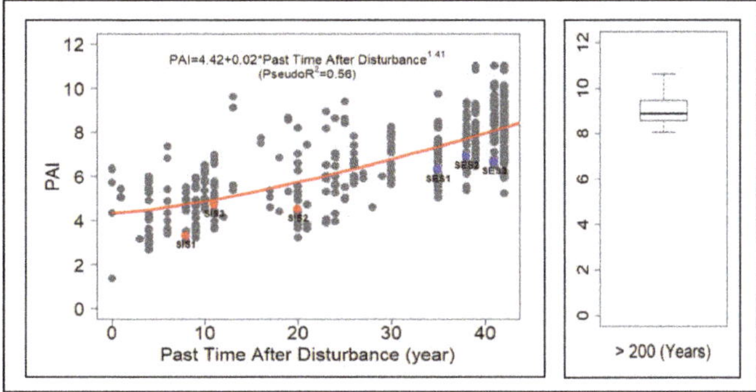

Figure 6. Relationship between the PAI estimated from the LiDAR and past time after disturbance (grey dots). The red line represents the fitted power model. Red dots represent the PAI of SIS in successional sample plots. Blue dots represent PAI of SES in successional sample plots. The box plots represent the PAI values of the OGS in Mo Singto plots with a forest age of ~ 200 years. The box edges indicate the 25th and 75th percentiles, the solid line indicates the median.

Topographic factors showed mixed effects on both the PAI values and residual values. As shown in Figure 7 and Table 2, a weak trend was observed in the relationship between both the PAI values and residual values, and topographic factors such as TWI, slope, and elevation. Results from the stepwise multiple regression analysis are shown in Table 2 and an interpretation is described below Table 2.

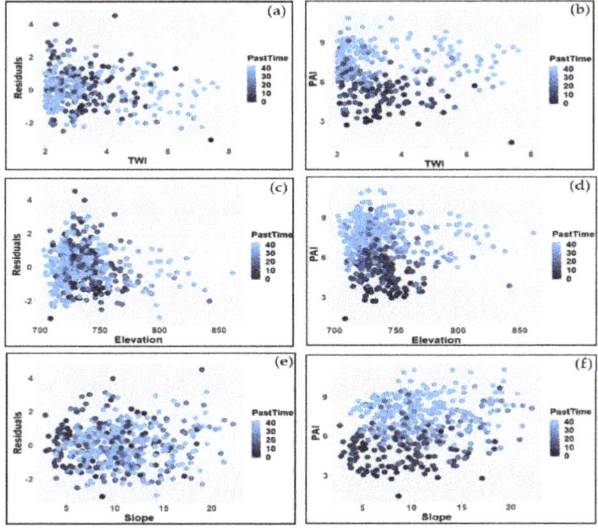

Figure 7. The left colmn showed the relationship between PAI residuals and (**a**) topographic wetness index (TWI), (**c**) elevation, and (**e**) slope. The right colmn showed the relationship between the PAI values and (**b**) TWI, (**d**) elevation, and (**f**) slope. The color grediants represent the the past time after disturbance.

Table 2. Effects of topographic factors on PAI values and residual values using stepwise multiple regression analysis.

Description	Coefficients	*p*-Value	S.E.	*t* Stat
Depedent PAI value				
Slope	0.10156 **	0.000	0.02096	4.85
Elevation	−0.30905 *	0.017	0.12877	−2.40
Elevation^2	0.00020 *	0.020	0.00008	2.33
Intercept	126.02670 *	0.010	48.82904	2.58
Dependent PAI residual value				
Slope	−0.11043 *	0.016	0.04588	−2.41
Elevation	0.15834	0.070	0.08728	1.81
Elevation^2	−0.00011	0.057	0.00006	−1.91
Intercept	−56.76872	0.088	33.20418	−1.71

Note: * Significant coefficients at the 95% confidence level; ** Significant coefficients at the 99% confidence level. S.E. = Standard error.

By the backward stepwise elimination method, the variable TWI was dropped from the PAI model, suggesting that TWI did not significantly influence PAI value. On the contrary, the slope had a significant and positive linear effect on the PAI value. When the slope increases by one unit, PAI is predicted to increase by 0.1016 unit on average, while controlling elevation.

Elevation had a significant quadratic effect on PAI (Table 2), where the effect of elevation on PAI varies by the level of elevation. Using differential calculus, the effect of elevation on PAI was expressed by the following equation:

$$\text{Marginal effect of elevation on PAI} = 2 \times 0.0001977 \times \text{Elevation} - 0.3090521 \tag{6}$$

This implies that at elevation = 700, the effect of elevation was −0.0323, indicating that when elevation increases from 700 to 701, PAI is expected to decrease by 0.0323 on average, while controlling for slope. Likewise, we obtained the effect of elevation as −0.0125 at elevation = 750, 0.0073 at elevation = 800, 0.0270 at elevation = 850, and 0.0468 at elevation = 900. As elevation increases from 700, the effect on PAI is initially negative, but increases and reaches zero when elevation is between 780 and 790. Thereafter, the effect turns positive and continues increasing. In other words, the PAI value took its minimum when elevation was between 780 and 790.

Likewise, the backward stepwise elimination method was used to analyze the PAI residuals after the time variable was accounted for. In this analysis, the slope variable was dropped, suggesting that slope did not significantly influence PAI residual. This also implies that time and slope were correlated.

TWI had a significant and negative linear effect on PAI residual. When TWI increases by one unit, PAI residual is predicted to decrease by 0.1104 unit on average, while controlling elevation. Elevation was not statistically significant at 5% significance level. This can mean that elevation had no significant effect on PAI residual. If we consider a 10% significance level, then quadratic effects were identified, where the effect of elevation on PAI residual varies by the level of evelation. Using differential calculus, the effect of elevation on PAI residual was expressed by the following equation:

$$\text{Marginal effect of elevation on PAI} = 2 \times (-0.0001096) \times \text{Elevation} + 0.1583402 \tag{7}$$

This implies that at elevation = 700, the effect of elevation was 0.0049, indicating that when elevation increases from 700 to 701, PAI residual is expected to decrease by 0.0049 on average, while controlling for TWI. Likewise, we obtained the effect of elevation as -0.0061 at elevation = 750, −0.0170 at elevation = 800, −0.0280 at elevation = 850, and −0.0389 at elevation = 900. As elevation increases from 700, the effect on PAI residual is initially positive, but decreases and reaches zero when elevation is between 720 and 730. Thereafter, the effect turns negative and continues increasing. In other words, PAI residual took its maximum when elevation was between 720 and 730.

4. Discussion

4.1. Spatial PAI Variability

The spatial variability in the PAI was derived from the LiDAR-derived PAI for all sample plots. For successional sample plots, we found that the PAI values differed greatly among the three successional stages, which related to canopy height and stand age. The PAI values were higher in old-growth plot than in stem exclusion and stand initiation stage plots (Table 1). The old-growth stage included a larger number of tall trees with random gaps, while the stem exclusion stage included trees with a smaller crown size and lower height, and the stand initiation stage included the smallest and lowest number of trees (Figure 4). This result indicates that differences in forest structure have a strong impact on the PAI value. For the LTS plots, the results also showed that the older areas had higher PAI values (Figure 8a,b). For all 510 plots, the PAI varied from 2.69 to 11.02 $m^2\,m^{-2}$ (mean = 7.07). The coefficient of variation was 26.12%, indicating a high spatial variability of the PAI at the investigated sites. Compared to other studies, our PAI values were in the range (2.66–12.94) reported by Clark et al. [36], who conducted the first direct measurements in the heterogeneous landscape of the tropical forest, and Vincent et al. [27] and Tang et al. [22], who both obtained PAI via LiDAR technology. With respect to the tree height, the relationships between the PAI and CHM were significant (R^2 = 0.75, p-value = 0.001) (Figure 5), indicating that the stand height is a key variable in understanding the spatial distribution and variability of the PAI over large areas.

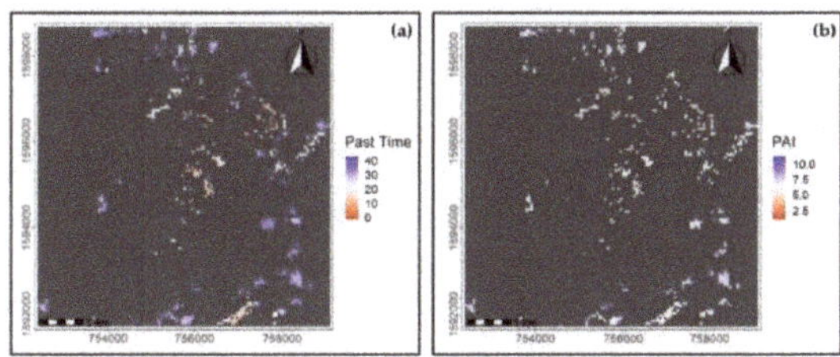

Figure 8. (a) Map showing the past time after disturbance of LTS sample plots. (b) Map showing the PAI values of LTS sample plots.

4.2. Long-Term PAI Accumulation Through Succession

The LAI is considered to be one of the most valuable determinants of forest growth and productivity. At the same time, long-term monitoring during the succession of the LAI in tropical forests is very scarce. We demonstrated that the combination of LiDAR and data of the classified forest age obtained from time-series Landsat data can quantify the long-term PAI accumulation over 42 years of succession in a tropical moist forest in Khao Yai National Park, Thailand. As a result, our analyses showed that a non-linear power model with an exponent greater than 1 was the best fit to our data, indicating an increase in the PAI during the first stage over the 42 years of succession (Figure 6). In agreement with the model, the estimated PAI of the successional field plots, which were used for validating our model, also showed that the PAI increment following the succession stages and their values were close to our predicted PAI (Figure 6). In addition, our study showed that the PAI increased gradually in the first 20 years and continued to increase twofold after 40 years. Compared to the PAI estimated for adjacent old-growth forests (mean = 8.92 $m^2\,m^{-2}$, more than 200 years), the average PAI (mean = 8.07 $m^2\,m^{-2}$) after 42 years of succession decreased by 0.85 $m^2\,m^{-2}$. These differences were very small when compared to the deviation in forest age. Accordingly, our results suggest that the PAI would increase rapidly during the first 42 years of forest succession, while it would only slightly increase thereafter and remain constant afterwards. In tropical forests, LAI predictions along successional gradients, assuming the distinct structure of successional stages or ages, are limited and were only reported on the plot scale (mostly ~0.5–1 ha). Chanthorn et al. [7] reported that the estimated LAI differed considerably between successional stages, i.e., 3.83, 5.47, and 3.81–5.74 m^2 m^{-2} for very young forests, smooth young forests, and old-growth forest, respectively. Although we conducted our study in the same study area, the LAI is very different due to the different methodology and sensor used. Tang et al. [22] reported that LAI values were estimated to be 1.74, 5.20, and 5.62 m^2 m^{-2} for open pasture, secondary forest, and old-growth forests, respectively. These studies showed a similar pattern when compared to our study, i.e., the LAI increased rapidly during successional stand development, and, then, the increase weakened during the transformation of secondary to mature forests. However, the estimated LAI values of these studies differs greatly from that of our study (Table 1), and the rate of recovery remains unclear because the LAI values were estimated with a different methodology, successional stage definition, and in a different study site.

4.3. Topographic Factors Influence the Variation in the PAI Increase

Topographic factors were found to be critical factors in determining the variation in the PAI across the landscape in several previous studies [37]. In those studies, topographic factors were proposed to be the most important factors contributing to tree growth and LAI variation in tropical forests. Moser et al. [38] found that the LAI decreased by 40%–60% in the elevation range of 1000–300 m. Unger et al. [39] reported that the LAI significantly decreased in the elevation range of 500–2000 m by about 1.1 per 1000 m. Liu et al. [40] reported a positive correlation between the LAI and soil moisture. However, this positive correlation was only found in clay-rich soil, which indicates that soil texture is one of the relevant factors determining the soil moisture-LAI relationship [41]. At our site, all three topographic factors, TWI, slope, and elevation showed week relationship with the PAI residual values and PAI values, indicating that the variability in the PAI in this study area was less influenced by these factors (Figure 7, Table 2). It has been hypothesized that the variability in the PAI is due to the variation in soil nutrients and properties during succession. In addition, our study site is a special case in that it is a natural ecosystem, affected by anthropogenic activities, such as agriculture, before the establishment of the park, and maintained as pasture land by fire. The study area comprises a mosaic of secondary forest patches, with different age and surrounded with old growth forest. Therefore, some of the sample plots are in the transition zone and/or large trees were partly left in the secondary forests, which probably affected the PAI variability in the recovery model.

5. Conclusions

Study on PAI patterns in the successional forests after disturbance is rare in the tropics. This study demonstrated the combined use of LiDAR data and stand age data derived from Landsat time-series to determine PAI patterns in a forest succession in Khao Yai National Park, Thailand. The PAI values non-linearly increased following forest successional stages with a rapid increase during the first 20 years, twofold increase during the first 40 years or, so but a constant slow increase after 200 years onward. Effects of the topographic factors on PAI values and residuals are less significant or weak in our study.

These findings provide an information of the long-term PAI recovery patterns during successional processes and the spatial variation in the PAI in heterogeneous tropical moist forests following disturbance. This information is very important for understanding the vegetation regrowth and the rate of change in a disturbed area as the large areas of tropical forests have experienced overexploitation, forest fires, and forest degradation. Our study findings can also provide the information on PAI values in tropical forests in Southeast Asia, where such values are still limited. The information on the spatial variation of the PAI in in the successional forests as a consequence of forest disturbance is critically needed for the development of ecological surveys and ground-based sampling strategies for ecological research and conservation. Further study on the effects of soil conditions and seasonal variations of vegetation on the PAI values could improve our understanding of the PAI patterns under various environmental and topographic factors.

Author Contributions: Conceptualization, S.P. and N.K.T.; methodology, S.P.; validation, S.P., N.K.T., S.N. and N.S.; formal analysis, S.P. and N.S.; investigation, S.P., N.K.T., S.N. and N.S.; resources, S.P. and N.K.T.; data curation, S.P. and N.K.T.; writing—original draft preparation, S.P.; writing—review and editing, S.P., N.K.T., S.N. and N.S.; visualization, S.P.; supervision, N.K.T.; project administration, S.P. All authors have read and agreed to the published version of the manuscript.

Funding: This research received no external funding.

Acknowledgments: The authors gratefully thank the support of the National Science and Technology Development Agency (Thailand), Kasetsart University, the Department of National Parks, Wildlife and Plant Conservation (DNP), and the Institute of Research for Development (IRD). Takuji W. Tsusaka of the Natural Resource Management program at the Asian Institute of Technology is acknowledged for helping with the statistical analysis.

Conflicts of Interest: The authors declare no conflict of interest.

References

1. Estoque, R.C.; Ooba, M.; Avitabile, V.; Hijioka, Y.; Dasgupta, R.; Togawa, T.; Murayama, Y. The future of Southeast Asia's forests. *Nat. Commun.* **2019**, *10*, 1829. [CrossRef] [PubMed]
2. Stibig, H.-J.; Achard, F.; Carboni, S.; Raši, R.; Miettinen, J. Change in tropical forest cover of Southeast Asia from 1990 to 2010. *Biogeosciences* **2014**, *11*, 247–258. [CrossRef]
3. Chazdon, R.L. Change and determinism in tropical forest succession. In *Tropical Forest Community Ecology*; Wiley-Blackwell: Oxford, UK, 2008; pp. 384–408.
4. Lu, D.; Mausel, P.; Brondízio, E.; Moran, E.F. Classification of successional forest stages in the Brazilian Amazon basin. *For. Ecol. Manag.* **2003**, *181*, 301–312. [CrossRef]
5. Pena-Claros, M. Changes in Forest Structure and Species Composition during Secondary Forest Succession in the Bolivian Amazon. *BIOTROPICA* **2003**, *35*, 450–461. [CrossRef]
6. Brown, S.; Lugo, A.E. Tropical secondary forests. *J. Trop. Ecol.* **1990**, *6*, 1–32. [CrossRef]
7. Chanthorn, W.; Ratanapongsai, Y.; Brockelman, W.Y.; Allen, M.A.; Favier, C.; Dubois, M.A. Viewing tropical forest succession as a three-dimensional dynamical system. *Theor. Ecol.* **2015**, *9*, 163–172. [CrossRef]
8. Asner, G.P.; Scurlock, J.M.O.; Hicke, J.A. Global synthesis of leaf area index observations: Implications for ecological and remote sensing studies. *Glob. Ecol. Biogeogr.* **2003**, *12*, 191–205. [CrossRef]
9. Chen, J.M.; Cihlar, J. Retrieving Leaf Area Index of boreal conifer forests using Landsat TM images. *Remote Sens. Environ.* **1996**, *55*, 153–162. [CrossRef]
10. Zheng, G.; Moskal, L.M. Retrieving Leaf Area Index (LAI) Using Remote Sensing: Theories, Methods and Sensors. *Sensors* **2009**, *9*, 2719–2745. [CrossRef]
11. Gupta, R.; Sharma, L.K. The process-based forest growth model 3-PG for use in forest management: A review. *Ecol. Model.* **2019**, *397*, 55–73. [CrossRef]
12. Zheng, D.; Wang, Y.; Shao, Y.; Wang, L. The Vegetation Dynamics and Climate Change Responses by Leaf Area Index in the Mu Us Desert. *Sustainability* **2019**, *11*, 3151. [CrossRef]
13. Olivas, P.C.; Oberbauer, S.F.; Clark, D.B.; Clark, D.; Ryan, M.G.; O'Brien, J.J.; Ordoñez, H. Comparison of direct and indirect methods for assessing leaf area index across a tropical rain forest landscape. *Agric. For. Meteorol.* **2013**, *177*, 110–116. [CrossRef]
14. Bréda, N. Ground-based measurements of leaf area index: A review of methods, instruments and current controversies. *J. Exp. Bot.* **2003**, *54*, 2403–2417. [CrossRef]
15. Jonckheere, I.; Fleck, S.; Nackaerts, K.; Muys, B.; Coppin, P.; Weiss, M. Review of methods for in situ leaf area index determination Part I. Theories, sensors and hemispherical photography. *Agric. For. Meteorol.* **2004**, *121*, 19–35. [CrossRef]
16. Zhu, X.; Skidmore, A.K.; Wang, T.; Liu, J.; Darvishzadeh, R.; Shi, Y.; Premier, J.; Heurich, M. Improving leaf area index (LAI) estimation by correcting for clumping and woody effects using terrestrial laser scanning. *Agric. For. Meteorol.* **2018**, *263*, 276–286. [CrossRef]
17. Blinn, C.E.; House, M.N.; Wynne, R.; Thomas, V.; Fox, T.R.; Sumnall, M.J. Landsat 8 Based Leaf Area Index Estimation in Loblolly Pine Plantations. *Forests* **2019**, *10*, 222. [CrossRef]
18. Davi, H.; Soudani, K.; Deckx, T.; Dufrene, E.; Le Dantec, V.; François, C. Estimation of forest leaf area index from SPOT imagery using NDVI distribution over forest stands. *Int. J. Remote Sens.* **2006**, *27*, 885–902. [CrossRef]

19. She-Zhou, L.; Wang, C.; Gui-Bin, Z.; Xiao-Huan, X.; Gui-Cai, L. Forest Leaf Area Index (LAI) Estimation Using Airborne Discrete-Return Lidar Data. *Chin. J. Geophys.* **2013**, *56*, 233–242. [CrossRef]

20. Van Ewijk, K.Y.; Treitz, P.; Scott, N.A. Characterizing Forest Succession in Central Ontario using Lidar-derived Indices. *Photogramm. Eng. Remote Sens.* **2011**, *77*, 261–269. [CrossRef]

21. Falkowski, M.J.; Evans, J.; Martinuzzi, S.; Gessler, P.E.; Hudak, A.T. Characterizing forest succession with lidar data: An evaluation for the Inland Northwest, USA. *Remote Sens. Environ.* **2009**, *113*, 946–956. [CrossRef]

22. Tang, H.; Dubayah, R.; Swantaran, A.; Hofton, M.A.; Sheldon, S.; Clark, D.B.; Blair, B. Retrieval of vertical LAI profiles over tropical rain forests using waveform lidar at La Selva, Costa Rica. *Remote Sens. Environ.* **2012**, *124*, 242–250. [CrossRef]

23. Jensen, J.L.; Humes, K.S.; Vierling, L.A.; Hudak, A.T. Discrete return lidar-based prediction of leaf area index in two conifer forests. *Remote Sens. Environ.* **2008**, *112*, 3947–3957. [CrossRef]

24. Morsdorf, F.; Kötz, B.; Meier, E.; Itten, K.; Allgöwer, B. Estimation of LAI and fractional cover from small footprint airborne laser scanning data based on gap fraction. *Remote Sens. Environ.* **2006**, *104*, 50–61. [CrossRef]

25. Solberg, S.; Brunner, A.; Hanssen, K.H.; Lange, H.; Næsset, E.; Rautiainen, M.; Stenberg, P. Mapping LAI in a Norway spruce forest using airborne laser scanning. *Remote Sens. Environ.* **2009**, *113*, 2317–2327. [CrossRef]

26. Kucharik, C.J.; Norman, J.M.; Gower, S.T. Measurements of branch area and adjusting leaf area index indirect measurements. *Agric. For. Meteorol.* **1998**, *91*, 69–88. [CrossRef]

27. Vincent, G.; Antin, C.; Laurans, M.; Heurtebize, J.; Durrieu, S.; LaValley, C.; Dauzat, J. Mapping plant area index of tropical evergreen forest by airborne laser scanning. A cross-validation study using LAI2200 optical sensor. *Remote Sens. Environ.* **2017**, *198*, 254–266. [CrossRef]

28. Badreldin, N.; Sánchez-Azofeifa, G.A. Estimating Forest Biomass Dynamics by Integrating Multi-Temporal Landsat Satellite Images with Ground and Airborne LiDAR Data in the Coal Valley Mine, Alberta, Canada. *Remote Sens.* **2015**, *7*, 2832–2849. [CrossRef]

29. Deo, R.K.; Russell, M.B.; Domke, G.M.; Woodall, C.; Falkowski, M.J.; Cohen, W.B. Using Landsat Time-series and LiDAR to Inform Aboveground Forest Biomass Baselines in Northern Minnesota, USA. *Can. J. Remote Sens.* **2016**, *43*, 28–47. [CrossRef]

30. Jha, N.; Tripathi, N.K.; Chanthorn, W.; Brockelman, W.; Nathalang, A.; Pélissier, R.; Pimmasarn, S.; Ploton, P.; Sasaki, N.; Virdis, S.G.P.; et al. Forest aboveground biomass stock and resilience in a tropical landscape of Thailand. *Biogeosciences* **2020**, *17*, 121–134. [CrossRef]

31. Brockelman, W.Y.; Nathalang, A.; Gale, G.A. The Mo Singto forest dynamics plot, Khao Yai National Park, Thailand. *Nat. Hist. Bull. Siam Soc.* **2011**, *57*, 35–55.

32. Chanthorn, W.; Hartig, F.; Brockelman, W.Y. Structure and community composition in a tropical forest suggest a change of ecological processes during stand development. *For. Ecol. Manag.* **2017**, *404*, 100–107. [CrossRef]

33. Tang, J.; Pilesjö, P. Estimating slope from raster data: A test of eight different algorithms in flat, undulating and steep terrain. *River Basin Manag. VI* **2011**, *146*, 143–154. [CrossRef]

34. Mattivi, P.; Franci, F.; Lambertini, A.; Bitelli, G. TWI computation: A comparison of different open source GISs. *Open Geospat. Data Softw. Stand.* **2019**, *4*. [CrossRef]

35. FAO. *The State of Food and Agriculture 2012*; FAO publications: Rome, Italy, 2012.

36. Clark, D.B.; Olivas, P.C.; Oberbauer, S.F.; Clark, D.; Ryan, M.G. First direct landscape-scale measurement of tropical rain forest Leaf Area Index, a key driver of global primary productivity. *Ecol. Lett.* **2007**, *11*, 163–172. [CrossRef] [PubMed]

37. Aragão, L.E.O.; Shimabukuro, Y.E.; Espirito-Santo, F.; Williams, M. Landscape pattern and spatial variability of leaf area index in Eastern Amazonia. *For. Ecol. Manag.* **2005**, *211*, 240–256. [CrossRef]

38. Moser, G.; Hertel, D.; Leuschner, C. Altitudinal Change in LAI and Stand Leaf Biomass in Tropical Montane Forests: A Transect Study in Ecuador and a Pan-Tropical Meta-Analysis. *Ecosystems* **2007**, *10*, 924–935. [CrossRef]

39. Unger, M.; Homeier, J.; Leuschner, C. Relationships among leaf area index, below-canopy light availability and tree diversity along a transect from tropical lowland to montane forests in NE Ecuador. *Trop. Ecol.* **2013**, *54*, 33–45.

40. Liu, L.; Zhang, R.; Zuo, Z. The Relationship between Soil Moisture and LAI in Different Types of Soil in Central Eastern China. *J. Hydrometeorol.* **2016**, *17*, 2733–2742. [CrossRef]

41. Zhang, W.; Hu, B.; Woods, M.; Brown, G. Characterizing Forest Succession Stages for Wildlife Habitat Assessment Using Multispectral Airborne Imagery. *Forests* **2017**, *8*, 234. [CrossRef]

Article

Shrub Biomass Estimates in Former Burnt Areas Using Sentinel 2 Images Processing and Classification

José Aranha [1,2,*], Teresa Enes [1,2], Ana Calvão [3] and Hélder Viana [1,4]

1   Centre for the Research and Technology of Agro-Environmental and Biological Sciences (CITAB), University of Trás-os-Montes and Alto Douro, 5001-801 Vila Real, Portugal; tenes@utad.pt (T.E.); hviana@esav.ipv.pt (H.V.)
2   Department of Forestry Sciences and Landscape Architecture (CIFAP), University of Trás-os-Montes and Alto Douro, 5001-801 Vila Real, Portugal
3   Águeda School of Technology and Management, University of Aveiro (ESTGA-UA), 3754-909 Águeda, Portugal; arc@ua.pt
4   Agricultural High School, Polytechnic Institute of Viseu, 3500-606 Viseu, Portugal
*   Correspondence: j_aranha@utad.pt; Tel.: +351-259-350-986; Fax: +351-259-350-480

Received: 1 April 2020; Accepted: 12 May 2020; Published: 14 May 2020

Abstract: Shrubs growing in former burnt areas play two diametrically opposed roles. On the one hand, they protect the soil against erosion, promote rainwater infiltration, carbon sequestration and support animal life. On the other hand, after the shrubs' density reaches a particular size for the canopy to touch and the shrubs' biomass accumulates more than 10 Mg ha^{-1}, they create the necessary conditions for severe wild fires to occur and spread. The creation of a methodology suitable to identify former burnt areas and to track shrubs' regrowth within these areas in a regular and a multi temporal basis would be beneficial. The combined use of geographical information systems (GIS) and remote sensing (RS) supported by dedicated land survey and field work for data collection has been identified as a suitable method to manage these tasks. The free access to Sentinel images constitutes a valuable tool for updating the GIS project and for the monitoring of regular shrubs' accumulated biomass. Sentinel 2 VIS-NIR images are suitable to classify rural areas (overall accuracy = 79.6% and Cohen's K = 0.754) and to create normalized difference vegetation index (NDVI) images to be used in association to allometric equations for the shrubs' biomass estimation (R^2 = 0.8984, *p*-value < 0.05 and RMSE = 4.46 Mg ha^{-1}). Five to six years after a forest fire occurrence, almost all the former burnt area is covered by shrubs. Up to 10 years after a fire, the accumulated shrubs' biomass surpasses 14 Mg ha^{-1}. The results described in this paper demonstrate that Northwest Portugal presents larger shrubland areas and greater shrub biomass accumulation (average 18.3 Mg ha^{-1}) than the Northeast (average 7.7 Mg ha^{-1}) of the country.

Keywords: sentinel 2; landsat; remote sensing; GIS; shrubs biomass; bioenergy; vegetation indices

---

1. Introduction

Portugal is an European country with a constituent land mass and 4 separate archipelagos. The former is located in the east of the Iberian Peninsula with an area of approximately 90,000 km^2. From the mainland area (52%) there are: forest stands (39%), dense shrubland (12%), and sparse shrubland (1%) [1,2]. Between the mid-1980s and 2020, due to increasing human rural abandonment and edaphoclimatic conditions, a large number of forest fires occurred in mainland Portugal during the summer. The intensity of these fires increased dramatically each decade [3–7]. The same edaphoclimatic conditions that make the territory prone to wildfire occurrences, however, also create suitable ecological conditions for shrub regrowth after the fires. Previously published results [8–12] demonstrate that, five years after wildfire occurrence, the fire scars are no long visible because they have been covered by shrubs as well as with the growth of scattered trees from self seedling processes.

Shrublands assume several diametrically opposed roles. On the one hand, they constitute the vegetable fuel that will eventually burn, and a social-economic problem. Portuguese law [13], specifies that it is mandatory to cut shrubs 10 m alongside the road network and 50–100 m beside other man-made structures on a regular basis. This has led to the use of fire as a means to eliminate the remaining cut area. On the other hand, these shrubs are also a source of carbon [12,14,15], they promote water and nutrients circulation in forested areas [8,16,17], protect the soil from erosion processes [16–19], support animal life, and promote biodiversity [8,19]. In essence, the ecological benefits to these shrublands could be summarised in two words: ecosystem services [20–22]. This is a central measure within European Community [22] and used as a way to assess forested areas' values. Thus such shrublands could take on an unanticipated new economic role potentially generating biofuel for power plants [23,24].

Analysis of shrubland location and its biomass accumulation is therefore important as it could influence the working processes for several stakeholders: forest management [25,26], wildfire hazard reduction, ecosystem surveying and biomass harvesting for energy production.

The calculation of forest biomass can be achieved using destructive processes, such as cutting and weighing vegetation in sampling plots. Subsequently, the results obtained can be analysed using appropriate geostatistics processes [23] that generate indicative biomass maps.

The results obtained through these destructive processes can later be used to adjust allometric equations enabling the estimation of biomass weight based on the volume collected. This is a non-destructive process. The results can also be used in conjunction with relevant satellite images. This can also support the calculation used in the allometric models for accumulated biomass estimation. This is achieved by comparing the relevant bands from the satellite images as well as analyzing the vegetation indices calculated by means of those same bands. This is a good example of a non-destructive method of estimating biomass.

It is important to note, however, that the first phase of a process to calculate and estimate accumulated biomass always begins with a destructive method.

The assessment of land cover dynamics in former burnt areas of forest as well as shrubs' regrowth must be carried out over vast areas of territory, and on an annual basis. It also requires the use of appropriate computer technology. Firstly, remote sensing techniques (RS) can be used for image processing and classification to create updated land cover maps. Secondly, geographic information systems (GIS) can be employed to record, manipulate, and present data. In addition, GIS allows the combination of multiple data sources enabling spatial analysis and can enhance the sampling process too. It is recognized that annual fieldwork for data collection is very expensive and time consuming, thus the use of RS and GIS provides a cheaper and appropriate sampling mechanism. Previous research in this subject area has generated several RS based approaches that used multitemporal satellite image classification and comparison.

A literature review on the use of GIS, RS, and combined RS/GIS for forest biomass and shrubs' biomass [25–29] enabled the consideration of different approaches to biomass estimation and mapping on given dates. This resulted in the use of particular regression models that were based on specific vegetation indices [30–34], and are presented in Table 1.

It should be noted that almost all of these previously presented data were not based on a shrubs' biomass time series sampling system thus do not account for any variation due to elapsed time over the former burnt areas. The use of allometric equations adjusted for a given geographic area requires local validation before it is used elsewhere, i.e., necessitating field work for data collection and the use of mathematical models for data analysis.

Table 1. Biomass regression models based on vegetation indices.

Allometric Model	R^2 (Adj)	Ref
Trees, shrubs and herbaceous $y = 73{,}709.9241 - 48{,}420.44\,\chi^1 + 67{,}242.43\,\chi^2$ where, y = Biomass (kg), χ^1 = NDVI value, χ^2 = NDVI MIR index value	0.70	[30]
Trees, shrubs and herbaceous $Log10\ y = 3.7163 - 0.01078\,\chi^1 + 0.007065\,\chi^2$ where, y = Biomass (kg), χ^1 = Brightness value, χ^2 = Wetness value	0.66	[31]
Shrubs $y = 46{:}678\,\chi^1 + 7{:}929\,\chi^2 + 32{:}565$ where, y = Biomass (kg), χ^1 = Brightness value, χ^2 = RVI (ratio vegetation index)	0.70	[31]
Total biomass AGB prediction = 3.35 + 3.13 VV + 0.21 VH + 1.53 NDVI where: VV—the backscatter coefficients for a specific polarization; VH—the backscatter coefficients for a specific polarization; NDVI—normalized difference vegetation index.	0.66	[32]
Shrubs Biomass y = 0.18363 + 0.85669 NDVI where, y = Biomass (Mg), NDVI—normalized difference vegetation index	0.74	[33]
Fractional green vegetation cover (fc) fc = 0.114 + 1.284 NDVI (R^2 = 0.89)	0.89	[34]

After 18 years of carrying out fieldwork to measure the volume and the weight of shrublands, the authors' main aim now is to present a suitable methodology that enables the estimation of the accumulated shrubs' biomass. This process takes into account the elapsed time after wildfires, based on satellite imagery processing and classification. The methodology is non-destructive and does not require fieldwork for data collection, thus allows accurate estimates when used in conjunction with RS techniques. It also enables stakeholders to perform dynamic analysis using satellite images in time series processing. To achieve this main aim, the authors adopted a methodology using Sentinel 2 images processing and classification as a way to identify former burnt areas, shrubland and to adjust an allometric equation that enables to estimate shrubs' biomass through using NDVI images.

This methodology also incorporates the elapsed time after identifying any wildfire occurrence effect on the shrubs' regrowth as an estimate. This then enables the creation of accurate maps related to the shrubs' biomass accumulation. It also established that, if the growth rate in the Northwest area of Portugal is different from that of the Northeast area (comparison was using annual satellite images), then this may have an influence on the adjustment of allometric equations. In addition this methodology has used dynamic models to identify forest fire hazards and also used design logistics models for biomass harvesting for energy purposes. These can indicate the areas where the intervention of forest managers (necessary to comply with the law) has taken place. Another unexpected result of this methodology was to help design post fire ecosystem recovery actions.

2. Materials and Methods

2.1. Study Area Characteristics and Sampling Plots Location

The study area is located in North Portugal (Figure 1) and comprises a forested area of 813,846 ha (432,000 ha forest stands and 381,846 ha shrubland).

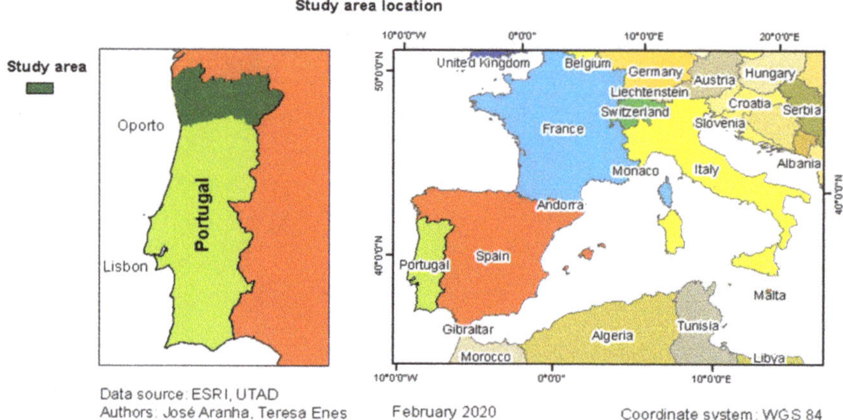

Figure 1. Study area (North Portugal) and Portugal world geographical location.

This is a very fragmented landscape and the small forest areas are often side by side with shrubland and agricultural areas. It should be noted that, for cultural reasons, the rural population often uses fire as an instrument for pruning as well as a method for removing any residues. The result of this approach is that a higher number of rural fires every year occurs than is actually appropriate for this type of landscape.

The study area in Northern Portugal includes many morphological and edaphoclimatic conditions typical of this region. Altitude ranges from sea level (0 m) to 1546 m in the Gerês mountains (Figure 2). Mean annual accumulated precipitation ranges from 1000 to 2400 mm in the Northwest areas and from 600 to 1200 mm in the Northeast areas. Mean annual temperature ranges from 12.5 °C to 15.0 °C in the Northwest areas and from 7.5 °C to 12.5 °C in the Northeast areas.

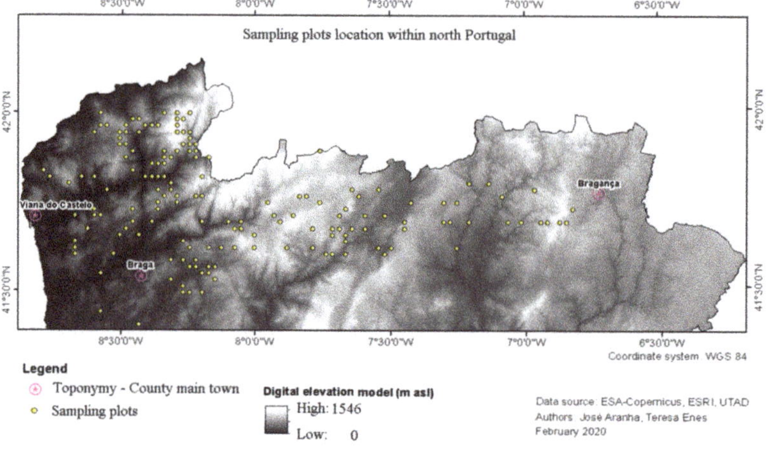

Figure 2. Sampling plots location.

2.2. Data Sources

2.2.1. Using GIS in the Project

The project aims were to analyse former burnt areas, assess potential vegetation regrowth and to estimate the shrubs' biomass taking into account the elapsed time since any wildfire took place.

The initial process used the burnt areas data. It is possible to download a set of vector files representing the burnt areas' perimeter by year of occurrence from the Portuguese Forest Services website [35] or the European Forest Fire Information System [36].This information has been used to establish one of the layers within the GIS project since 2000. Every year, the burnt areas' vector file is updated with new burnt areas boundaries. Spatial analysis then enabled new calculations to be made such as identifying fire recurrence areas as well as assessing the time since the last occurrence of a fire in the same areas.

For the period between 2000 and 2016, 10 sampling plots per year after the last fire were selected. This resulted in 170 sampling points, dispersed throughout North Portugal. In 2017 and 2018, after the severe rural fires that occurred in those years, the GIS project was updated and new sampling plots were added, increasing the number of samples to 234.

These sample points (Figure 2) were then used to create a survey GIS project (e.g., ArcPad, Survey 123, QField) that was transferred to a DGPS receiver (Trimble Inc., Sunnyvale, CA, USA). All the sample points were also marked (based on the GIS layout) on the Topographic Plan of Portugal on a 1/25000 scale. These were then printed to support fieldwork in areas with no GNSS signal.

Thus the data generated from 2000 to 2018 were used to identify the sampling points on the ground enabling us to record any shrub regrowth since the last known fire occurrence.

Circular 500 m^2 sampling plots (12.62 m in radius) were used along with the cross transect method. Two 25.24 m fiberglass tape measures were stretched perpendicularly across each sampling plot. Then, the shrubs intersecting each fiberglass tape were measured in 3 dimensions: length, width, and height. This enabled a calculation of the volume, assuming that the shrubs canopy was a sphere, and by using Equation (1).

$$Vsh = 1/6 \, \pi \, L \, W \, H \tag{1}$$

where Vsh = shrubs canopy volume (m^3); L = length (m), W = width (m), H = height (m) (Formula demonstration in Appendix A).

After measuring the total shrubs' volume along the 2 transects within the sampling plot, 10 shrub plants, 5 per transect (at the edges of the plot, halfway from the center and in the center) were cut in order to be weighed. They were then placed in plastic bags, brought to the lab, put to dry in the shade and weighed after reaching 30% moisture. The achieved results for the 500 m^2 circular sampling plots were then extrapolated to an area of 1 hectare and the amount of shrub biomass per plot was estimated using Equation (2).

$$Biomass = V \, W \, 200/Mxw \tag{2}$$

where Biomass in Mg ha^{-1}, V = total shrubs volume along the 2 transects (m^3), W = average shrubs weight (Mg m^{-3} at 30% moisture), Mxw = maximum width measured along both transects (Formula demonstration in Appendix A).

2.2.2. The Processing and Classification of Sentinel 2 Images

Sentinel 2 images were freely downloaded either from the Copernicus Hub website [37] or from the Glovis website [38].

As the study area is not covered by a single image, eight Sentinel 2 images were used, two per year, for the years of 2016, 2017, 2018, and 2019. All of them were recorded in the summer season.

It was not possible to download Sentinel-2S2A images for the years 2017, 2018 and 2019. Only Sentinel-2L1C were available. When processing these images for the various dates, it was noticed, after computing the NDVI images, that the calculated values for the water surfaces (e.g., dams) were not consistent. This situation led to performing Sentinel-2 SEN2COR280 Processor (7.0.0) analysis by SNAP. For this reason, an images atmospheric calibration was performed, based on the spectral signature of the water collected by the research team with a spectra radiometer Ocean Optics VIS NIR (Spectrecology, Inc., St Petersburg, FL, USA) and on the atmospheric scattered model proposed by Chavez (1988) [39]. This procedure was carried out in order to have the same spectral signature for all water surfaces. Each image was also submitted for atmospheric correction because solar elevation and the state of the atmosphere introduce differences in the radiation detected by the sensor for the same area on different dates [40–43].

To ensure that differences in reflectance are due to changes in land cover, and not caused by radiometric distortions, it was also necessary to apply a radiometric correction. One of the most used models for atmospheric correction is a process called dark object subtraction (DOS) proposed by reference [39]. This process is based on an atmospheric scattering model and reduces the haze effect by calculating the expected minimum for a given band after atmospheric correction. This was carried out in relation to the following criteria: at-satellite radiances were converted to surface reflectance by correcting for both solar and atmospheric effects. Then, at-satellite radiance values were converted into surface reflectance using a DOS approach [39]. This assumes no atmospheric transmittance loss and no downward diffuse radiation. The surface reflectance of the dark object was assumed to be 1%, and thus the path radiance was assumed to be the dark-object radiance minus the radiance contributed by 1% surface reflectance [39–43].

Spectral reference signatures, such as water, bare soil and dense shrubland, were created after dedicated work using an Ocean Optics VIS NIR spectra radiometer.

After the images processing operations, a RGB false colour composition image and an NDVI image was created for all dates. The Sentinel 2 RGB482 composition was used, because it employs the near infrared band in the green channel which allows the highlighting of vegetation thus enabling an identification of the burnt areas, both recent and old.

Based on the analysis of these new images, and with the support of the GIS project and the orthophotomaps made available by Bing Maps, spectral signatures were created to support the images supervised classification. In a second stage, spectral signatures for the main rural and forest land cover classes were created, namely:

- Agriculture
- Bare soil
- Deciduous
- Burnt areas
- Coniferous
- Grass
- Rocky areas with shrubs
- Shrubs
- Urban areas
- Water

These spectral signatures were then used to perform supervised classification techniques using the 10 m Sentinel 2 bands: B2, B3, B4, and B8. The minimum distance, maximum likelihood, and random forest were tested.

After the supervised classification process completed, a new image was created indicating the burnt areas and shrubland. This is a necessary step whereby a raster mask is created then applied to the NDVI image in order to estimate the shrubs' biomass that has regrown on former burnt areas.

This raster mask is required because the vector files represent all of the annual burnt areas created. When placed over the relevant satellite images showing the burnt areas boundary or perimeter, they do not consider the unburnt 'islands', nor the rocky areas [44–46]. Thus, to calculate biomass estimates by means of a vector mask may lead to overestimations. In order to be able to calculate the amount of shrub biomass that regenerated in the former burnt areas, it is necessary to create a raster mask that represents only the areas that were actually burnt.

Finally, the sampling points attribute table in the GIS was updated with the NDVI values using a known technique that enables extracting raster values to point-type vector files. Lastly, this attribute table was exported in dBase format and processed in Excel. This way, it was possible to analyse the relationship that could be established between the measured shrubs' biomass, and that which was calculated using dedicated field work. In addition an NDVI value was also calculated using satellite images processing techniques.

Figure 3 outlines a summary of all the working stages used in this research.

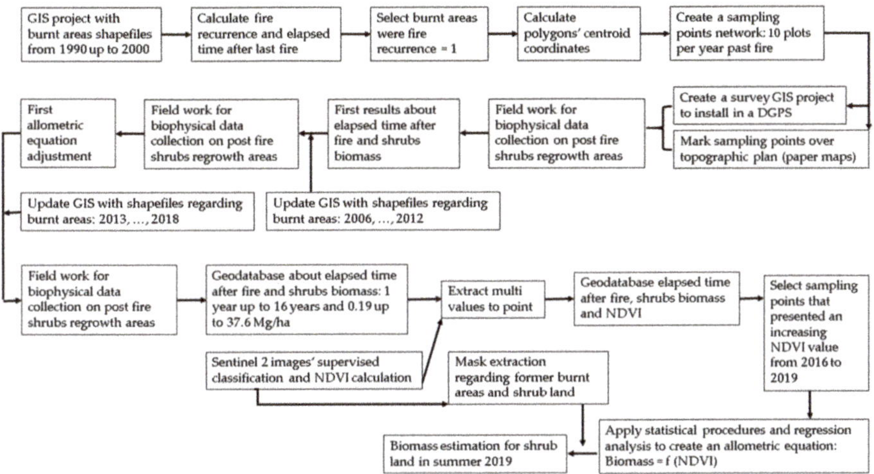

Figure 3. Work flow chart.

3. Results

3.1. Landcover Characterization

The landcover characterization work carried out was based on false colour RGB482 composition visual analysis (Figure 4) and used NDVI images (Figure 5) for interpretation (all dates). The ensuing results for 2019 are shown in Figures 4 and 5.

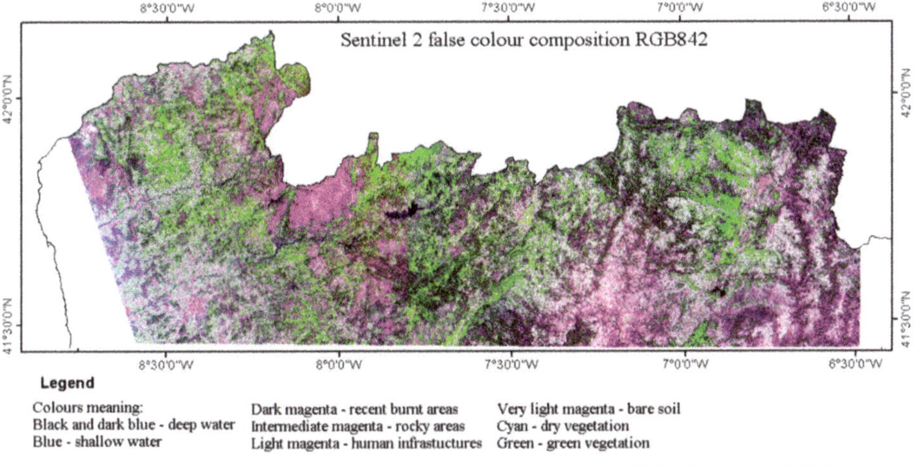

Figure 4. Sentinel 2 images (summer 2019) for the study area using false colour composition RGB482.

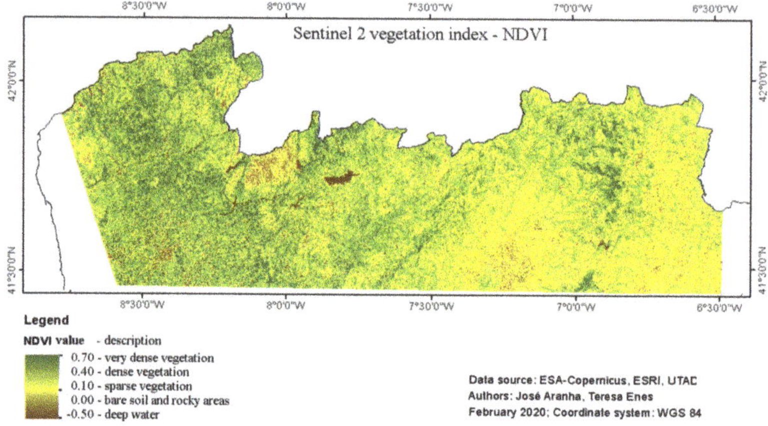

Figure 5. Sentinel 2 images (summer 2019) for the study area using an NDVI calculation.

As healthy vegetation has its maximum spectral reflectance in near-infrared wavelength, the Sentinel 2 band 8 (near-infrared) was used and coloured in green. Thus, green tonalities depicted in the RGB482 images indicated vegetation density and consequently the darker the green colour indicated the denser the vegetation.

The vegetation index NDVI was calculated using the normalized difference between the near-infrared and the red images. As the vegetation red reflectance is always lower to the near-infrared reflectance, the positive NDVI achieved values could also indicate vegetation density. Thus, the higher the NDVI values the denser the vegetation. Each of the dots depicted in Figure 2 represent a sampling point within a former burnt area. Analysing the NDVI image, Figure 5, it appears that the old burnt areas are in various states of vegetation recovery. It seems, therefore, that the Northwest area of Portugal has more dense vegetation and less burnt areas scars than the Northeast area.

It appears, however, that neither the green colour intensity shown in the RGB482 images (Figure 4) nor the NDVI values (Figure 5) enable a classification of the vegetation type (e.g., burnt areas, forest land, shrubland). These two images alone only infer the potential density of the vegetation cover. Thus, it was necessary to carry out further analysis using Sentinel 2 images and an assisted classification process to classify the land cover in classes.

Using the results from the Sentinel 2 images supervised classification process as well as the Minimum Distance Classifier, enabled us to state that the main land cover features are suitable to be classified accurately as presented in Table 2. This classification accuracy is particularly high for forest features. For example, burnt areas and shrubland was easy to identify and classify with an accuracy over 80%.

It was possible, therefore, to create a mask image for use with an associated NDVI image in order to isolate burnt areas and shrubland. This mask was later used to 'cut' the NDVI image enabling a shrub biomass calculation to be carried out along with the allometric equation application.

With an overall accuracy of 79.6% and a Cohen's K coefficient of 0.754, it can be stated that the Sentinel 2 images were found to be suitable for use in forestry applications as well as in the dynamic analysis of former burnt areas too. The Sentinel 2 classified images created a raster mask that was used subsequently to isolate the burnt areas and the shrubs areas.

Table 2. Achieved results after confusion matrix for classification accuracy from Sentinel-2 images.

Sample Class	N	Pa (%)	Ua (%)	Ce (%)	Oe (%)
Agriculture	77	46	80	54	20
Bare soil	35	80	48	20	52
Deciduous	31	87	68	13	32
Burnt areas	16	100	89	0	11
Coniferous	161	96	96	4	4
Grass	13	69	69	31	31
Rocky and shrubs	46	83	64	17	36
Shrubs	67	84	92	16	8
Urban areas	31	48	65	52	35
Water	9	89	100	11	0

N: Number of ground control points; Pa: Producer's accuracy; Ua: User's accuracy; Ce: Commission error; Oe: Omission error.

3.2. Allometric Model for Shrub Biomass Estimation

For the allometric equation adjustment, 110 pairs (NDVI, shrub biomass) were used: 46 extracted from 2016 image, 33 from 2017 image and 31 from 2018 image. In a first approach to data processing, descriptive statistics were calculated for each date and region, as presented in Table 3.

Table 3. Descriptive statistic for the sub-samples.

	2016		2017		2018	
	NW	NE	NW	NE	NW	NE
NDVI						
Count	28	30	21	12	23	8
Minimum	0.388	0.378	0.280	0.136	0.048	0.120
Maximum	0.700	0.700	0.696	0.655	0.694	0.688
Average	0.590	0.580	0.552	0.345	0.521	0.390
Standard deviation	0.090	0.100	0.144	0.193	0.219	0.259
Age						
Count	28	30	21	12	23	8
Minimum	5	5	3	3	1	2
Maximum	15	15	15	11	15	14
Average	8.7	8.7	8.6	4.8	7.8	5.9
Standard deviation	3.4	3.4	4.1	2.6	4.2	4.5
Shrub biomass						
Count	28	30	21	12	23	8
Minimum	3.49	4.80	1.73	0.46	0.19	0.67
Maximum	34.48	37.60	34.48	27.90	30.82	37.60
Average	17.04	18.76	16.46	6.50	15.97	12.77
Standard deviation	8.24	10.37	11.62	8.69	10.26	14.58

Subsequently, student-t tests were performed in order to verify that the sample points for the NW area of Portugal are different to those of the sample points for the NE area. As no significant differences were found, all the sampling points for each year were then merged into a single sample file.

The NDVI and shrub biomass values were centered and reduced in order to verify if this composite sample had a normal distribution. Descriptive statistic and accumulated probability values were calculated. Initially, the descriptive statistic for NDVI and shrubs biomass was calculated, shown in Table 4. Then, accumulated probability values were calculated as depicted in Figure 6.

Table 4. Descriptive statistic for the total sample.

	NDVI (Dimension Less)	Shrubs Biomass (Mg ha^{-1})
Count	110	110
Minimum	0.048	0.186
Maximum	0.700	37.596
Average	0.525	15.494
Standard deviation	0.179	10.781
Standard error	0.342	0.696
Median	0.596	15.291

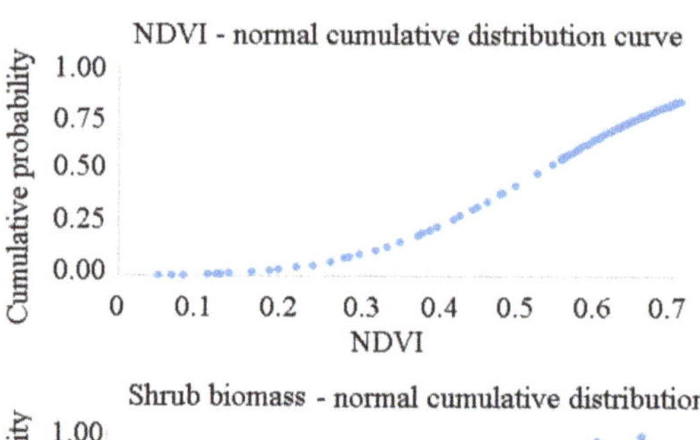

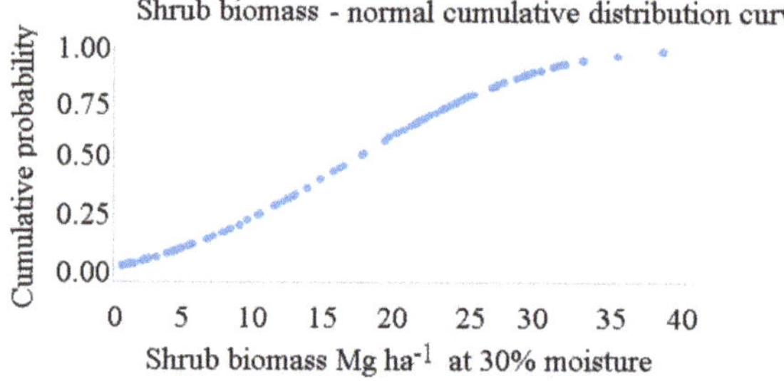

Figure 6. Normal cumulative curves to NDVI (top) and to shrubs biomass (bottom).

The achieved results show that both distributions presented a normal distribution (Figure 6).

In the third stage of the process, a XY graphical representation was used in order to analyse the relationship that could be established between NDVI values and shrub biomass (Figure 7). The resulting graphic shows a narrow points cloud on the left for the minimum values and a scattered cloud on the right for the maximum values. This indicates that there were constraints in the regression analysis for the allometric equation calculation, possibly suggesting that there were too few options available.

During the regression analysis processing, it was noted that the NDVI tends to saturate at 0.7, suggesting that the shrubs' growth process maybe asymptotic to approximately 50 Mg ha^{-1}. This maybe due to the nature of the plants and the space they occupy over periods of time. As a consequence of the approach adopted, the ensuing model led to better results than anticipated. It is presented in Equation (3).

$$\text{Shrub biomass} = 70.078 \text{ NDVI}^{2.8113} \tag{3}$$

$R^2 = 0.8984$ (*p*-value < 0.05) and RMSE $= 4.46$ Mg ha^{-1}

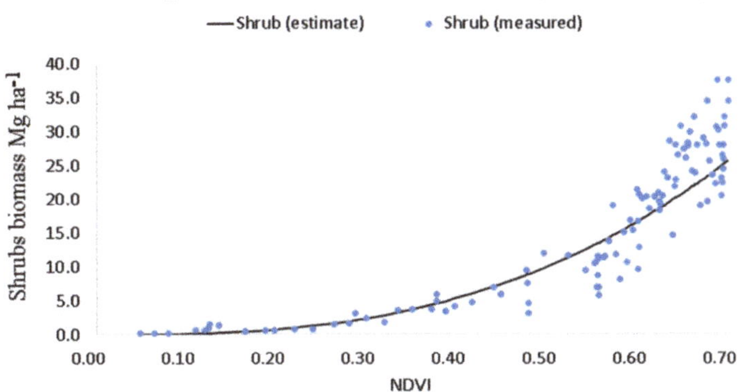

Figure 7. Relationship between NDVI and shrub biomass.

3.3. Shrub Biomass Estimation Using NDVI Image Processing

When the regression analysis completed, the adjusted allometric equation was used to generate the shrubs' biomass estimation using the NDVI. It was also applied to the 2019 Sentinel 2 images before the final calculation. General NDVI images were submitted to mask extraction in order to create new images indicating the former burnt and shrub areas, as depicted in Figure 8.

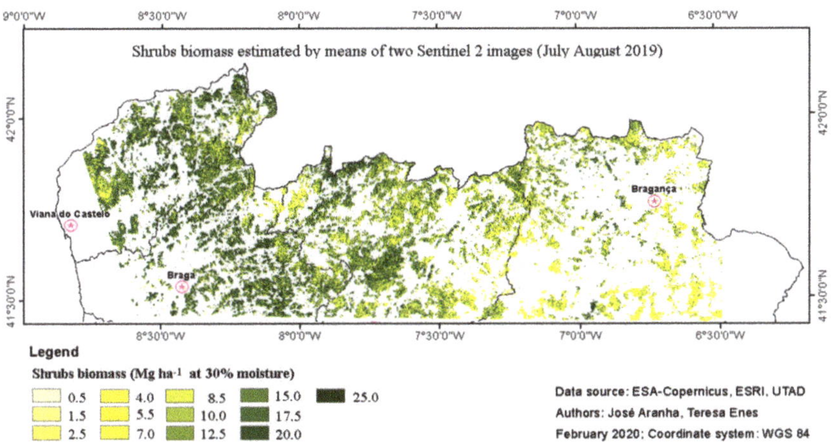

Figure 8. Accumulated shrubs' biomass in former burnt areas in the North of Portugal, estimated using an NDVI image summer 2019 and an allometric equation.

The results show that it is possible to account for an extent of some 172,022 ha and of 1,323,222 Mg of accumulated shrubs' biomass in the Northeast area. Likewise some 209,824 ha and 3,835,047 Mg can be identified in the Northwest area. On average, the Northeast area has approximately 7.7 Mg ha⁻¹ and the Northwest has 18.3 Mg ha⁻¹ of accumulated shrubs' biomass identified in the former burnt areas and shrubland.

4. Discussion

4.1. Sentinel 2 Images

The Sentinel 2 images, bands B2, B3, B4 and B8 (VIS–NIR), were deemed to be suitable for use in rural areas characterization and mapping, mainly in forested and shrubland areas, as demonstrated by the calculated overall accuracy (OA = 79.6) and the Cohen's K coefficient (k = 0.754) described earlier.

It was possible to classify, with good accuracy, the forest features in deciduous, coniferous and shrubs areas too. It was not possible, however, to classify mixed forest stands of deciduous and coniferous woodlands as the classification methods only isolate deciduous clusters from coniferous clusters. It was also not possible to classify forest stands by species.

Due to their spectral signatures, burnt areas were identified and classified with an users' accuracy of 89%. It was only possible to classify these former burnt areas, i.e., younger than a year and a half, after the fire because this timeframe indicates when a vegetative period has taken place and the shrubs were beginning to grow in these burnt areas. For example, if two full years have passed since the fire, the former burnt areas classification through satellite images starts to present results that confuse these areas with agricultural land and rocky areas with scattered vegetation.

The B4 and B8 bands enabled us to calculate NDVI images which proved to be adequate for the shrubs' biomass characterization and quantification. This was demonstrated using the statistical values of correlation between NDVI and shrubs biomass (r = 0.853) and the determination coefficient for the allometric equation (R^2 = 0.8984).

4.2. Shrubland Characterization

Vegetation, mainly shrubs, has a great potential to regrow on former burnt areas. The capability to colonize the space and to produce biomass is closely related to local morphology and edaphoclimatic conditions. As previously presented in Table 3, after calculating descriptive statistics for sub-samples and after adjusting allometric equations to each year, no statistically significant differences were found. In order to guarantee the accuracy of the estimates obtained by the general allometric equation now presented, however, it is necessary to verify if the growth rate of the shrubs is different in the two study areas. To achieve this, an allometric equation per area to the pairs was developed using: age and shrubs' biomass. This is presented as Equations (4) and (5).

$$\text{NW region: Shrub biomass} = -0.0062\ t^3 + 0.2089\ t^2 - 0.2738\ t + 2.279 \tag{4}$$

R^2 = 0.7349 (*p*-value < 0.05) and RMSE = 4.9 Mg ha^{-1}

$$\text{NE region: Shrub biomass} = -0.0072\ t^3 + 0.2211\ t^2 - 0.2738\ t + 2.043 \tag{5}$$

R^2 = 0.6921 (*p*-value < 0.05) and RMSE = 3.7 Mg ha^{-1}

The resulting shrubs' estimates, by means of these two equations, showed that there was no statistically significant differences found between regrowth rate in either of the study areas. This is shown in Figure 9.

The Northwest study area was found to have better morphological and edaphoclimatic conditions for shrubs' regrowth after fire than the Northeast area, possibly because the latter had retained burnt area scars for a longer time. After 30 years of forest fires, many of the mountainous areas had lost almost all vegetation and, as a consequence, top soil too. Figures 4 and 5 show that these areas are now characterized by small forested spaces or shrub zones surrounded by rocky extents with scattered shrubs.

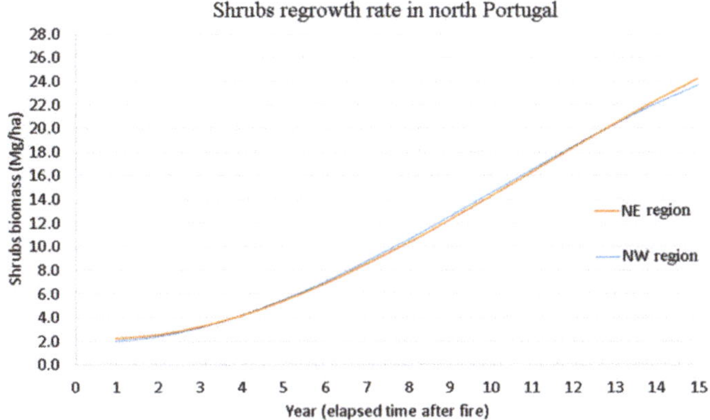

Figure 9. Estimations for shrubs regrowth rates in the study area.

The achieved results were classified regarding the average value of accumulate shrubs' biomass according to the elapsed time after the last fire event and the sampling date. This reclassification process enabled us to calculate the accumulated shrubs' biomass amount per class and also to create a histogram for the number of available hectares per class. The results are shown in Figure 10.

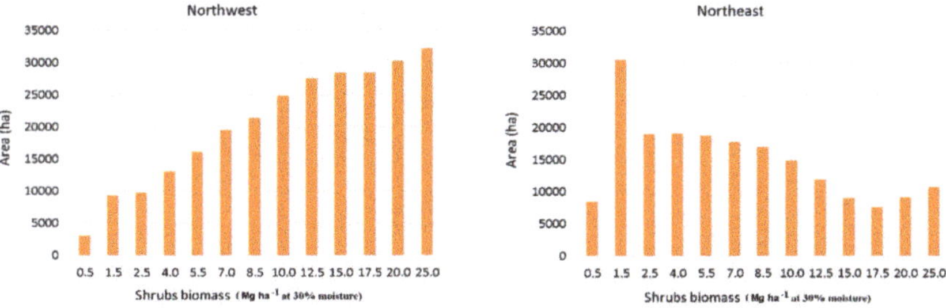

Figure 10. Accumulated shrubs biomass in summer 2019.

Although the growth rates in the two areas are not significantly different, it appears that the Northwest area is more densely vegetated than the Northeast. This was expected since this former area is facing the Atlantic Ocean, has a milder climate and greater rainfall than the Northeast landlocked territory.

It can be noted that for both of the study areas, within two years, the vegetation was capable to regrow enough to disguise the black landscape caused by fire. After five years, almost all the former burnt area was covered by vegetation and the accumulated shrubs' biomass grew up to 5 Mg ha^{-1} (30% moisture). Thus it can be demonstrated that in up to five years after a fire occurrence, the ecosystem will recover as evidenced by the fire hard index ranges from very low to medium. It appears that between five and 10 years after a fire, the accumulated shrubs' biomass can reach 14 to 18 Mg ha^{-1}. In terms of the ecosystem, the situation is favorable, but in terms of fire danger less so. Between 10 and 15 years after a fire, the shrubs' accumulated biomass can reach 26 Mg ha^{-1}. It must be noted, however, that this biomass already has a large wood structure that gives it properties suitable for its use as fuel for thermoelectric power plants. This means that this shrubs biomass has a potential economic value that could change very dense shrubs areas from a severe fire hazard issue into a green fuel source.

4.3. Allometric Equation for Shrub Biomass Estimation

The accumulated shrubs' biomass estimation was made using an allometric equation based on the elapsed time after a fire, as previously presented in Section 4.2 and also shown in Equations (4) and (5) as well as Figure 9. In addition the NDVI values derived from satellite image processing, also played a significant role (described earlier in this paper).

When working in a large area that presents different morphological and edaphoclimatic characteristics and which requires the use of multiple satellite scenes, it is appropriate to verify in advance if there are differences in the shrubs' growth rate and if there are differences in relation to the year under study. It was verified, as previously stated, that no statistically significant differences were found in relation to the shrubs' growth rate of the bush in the two study areas.

To verify the second hypothesis, an allometric equation for each date was adjusted. This is presented in Table 5.

Table 5. Allometric equations adjusted for the 3 years in analysis.

Date	Allometric Equation	R^2 (Adj)	RMSE (Mg/ha)
2016	66.383 NDVI $^{2.6073}$	0.894	4.08
2017	68.476 NDVI $^{2.6053}$	0.876	4.22
2018	58.139 NDVI $^{1.9541}$	0.855	4.95

Each of the allometric equations used the 110 sampling points and consequently the results generated were very similar with no significant differences found between estimates. These are shown in Figure 11.

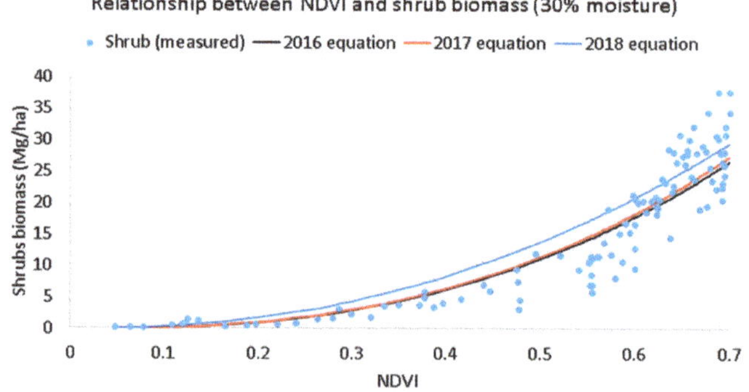

Figure 11. Relationship between NDVI and shrubs biomass estimates.

As no statistically significant differences were found between the shrubs' biomass estimations using the three previously presented equations, a general equation was developed using the 110 sampling points (Equation (3)).

The adjusted allometric equation described here proved to be suitable for assessing the shrubs' biomass estimation using NDVI values. Employing this method also made it possible to monitor the shrubs' biomass regrowth in the former burnt areas on a regular basis by means of Sentinel 2 image processing and classification.

As neither the shrubs' growth rates are significantly different for either study area or the use of the allometric equations (adjusted for each of the years: 2016, 2017, and 2018) led to statistically significant estimates between them and to the general equation, it can be state that the methodology was suitable to be used in this area for any year. This has been demonstrated by using accurate estimates of the accumulated shrubs' biomass (based on satellite imagery) and on a regular basis. It also enables the monitoring of the shrubs' biomass variation as well as calculating the biomass gains and losses. Biomass gains can be converted into sequestered carbon and used to analyse the ecosystem's state of health as well as its production capability [47,48]. It can also be used to update the fire hazard indices calculation and to identify the places most prone to be burnt by large fires and, therefore, requiring special attention. Biomass losses can be converted into carbon released into the atmosphere by forest fires [9] or used to calculate the intensity of forest fires. The difference between post-fire gains and losses can be used to calculate the fire severity [46,49] in any given spot.

This equation was adjusted for the North Portugal study areas and it is suggested that it could be used in other territories with similar morphological and edaphoclimatic conditions. It is important to remember, however, that such work requires a validation process for any area.

5. Conclusions

Free Sentinel 2 images were an asset to derive multi temporal and dynamic studies about land cover, for monitoring former burnt areas and to estimate the shrubs' biomass accumulation.

The achieved results enable us to state that the methodology presented in this manuscript proved to be robust and that the NDVI derived from Sentinel 2 images can be used to calculate accurate and dynamic estimates of accumulated shrubs biomass.

The allometric equation presented here also allowed us to estimate the shrubs' biomass using Sentinel 2 images without depending on the vector files provided by EFFIS or by ICNF (Portuguese Institute for Nature and Forest). Comparing the shrubs' biomass estimates achieved through the two allometric equations, using NDVI and using elapsed time after fire, demonstrated that no statistically significant differences were found. Thus, it can be stated that the allometric equation presented in this manuscript incorporated the effect of elapsed time after the fire.

The combined use of GIS and RS techniques, complemented by regression analysis proved to be useful for monitoring the shrubs' regrowth in the former burnt areas. It was also useful to analyse the land cover dynamics and also to quantify the accumulated shrubs' biomass. GIS supported data records, management, and the sampling system development was helpful whilst RS supported multitemporal land cover analysis and biomass estimation using associated satellite image processing. Classification also offered major benefits too.

Regression analysis and allometric equation adjustments were found to be suitable processes to assign the biophysical data that was collected via field work as well as the use of the freely available satellite images. This led to the calculation and estimatation of the shrubs' biomass.

Although the NDVI saturates only measured 0.7, it was still possible to obtain good estimates of the shrubs' biomass before the complete canopy closure which is when the accumulated values reach their maximum. It was discovered that the NDVI values were not, however, specific for all types of vegetation. Consequently it was always necessary to create a raster mask in advance that identified the type of vegetation or area under analysis in order to define the estimates for those particular places of interest.

From a forest management perspective, it was found that, after five years, the accumulated shrubs' biomass starts to be a fire hazard related issue as it creates a horizontal continuous coverture which encourages any fire to spread. If it was 10 years since the last fire occurrence, the amount of accumulated shrubs' biomass was found to be over 14 Mg ha^{-1} which led to a severe wild fires spread. This is commonly referred to as a 10 year of fire recurrence cycle in Portugal.

The methodology presented in this paper was found to be suitable for use in forest land management and also served a number of unexpected different purposes. From an ecological persepective it has been demonstrated that, over a two-year period, the vegetation was capable of enough regrowth to minimize erosion actions and to support animal life. In a five-year period, it appears that almost all the former burnt areas are covered by vegetation. From this point of view it may be possible to use the shrubs' biomass for energy purposes but it was found that only after a 10-year period that the amount of accumulated shrubs' biomass became economically valuable to cut and transport elsewhere. This was determined by the monetary biomass value for any potential power plant location, the man hours cost involved to capture it as well as the necessary transportation costs.

Author Contributions: Conceptualization, J.A.; performed the experiment and analysis, J.A., T.E., H.V. and A.C.; performed the statistical analysis, J.A., T.E., A.C. and H.V.; resources, H.V., A.C. and J.A.; writing—original draft preparation, T.E., J.A.; writing—review and editing, J.A., H.V.; project administration, J.A. All authors have read and agreed to the published version of the manuscript.

Funding: This research is supported by National Funds by FCT—Portuguese Foundation for Science and Technology, under the project UIDB/04033/2020.

Acknowledgments: The authors would like to express their sincere thanks for the critical review of the manuscript carried out by Teresa Connolly.

Conflicts of Interest: The authors declare no conflict of interest.

Appendix A

Additional information:

- Sentinel-2S2A_20160828T113040_20160828T164718_A006183_T29TPG_N02_04_01
- Sentinel-2S2A_20150804T113226_20160319T010337_A000606_T29TPN_N02_04_01
- Sentinel-2L1C_T29TPG_A010759_20170714T112114
- Sentinel-2L1C_T29TNG_A010759_20170714T112114
- Sentinel-2L1C_T29TPG_A015621_20180619T112602
- Sentinel-2L1C_T29TNG_A006784_20180624T112452
- Sentinel-2L1C_T29TNG_A021341_20190724T112448
- Sentinel-2L1C_T29TPG_A021484_20190803T112140

Formulae demonstration:

$$Vsh = 1/6 \, \pi \, L \, W \, H \tag{A1}$$

where Vsh = shrubs canopy volume (m^3), L = length (m), W = width (m), H = height (m), and the volume of a sphere = $4/3 \, \pi \, r^3$.

In order to use the shrubs dimensions measured along the transect, the equation can be rewritten as:

$$\text{Volume of shrub canopy} = 4/3 \, \pi \, \text{Length}/2 \, \text{Width}/2 \, \text{Height}/2$$

$$\text{Volume of shrub canopy} = 4/24 \, \pi \, \text{Length Width Height}$$

$$\text{Volume of shrub canopy} = 1/6 \, \pi \, \text{Length Width Height}$$

$$\text{Biomass} = V \, W \, 200/Mxw \tag{A2}$$

where Biomass in Mg/ha V = Total shrubs volume along the 2 transects (m^3) W = Average shrubs weight (Mg/m^3 at 30% moisture) Mxw = Maximum width measured along both transects.

In a 500 m^2 sampling plot, the plot radius is 12.62 m. This way, each transect has 25.24 m. The 2 transect account for 50.48 m.

When measuring the shrubs along these transects, the sum of shrubs width plus shrubs length define the area occupied by shrubs within the 2 cross section transects. Using the maximum measured shrubs width and the length of both transects is possible to calculate the maximum area for the 2 transects.

$$\text{Sampling area} = \text{Maximum width } 50.48 \text{ m}$$

For converting the measured shrubs biomass within the sampling area, it is necessary to convert this area to 1 hectare.

$$10,000/\text{Maximum width } 50.48 \approx 200/\text{Maximum width}$$

References

1. Caetano, M.; Igreja, C.; Marcelino, F.; Costa, H. Estatísticas e dinâmicas territoriais multiescala de Portugal Continental 1995–2007–2010 com base na Carta de Uso e Ocupação do Solo (COS). *Direcção Geral do Território* **2017**, 149.
2. Oliveira, S.L.J.; Pereira, J.M.C.; Carreiras, J.M.B. Fire frequency analysis in Portugal (1975–2005), using Landsat-based burnt area maps. *Int. J. Wildl. Fire* **2011**, *21*, 48–60. [CrossRef]
3. Marques, S.; Borges, J.G.; Garcia-Gonzalo, J.; Moreira, F.; Carreiras, J.M.B.; Oliveira, M.M.; Cantarinha, A.; Botequim, B.; Pereira, J.M.C. Characterization of wildfires in Portugal. *Eur. J. For. Res.* **2011**, *130*, 775–784. [CrossRef]
4. Parente, J.; Pereira, M.G.; Amraoui, M.; Tedim, F. Science of the Total Environment Negligent and intentional fires in Portugal: Spatial distribution characterization. *Sci. Total Environ.* **2018**, *624*, 424–437. [CrossRef] [PubMed]
5. Pereira, M.G.; Aranha, J.; Amraoui, M. Land cover fire proneness in Europe. *For. Syst.* **2014**, *23*, 598–610. [CrossRef]
6. Ferreira Leite, F.; Bento Gonçalves, A.; Vieira, A. The recurrence interval of forest fires in Cabeço da Vaca (Cabreira Mountain-northwest of Portugal). *Environ. Res.* **2011**, *111*, 215–221. [CrossRef] [PubMed]
7. Ursino, N.; Romano, N. Wild forest fire regime following land abandonment in the Mediterranean region. *Geophys. Res. Lett.* **2014**, *41*, 8359–8368. [CrossRef]
8. Cerdá, A.; Doerr, S.H. Influence of vegetation recovery on soil hydrology and erodibility following fire: An 11-year investigation. *Int. J. Wildl. Fire* **2005**, *14*, 423–437. [CrossRef]
9. Pereira da Silva, T.; Cardoso Pereira, J.M.; Paúl, P.J.C.; Teresa Santos, M.N.; José Vasconcelos, M.P.; de Investigação, B.; Catedrático, P.; Auxiliar, I. Estimativa de Emissões Atmosféricas Originadas por Fogos Rurais em Portugal Estimate of Atmospheric Emissions Originated by Wildfires in Portugal. *Silva Lusit.* **2006**, *14*, 239–263.
10. Viana, H. Modelling and Mapping Aboveground Biomass for energy Usage and Carbon Storage Assessment in Mediterranean Ecosystems. Ph.D. Thesis, Universidade de Trás os Montes e Alto Douro, Vila Real, Portugal, 2012.
11. Aranha, J.; Calvão, A.R.; Lopes, D.; Viana, H. Quantificação Da Biomassa Consumida Nos Últimos 20 Anos De Fogos Florestais No Norte Portugal. *Info* **2011**, *26*, 44–49.
12. Kazanis, D.; Xanthopoulos, G.; Arianoutsou, M. Understorey fuel load estimation along two post-fire chronosequences of Pinus halepensis Mill. forests in Central Greece. *J. For. Res.* **2012**, *17*, 105–109. [CrossRef]
13. DL 76/2017-5.ª alteração ao DL 124-2006. Diário da República n.º 158/2017, Série I de 2017-08-17. DR. *J. Portuguese Rep.* **2017**, 4734–4762.
14. Viana, H.; Fernandes, P.; Rocha, R.; Lopes, D.; Aranha, J. Alometria, Dinâmicas da Biomassa e do Carbono Fixado em Algumas Espécies Arbustivas de Portugal. *6.º Congr. Florest. Nac.* **2009**, 244–252.
15. Gratani, L.; Varone, L.; Ricotta, C.; Catoni, R. Mediterranean shrublands carbon sequestration: Environmental and economic benefits. *Mitig. Adapt. Strateg. Glob. Chang.* **2013**, *18*, 1167–1182. [CrossRef]
16. Cairns, M.A.; Lajtha, K.; Beedlow, P.A. Dissolved carbon and nitrogen losses from forests of the Oregon Cascades over a successional gradient. *Plant Soil* **2009**, *318*, 185–196. [CrossRef]
17. Zuazo, V.H.D.; Pleguezuelo, C.R.R. Soil-erosion and runoff prevention by plant covers: A review. *Agron. Sustain. Dev.* **2008**, *28*, 65–86. [CrossRef]
18. Imeson, A. *Desertification, Land Degradation and Sustainability*; John Wiley & Sons: West Sussex, UK, 2012; ISBN 1119978483.
19. Bochet, E.; Poesen, J.; Rubio, J.L. Runoff and soil loss under individual plants of a semi-arid Mediterranean shrubland: Influence of plant morphology and rainfall intensity. *Earth Surf. Process. Landforms* **2006**, *31*, 536–549. [CrossRef]
20. Mangas, J.G.; Lozano, J.; Cabezas-Díaz, S.; Virgós, E. The priority value of scrubland habitats for carnivore conservation in Mediterranean ecosystems. *Biodivers. Conserv.* **2008**, *17*, 43–51. [CrossRef]
21. Sandifer, P.A.; Sutton-Grier, A.E.; Ward, B.P. Exploring connections among nature, biodiversity, ecosystem services, and human health and well-being: Opportunities to enhance health and biodiversity conservation. *Ecosyst. Serv.* **2015**, *12*, 1–15. [CrossRef]

22. Helming, K.; Diehl, K.; Geneletti, D.; Wiggering, H. Mainstreaming ecosystem services in European policy impact assessment. *Environ. Impact Assess. Rev.* **2013**, *40*, 82–87. [CrossRef]
23. Viana, H.; Cohen, W.B.; Lopes, D.; Aranha, J. Assessment of forest biomass for use as energy. GIS-based analysis of geographical availability and locations of wood-fired power plants in Portugal. *Appl. Energy* **2010**, *87*, 2551–2560. [CrossRef]
24. Quinta-Nova, L.; Fernandez, P.; Pedro, N. GIS-Based Suitability Model for Assessment of Forest Biomass Energy Potential in a Region of Portugal. *IOP Conf. Ser. Earth Environ. Sci.* **2017**, *95*, 042059. [CrossRef]
25. Cohen, W.B.; Maiersperger, T.K.; Spies, T.A.; Oetter, D.R. Modelling forest cover attributes as continuous variables in a regional context with Thematic Mapper data. *Int. J. Remote Sens.* **2001**, *22*, 2279–2310. [CrossRef]
26. Botequim, B.; Zubizarreta-Gerendiain, A.; Garcia-Gonzalo, J.; Silva, A.; Marques, S.; Fernandes, P.M.; Pereira, J.M.C.; Tomé, M. A model of shrub biomass accumulation as a tool to support management of portuguese forests. *iForest* **2014**, *8*, 114–125. [CrossRef]
27. Nguyen, H.K.L.; Nguyen, B.N. Mapping biomass and carbon stock of forest by remote sensing and GIS technology at Bach Ma National Park, Thua Thien Hue province. *J. Viet. Env.* **2016**, *8*, 80–87.
28. Deng, S.; Shi, Y.; Jin, Y.; Wang, L. A GIS-based approach for quantifying and mapping carbon sink and stock values of forest ecosystem: A case study. *Energy Proc.* **2011**, *5*, 1535–1545. [CrossRef]
29. Wijaya, A.; Kusnadi, S.; Gloaguen, R.; Heilmeier, H. Improved strategy for estimating stem volume and forest biomass using moderate resolution remote sensing data and GIS. *J. For. Res.* **2010**, *21*, 1–12. [CrossRef]
30. Roy, P.S.; Ravan, S.A. Biomass estimation using satellite remote sensing data—An investigation on possible approaches for natural forest. *J. Biosci.* **1996**, *21*, 535–561. [CrossRef]
31. Chen, W.; Zhao, J.; Cao, C.; Tian, H. Shrub biomass estimation in semi-arid sandland ecosystem based on remote sensing technology. *Glob. Ecol. Conserv.* **2018**, *16*, e00479. [CrossRef]
32. Norovsuren, B.; Tseveen, B.; Batomunkuev, V. Estimation for Forest Biomass and Coverage Using Satellite Data in Small Scale Area, Mongolia. *IOP Conf. Ser. Eart Environ. Sci.* **2019**, *320*, 012019. [CrossRef]
33. Filella, I.; Peñuelas, J.; Llorens, L.; Estiarte, M. Reflectance assessment of seasonal and annual changes in biomass and CO_2 uptake of a Mediterranean shrubland submitted to experimental warming and drought. *Remote Sens. Environ.* **2004**, *90*, 308–318. [CrossRef]
34. Xiao, J.; Moody, A. A comparison of methods for estimating fractional green vegetation cover within a desert-to-upland transition zone in central New Mexico, USA. *Remote Sens. Environ.* **2005**, *98*, 237–250. [CrossRef]
35. ICNF. Available online: http://www2.icnf.pt/portal/florestas/dfci/inc (accessed on 8 January 2020).
36. EFFIS. Available online: https://effis.jrc.ec.europa.eu/applications/data-request-form (accessed on 8 January 2020).
37. Copernicus. Available online: https://scihub.copernicus.eu/dhus/#/home (accessed on 10 January 2020).
38. Glovis. Available online: https://glovis.usgs.gov/app (accessed on 10 January 2020).
39. Chavez, P.S. An improved dark-object subtraction technique for atmospheric scattering correction of multispectral data An Improved Dark-Object Subtraction Technique for Atmospheric Scattering Correction of Multispectral Data. *Remote Sens. Environ.* **1988**, *24*, 459–479. [CrossRef]
40. Teillet, P.M.; Staenz, K.; William, D.J. Effects of spectral, spatial, and radiometric characteristics on remote sensing vegetation indices of forested regions. *Remote Sens. Environ.* **1997**, *61*, 139–149. [CrossRef]
41. Louis, J.; Debaecker, V.; Pflug, B.; Main-knorn, M.; Bieniarz, J. Sentinel-2 Sen2cor: L2a Processor for Users. In Proceedings of the Living Planet Symposium, Prague, Czech Republic, 9–13 May 2016; Volume 2016, pp. 9–13.
42. Sola, I.; García-Martín, A.; Sandonís-Pozo, L.; Álvarez-Mozos, J.; Pérez-Cabello, F.; González-Audícana, M.; Llovería, R. Montorio Assessment of atmospheric correction methods for Sentinel-2 images in Mediterranean landscapes. *Int. J. Appl. Earth Obs. Geoinf.* **2018**, *73*, 63–76. [CrossRef]
43. Sentinel Online. Available online: https://sentinel.esa.int/web/sentinel/user-guides/sentinel-2-msi/processing-levels/level-2 (accessed on 9 January 2020).
44. Meddens, A.J.H.; Kolden, C.A.; Lutz, J.A.; Abatzoglou, J.T.; Hudak, A.T. Spatiotemporal patterns of unburned areas within fire perimeters in the northwestern United States from 1984 to 2014. *Ecosphere* **2018**, *9*, 1–16. [CrossRef]

45. Tedim, F.; Royé, D.; Bouillon, C.; Fernando, J.M.C.; Leone, V. Understanding unburned patches patterns in extreme wildfire events: A new approach. In Proceedings of the Advances in forest fire research—VIII International Conference on Forest Fire Research, Coimbra, Portugal, 9–16 November 2018; Viegas, D.X., Ed.; pp. 700–715. [CrossRef]

46. Picos, J.; Alonso, L.; Bastos, G.; Armesto, J. Event-based integrated assessment of environmental variables and wildfire severity through Sentinel-2 Data. *Forests* **2019**, *10*, 1021. [CrossRef]

47. Iqbal, K.; Hussain, A.; Negi, A.K. An overview of Biomass Estimation methods. *Res. J. Soc. Sci. Manag.* **2014**, *4*, 42–57.

48. Banti, M.A.; Kiachidis, K.; Gemitzi, A. Estimation of spatio-temporal vegetation trends in different land use environments across Greece. *J. Land Use Sci.* **2019**, *14*, 21–36. [CrossRef]

49. Díaz-Delgado, R.; Lloret, F.; Pons, X. Influence of fire severity on plant regeneration by means of remote sensing imagery. *Int. J. Remote Sens.* **2003**, *24*, 1751–1763. [CrossRef]

Article

Evaluation of Different Algorithms for Estimating the Growing Stock Volume of *Pinus massoniana* Plantations Using Spectral and Spatial Information from a SPOT6 Image

Jingjing Zhou [1,2], Zhixiang Zhou [1,2], Qingxia Zhao [3], Zemin Han [1,2], Pengcheng Wang [1,2], Jie Xu [4] and Yuanyong Dian [1,2,*]

[1] College of Horticulture & Forestry Sciences, Huazhong Agricultural University, Wuhan 430070, China; hupodingxiangyu@mail.hzau.edu.cn (J.Z.); whzhouzx@mail.hzau.edu.cn (Z.Z.); swxfhzm@163.com (Z.H.); pengchengwang@163.com (P.W.)
[2] Hubei Engineering Technology Research Center for Forestry Information, Wuhan 430070, China
[3] Shaanxi Institute of Zoology, Xi'an 710032, China; zhaoqingxia@nwafu.edu.cn
[4] Hubei Forestry Survey and Design Institute (Forestry Branch), Wuhan 430079, China; xuejie@webmail.hzau.edu.cn
* Correspondence: dianyuanyong@mail.hzau.edu.cn; Tel./Fax: +86-027-87281827

Received: 28 March 2020; Accepted: 9 May 2020; Published: 12 May 2020

Abstract: Precise growing stock volume (GSV) estimation is essential for monitoring forest carbon dynamics, determining forest productivity, assessing ecosystem forest services, and evaluating forest quality. We evaluated four machine learning methods: classification and regression trees (CART), support vector machines (SVM), artificial neural networks (ANN), and random forests (RF), for their reliability in the estimation of the GSV of *Pinus massoniana* plantations in China's northern subtropical regions, using remote sensing data. For all four methods, models were generated using data derived from a SPOT6 image, namely the spectral vegetation indices (SVIs), texture parameters, or both. In addition, the effects of varying the size of the moving window on estimation precision were investigated. RF almost always yielded the greatest precision independently of the choice of input. ANN had the best performance when SVIs were used alone to estimate GSV. When using texture indices alone with window sizes of $3 \times 5 \times 5$ or 9×9, RF achieved the best results. For CART, SVM, and RF, R^2 decreased as the moving window size increased: the highest R^2 values were achieved with 3×3 or 5×5 windows. When using textural parameters together with SVIs as the model input, RF achieved the highest precision, followed by SVM and CART. Models using both SVI and textural parameters as inputs had better estimating precision than those using spectral data alone but did not appreciably outperform those using textural parameters alone.

Keywords: machine learning algorithms; forest growing stock volume; SPOT6 imagery; *Pinus massoniana* plantations

---

1. Introduction

Forest growing stock volume (GSV) is one of the most important forest characteristics, both economically and environmentally, because it is a key determinant of forest productivity [1]. Precisely estimating forest GSVs on large scales is crucial for monitoring forest carbon dynamics, assessing forest ecosystem services, and evaluating forest quality [2–5]. Traditional ground-based GSV estimation strategies that rely on field measurements of tree height and diameters at breast height (DBH) are time-consuming and labor-intensive [1]. GSV is estimated over large areas almost everywhere in the world, and estimates are often precise. The (additional) benefits of including remote sensing (RS) data in this process are that spatially explicit information can be produced (i.e., maps) and that the precision of population parameter estimates can be improved.

Several modelling approaches, both parametric and non-parametric, have proven capable of precisely estimating forest variables such as the leaf area index, canopy cover, height, basal area, and stock volume based on RS data [4,6–10]. The most popular parametric regression methods for modelling forest attributes are simple or multiple linear regression models. Commonly used non-parametric models include classification and regression trees (CART), k-nearest neighbor (k-NN), random forest (RF), and support vector machine (SVM). Although many researchers have performed comparative evaluations of different machine learning algorithms (MLAs), no single technique has been revealed as universally superior for predicting forest inventory attributes [8,11,12]. Different machine learning algorithms (MLAs) have different estimating precisions; for example, CART achieves the lowest precision, SVM and ANN tend to achieve moderate performance, and RF has achieved the highest precision and lowest error for the forest variables under consideration [4]. Zhang et al. [12] found that ANN and SVM were similarly effective at estimating sawgrass aboveground biomass (AGB), with correlation coefficients (r values) exceeding 0.9. However, ANN offered the most precise total biomass estimates (r = 0.94). Wang et al. [13] found that the MLR method generally outperformed SVM and RF at predicting site-level AGB. The GSV of *Pinus massoniana* plantations has not been estimated extensively in China's northern subtropical regions. Therefore, there is a need to evaluate and compare the performance of different MLAs in this specific task.

The spectral information in optical remote sensing images generally provides insufficient information to precisely assess the state of vegetation in dense forests with complex structures. Therefore, texture parameters derived from remote sensing images are commonly analyzed to assess the spatial distribution of image tone variance and thereby acquire spatial information on vegetation cover, especially in complex and dense forests [14]. Several recent studies have concluded that textural parameters from high-resolution satellite images are more useful than spectral information for estimating forest variables because the interpretation of spectral data is complicated by saturation when studying multi-layer forests with full canopy cover [6,12,14–19]. However, it is not yet clear whether combining spectral and texture-based features can enable more precise estimation of forest variables than analyses based on textural information alone. Chrysafis et al. [5] argued that combining spatial and textural information yielded at most marginal improvements in precision over texture-only models when estimating GSV in a Mediterranean forest. However, others have reported that forest variables such as the leaf area index (LAI) can be estimated more precisely by combining spectral and textural features than by relying solely on one type or the other [6,15,17]. Texture is a complex parameter and obtaining precise texture information over the entire forest area is challenging. This is because texture values are highly sensitive to texture measures, landscape type, physiological growth period, associated parameters (i.e., moving window size, direction), and the type of remote sensing image [5,20]. It is still worthwhile to explore whether a large or small moving window size should be used in various forest types, and with an inversion algorithm.

Pinus massoniana is a major coniferous tree that is widely distributed in the subtropical forests of South China and exhibits relatively high tolerance to acid rain, drought, and phosphorus deficiency [21,22]. It is central to forest ecosystems and is also an economically important source of timber and wood pulp [23]. Few efforts have been made to evaluate the capability of modern remote sensing techniques for mapping *Pinus massonian* GSV in the subtropical area of China [24,25]. A growing stock volume map of *Pinus massoniana* trees in in the subtropical area of China has never been generated and published. The specific objectives of this study were therefore to: (1) examine the ability of four MLAs (CART, SVM, ANN, and RF) to predict the GSV of *Pinus massoniana* plantations, (2) compare the effectiveness of spectral vegetation indices (SVIs) and texture data derived from SPOT6 images for forest GSV quantification, and (3) assess the influence of texture index window size on the precision of forest GSV estimation.

2. Materials and Methods

2.1. Study Area

This study focused on an approximately 7600 ha region of *Pinus massoniana* plantations near Taizi mountain in Jingshan County of China's Hubei Province (30°48′–31°02′ N and 112°48′–113°03′ E) (Figure 1). Topographically, the area is characterized by a hilly landscape with moderate slope and its elevation is between 40.3 and 467.4 m above sea level. It has a subtropical monsoon climate with a mean temperature of 16.4 °C. The average annual precipitation is 1094.6 mm, of which over 53% falls between April and August. The dominant soil type at the site is yellow-brown soil. Yellow-brown soil is a transitional soil between yellow, red, and brown soils, which are common in mixed subtropical evergreen broad-leaved and deciduous broad-leaved forests widely. The landscape of the study area is characterized by pure forests, mixed forests, shrub land, water, cultivations, and forest openings. The percentage of forest cover is 85%. The main tree species in the area are *Pinus massoniana* Lamb. and *Quercus acutissima* Carr. Less abundant species include *Cunninghamia lanceolata (Lamb.)* Hook., *Ilex chinensis* Sims, *Platycarya strobilacea* Sieb.et Zucc., *Liquidambar formosana* Hance, *Dalbergia hupeana* Hance, *Pistacia chinensis* Bunge., and *Celtis sinensis* Pers. [26].

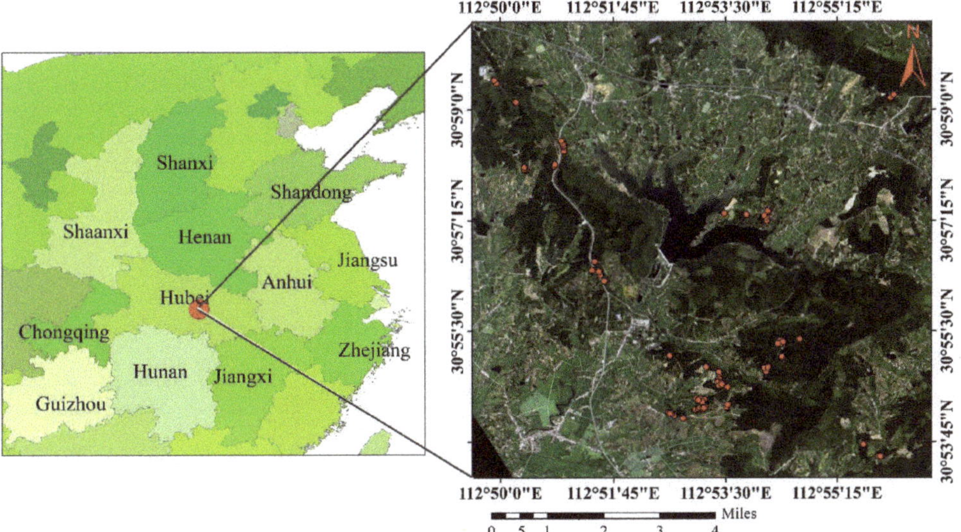

Figure 1. A portion of the study area and the location of the sample plots over a SPOT6 Multispectral Image (MSI) image acquired on 6 August 2015.

2.2. Data

2.2.1. Field Data and Stand Volume Estimation

The field sample plots were established in August 2015, August 2018, and November 2019. In total, 68 square plots of the size 20 × 20 m each were established in pure *Pinus massoniana* plantations by applying the method of combining stratified sampling and random sampling. Considering the standards for division of *Pinus massoniana* plantation age, there were 17 plots of young forest, 11 of middle-age forest, 15 of near-mature forest, 11 of mature forest, and 14 of over-mature forest. Sample plots in every group of forest ages were distributed randomly in terms of geography. In each plot, trees with DBH over 5 cm were tallied. DBH, tree height (H), and crown diameter (CD) of the individual tree in the plot were all measured. Trees with a DBH below 5 cm were excluded.

The LAI-2200 instrument (LI-COR Inc., Li-Cor, Lincoln, NE, USA, 2010) was used to indirectly measure the leaf area index (LAI) of each plot. Measurements were taken either under overcast conditions, or alternatively within two hours after sunrise or before sunset. At each site, two above-canopy and nine low-canopy readings were taken with an opaque, 180°, view-restricting cap placed over the sensor in order to mask out the operator [6]. After finding the total number of all trees in a plot, the stand density (Density) was estimated in trees per hectare (trees ha^{-1}). The survey also provided information about aspect, slope, and slope position. A differential global positioning unit (Trimble GeoXH6000 GPS units) was used to locate the center and the four corners of each plot, allowing the plots to be geo-referenced against satellite data. The summarized characteristics of the 68 plots are documented in Table 1.

The volumes of each individual growing trees were predicted by applying the allometric model (Equation (1)) based on DBH and H. These models yielded the following expression [27]:

$$V = 0.00006228789 \times DBH^{1.849839} \times H^{0.9843411} \tag{1}$$

where, V is the individual tree volume, DBH is the tree diameter at a height of 1.3 m, and H is the total tree height. The volumes estimated from all individual trees within a sample plot were summed to obtain the GSV of the sample plot, which was then up-scaled to per hectare. The GSV of the plot was measured in cubic meter per hectare (m^3 ha^{-1}). The field data revealed that the plot-level GSVs varied widely, ranging from 10.20 to 319.82 m^3 ha^{-1}, with a mean of 140.84 m^3 ha^{-1}. The inconsistency between the time of sample plot data collected in 2018 and 2019 and remote sensing (RS) image was an important factor which contributed to low precision of forest parameter estimation. We used the growth model of *Pinus massoniana* tree for conversing volume values of 2018 and 2019 to V values of 2015. The formula was as follows [28]:

$$V = 0.000025058 \times A^{2.5983243} \ (R = 0.996) \tag{2}$$

where, V is volume of a single tree and A is the stand age.

Table 1. The summarized characteristics of the 68 plots surveyed in August 2015, August 2018, and November 2019.

No. of Plots	Density (Stem ha^{-1})	Average DBH (cm)	Average Height (m)	GSV (m^3 ha^{-1})	LAI	Canopy Density	Aspect	Slope	Elevation
1	825	18.4	15.6	170.47	4.14	0.5	Shady	3°	95.0
2	775	17.2	13.1	119.98	3.40	0.5	Shady	4°	95.2
3	525	19.3	13.1	98.96	4.21	0.4	Shady	0°	83.1
4	2400	9.3	5.4	50.84	5.30	0.8	Sunny	2°	65.9
5	1375	10.3	5.6	35.95	5.24	0.8	Sunny	2°	73.9
6	2450	9.4	4.8	47.67	4.65	0.8	Sunny	4°	77.6
7	2175	10.2	5.9	59.00	5.25	0.8	Sunny	1°	83.4
8	2775	9.0	5.2	53.62	5.22	0.8	Sunny	3°	82.0
9	1350	22.3	18.0	234.93	3.33	0.4	Sunny	1°	79.1
10	2925	8.2	4.7	42.98	6.37	0.9	Sunny	2°	93.7
11	1850	20.2	15.6	249.51	5.53	0.6	Sunny	1°	110.4
12	900	13.9	9.5	70.46	5.23	0.7	Shady	2°	94.3
13	1200	8.4	5.2	10.20	5.50	0.8	Sunny	1°	87.7
14	1400	6.7	4.5	10.38	7.08	0.9	Sunny	3°	110.2
15	900	19.0	11.8	185.77	3.98	0.5	Sunny	3°	120.5
16	900	19.3	12.6	209.23	3.39	0.5	Sunny	3°	109.6
17	425	27.3	14.9	175.58	4.51	0.5	Shady	2°	91.2
18	800	17.4	10.1	143.82	5.46	0.7	Shady	1°	88.8
19	375	26.9	14.3	185.99	4.24	0.6	Shady	4°	95.1
20	600	20.3	13.7	139.56	2.91	0.4	Shady	3°	116.7
21	1300	26.2	16.2	270.60	6.13	0.8	Shady	2°	93.7
22	650	20.4	15.4	164.76	2.58	0.4	Shady	5°	128.1
23	625	19.5	15.8	150.89	2.35	0.4	Shady	4°	124.1

Table 1. *Cont.*

No. of Plots	Density (Stem ha^{-1})	Average DBH (cm)	Average Height (m)	GSV (m^3 ha^{-1})	LAI	Canopy Density	Aspect	Slope	Elevation
24	500	21.9	14.2	137.85	5.08	0.7	Shady	1°	88.6
25	550	22.0	14.4	199.61	4.51	0.7	Shady	2°	91.8
26	675	28.3	16.8	162.44	5.81	0.7	Shady	0°	104.6
27	435	27.1	18.2	225.16	3.96	0.6	Shady	2°	88.8
28	450	22.1	16.5	133.50	2.52	0.4	Sunny	8°	91.0
29	425	25.8	17.6	172.23	2.84	0.4	Sunny	5°	99.7
30	475	18.7	14.7	140.05	4.52	0.6	Sunny	0°	101.8
31	925	18.2	15.8	222.18	3.47	0.5	Sunny	2°	114.7
32	450	25.4	16.6	155.50	2.58	0.4	Sunny	0°	111.0
33	375	24.5	19.5	29.23	3.66	0.5	Shady	2°	115.4
34	425	34.2	20.4	290.30	3.05	0.6	Shady	5°	121.4
35	275	29.2	22.8	193.28	3.54	0.5	Shady	1°	115.5
36	350	31.6	21.1	175.27	3.89	0.6	Shady	2°	121.4
37	375	27.1	20.4	212.70	5.55	0.7	Shady	1°	108.0
38	300	32.1	20.1	168.31	3.61	0.5	Shady	1°	94.5
39	1875	13.7	10.1	54.78	6.08	0.8	Sunny	3°	190.1
40	725	22.0	16.9	219.12	3.16	0.6	Shady	0°	145.7
41	725	21.8	17.5	221.66	3.59	0.6	Sunny	0°	150.2
42	525	23.7	18.4	106.95	5.31	0.7	Sunny	5°	164.9
43	400	31.7	18.5	193.38	2.24	0.4	Sunny	7°	99.4
44	325	30.6	18.1	208.49	3.31	0.6	Shady	2°	95.0
45	500	26.9	16.5	222.52	4.58	0.6	Shady	5°	86.8
46	1000	5.2	5.2	38.31	5.30	0.9	Sunny	3°	97.7
47	600	20.5	16.2	165.05	6.28	0.9	Shady	4°	110.3
48	350	27.2	16.9	166.20	4.24	0.7	Shady	2°	144.4
49	900	19.3	12.6	209.24	3.39	0.5	Sunny	3°	109.6
50	650	8.2	5.2	41.41	5.18	0.8	Sunny	2°	98.5
51	775	8.7	5.4	57.26	5.97	0.8	Sunny	1°	91.1
52	825	9.0	4.0	52.00	7.34	0.9	Shady	3°	98.5
53	750	9.5	5.0	61.91	7.60	0.8	Shady	1°	107.9
54	775	7.8	5.5	52.97	5.50	0.8	Shady	7°	96.8
55	1000	22.1	16.5	250.98	6.08	0.8	Shady	1°	110.6
56	575	7.7	5.2	35.46	7.34	0.9	Shady	1°	100.1
57	1450	11.0	8.2	69.46	3.33	0.4	Sunny	6°	121.0
58	400	20.5	14.7	163.53	1.42	0.4	Sunny	5°	100.0
59	400	26.4	17.1	178.21	1.95	0.5	Sunny	0°	92.5
60	1850	10.6	9.6	89.74	4.24	0.6	Shady	4°	90.5
61	2075	9.6	9.1	81.48	1.24	0.4	Sunny	4°	115.0
62	875	20.6	21.4	319.82	2.17	0.5	Sunny	4°	102.5
63	400	15.6	8.2	26.41	2.04	0.4	Sunny	2°	95.6
64	750	13.4	8.1	33.16	2.08	0.4	Shady	3°	97.8
65	1000	19.2	11.2	184.09	2.20	0.5	Sunny	0°	96.5
66	375	23.5	19.4	153.87	3.66	0.6	Shady	3°	97.6
67	675	17.1	17.5	145.46	3.81	0.7	Shady	2°	95.5
68	925	28.9	24.9	275.53	2.03	0.8	Sunny	2°	98.0

2.2.2. SPOT 6 Image and Processing

This study was based on a single SPOT 6 panchromatic-multispectral image that was acquired on 6 August 2015 under clear sky conditions. The panchromatic image had a spatial resolution of 1.5 m while the resolution of the multispectral image was 6 m. The panchromatic image was orthorectified using ground control points and digital elevation model (DEM) data. Then, the corrected panchromatic image was used to rectify the multispectral data. The raw digital number values for the multispectral data were converted to spectral radiance values and then into top of atmosphere (TOA) reflectance. Atmospheric correction was performed using the Fast Line-of-sight Atmospheric Analysis of Spectral Hypercubes (FLAASH) approach [29]. All image processing was performed using the ENVI 5.1 software package.

2.2.3. Spectral Vegetation Indices and Texture Parameters

The average surface reflectance values from the multispectral data were used to compute eight vegetation indices (VIs) (Table 2) that are widely used in forest studies: the Atmospherically Resistant Vegetation Index (ARVI), Difference Vegetation Index (DVI), Enhanced Vegetation Index (EVI), Modified Soil Adjusted Vegetation Index (MSAVI), Normalized Difference Vegetation Index (NDVI), Non-Linear Vegetation Index (NLI), Soil Adjusted Vegetation Index (SAVI), and Simple Ratio (SR). Eight texture parameters—the Mean (MEAN), Homogeneity (HOM), Contrast (CON), Dissimilarity (DIS), Entropy (ENT), Variance (VAR), Angular Second Moment (ASM), and Correlation (COR)—were calculated (see Table 3) based on the SPOT6 panchromatic band at the maximum spatial resolution of 1.5 m using the GLCM method [30].

Table 2. Spectral Vegetation index and its calculation formula.

Spectral Vegetation Indices (SVIs)	Formula
1. Atmospherically Resistant Vegetation Index (ARVI) [31]	$\mathrm{ARVI} = \frac{\mathrm{NIR-RB}}{\mathrm{NIR+RB}},\ \mathrm{RB} = \mathrm{R} - \gamma(\mathrm{B}-\mathrm{R})$
2. Difference Vegetation Index (DVI) [32]	$\mathrm{DVI} = \mathrm{NIR} - \mathrm{R}$
3. Enhanced Vegetation Index (EVI) [33]	$\mathrm{EVI} = \mathrm{G}\frac{\mathrm{NIR-R}}{\mathrm{NIR}+\mathrm{C}_1\mathrm{R}-\mathrm{C}_2\mathrm{B}+\mathrm{L}}$
4. Modified Soil Adjusted Vegetation Index (MSAVI) [34]	$\mathrm{MSAVI} = \frac{2\mathrm{NIR}+1-\sqrt{(2\mathrm{NIR}+1)^2-8(\mathrm{NIR-R})}}{2}$
5. Normalized Difference Vegetation Index (NDVI) [35]	$\mathrm{NDVI} = \frac{\mathrm{NIR-R}}{\mathrm{NIR+R}}$
6. Non-linear Vegetation Index (NLI) [36]	$\mathrm{NLI} = \frac{\mathrm{NIR}^2-\mathrm{R}}{\mathrm{NIR}^2+\mathrm{R}}$
7. Soil Adjusted Vegetation Index (SAVI) [37]	$\mathrm{SAVI} = (1+\mathrm{L})\frac{\mathrm{NIR-R}}{\mathrm{NIR+R+L}}$
8. Simple Ratio (SR) [38]	$\mathrm{SR} = \frac{\mathrm{NIR}}{\mathrm{R}}$

Notes: B, R, and NIR represent SPOT6 reflectance in the blue, red, and near-infrared wavelengths, respectively. Parameters L and γ represent the SAVI term (set to 0.5) and the ARVI term (set to 1), respectively. The coefficients used in the EVI algorithm are C1 = 6.0, C2 = 7.5, C3 = 1, and G = 2.5 [17].

Table 3. Formula of texture measurements used in this study [30].

Grey Level Co-Occurrence Matrix Based Texture Parameter Estimation	Formula
1. Mean (MEAN)	$\mathrm{MEAN} = \frac{1}{N^2} \sum\limits_{i,j=0}^{N-1} P_{i,j}$
2. Homogeneity (HOM)	$\mathrm{HOM} = \sum\limits_{i,j=0}^{N-1} i\frac{P_{i,j}}{1+(i-j)^2}$
3. Contrast (CON)	$\mathrm{CON} = \sum\limits_{i,j=0}^{N-1} iP_{i,j}(1-j)^2$
4. Dissimilarity (DIS)	$\mathrm{DIS} = \sum\limits_{i,j=0}^{N-1} iP_{i,j}\lvert 1-j\rvert$
5. Entropy (ENT)	$\mathrm{ENT} = \sum\limits_{i,j=0}^{N-1} iP_{i,j}\left(-\ln P_{i,j}\right)$
6. Variance (VAR)	$\mathrm{VAR} = \frac{\sum_{i,j}(X_{ij}-\mu)^2}{n-1}$
7. Angular Second Moment (ASM)	$\mathrm{ASM} = \sum\limits_{i,j=0}^{N-1} iP_{i,j}{}^2$
8. Correlation (COR)	$\mathrm{COR} = \frac{\sum_{i,j=0}^{N-1} iP_{i,j}-\mu_1\mu_2}{\sigma_1^2\sigma_2^2}$ $\mu_1 = \sum\limits_{i=0}^{N-1} i \sum\limits_{j=0}^{N-1} P_{i,j}$ $\mu_2 = \sum\limits_{j=0}^{N-1} j \sum\limits_{j=0}^{N-1} P_{i,j}$ $\sigma_1^2 = \sum\limits_{i=0}^{N-1} (i-\mu_1)^2 \sum\limits_{j=0}^{N-1} P_{i,j}$ $\sigma_2^2 = \sum\limits_{j=0}^{N-1} (j-\mu_2)^2 \sum\limits_{j=0}^{N-1} P_{i,j}$

Here, i and j are the row and column numbers. N is the number of pixels that are summed. μ_i, μ_j, σ_i^2, and σ_j^2 are the means and standard deviations of P_i and P_j. $P(i, j)$ is the normalized cooccurrence matrix.

2.2.4. Optimum Window Selection

Image texture is a function of the image's spatial resolution. Texture parameters derived with the GLCM method are highly sensitive to the moving window size [20]. Small window sizes are known to exaggerate differences within the window but retain a high spatial resolution, whereas larger windows may cause inefficient extraction of texture information due to over-smoothing of textural variations [39]. There is no consensus as to what moving window size is optimal, so it is necessary to test a range of window sizes to determine which provides the best speed and precision when estimating GSV. Therefore, seven window sizes were used to estimate the stand volume in this work: $3 \times 3, 5 \times 5, 7 \times 7,$ $9 \times 9, 11 \times 11, 13 \times 13,$ and 15×15.

2.3. Machine Learning Algorithms (MLAs)

2.3.1. Classification and Regression Tree (CART)

The CART algorithm is a basic machine learning method proposed by Breiman et al. [40]. The regression tree algorithm was used to build models because the response variable in this study was the forest GSV. The CART algorithm was implemented with the "rpart" in the R software package (version 3.4.3; R Core Team, Viennna, Austria, 2017). Feature selection and pruning are the two core issues in decision tree algorithms. The Gini index and post-pruning were used as the binary split criterion for the selection of split features and to modify the tree, respectively.

2.3.2. Support Vector Machine (SVM)

Support vector machines (SVM) use a nonlinear kernel function to project input data onto a high-dimensional feature space, where complex non-linear patterns can be simply represented [12,41]. In the new hyperspace, the SVM aims to identify an optimal hyperplane that fits the data and minimizes the training error and the complexity of the model [42]. The Gaussian radial basis kernel function of the form was used in this study [43,44]. The best regularization and bandwidth parameter were determined using the training data. The SVM model was implemented with "e1071" packages in the R software package.

2.3.3. Artificial Neural Network (ANN)

A multi-layer neural network with a back-propagation algorithm, which may be the most popular type, was used in this study [42,45]. In this context, the neurons in the input layer were the predicted variables selected by the Boruta algorithm, while the number of neurons in the hidden layer, which carried weights representing the linkages between the predictors and GSV data, was determined using both the training and validation data. The output layer was a single neuron that represented the output values of the forest GSV. We examined the number of neurons ranging from 1 to 10 in the hidden layer to build different models. The ANN algorithm was implemented with the "neuralnet" package in the R software package.

2.3.4. Random Forest (RF)

Ntree (i.e., to the number of variables) and Mtry (i.e., to the number of variables to randomly sample as candidates at each split) parameters are highly sensitive to prediction robustness of RF algorithms [12]. Wang et al. [46], Breiman [47], and Zhao et al. [4] have all indicated that the default values are often a good choice, and that the influence of user-defined parameters on the sensitivity of prediction is very small. Thus, the Ntree value was set to 500 and the Mtry value was set to one-third of the predictive variables. The RF algorithms were implemented with the "randomForest" package in the R software package.

2.4. Model Testing and Comparison

The field data was randomly divided into training (70%) and testing (30%) data. The testing data ($n = 20$) was used to validate the CART, SVM, ANN, and RF models. A random selection of training samples was repeated 100 times to reduce variability. Three common statistical parameters, namely the coefficient of determination (R^2), root mean square error (RMSE), and relative RMSE (rRMSE) were calculated to evaluate the models' performances.

$$\text{RMSE} = \sqrt{\frac{\sum_{i=1}^{n} (\hat{y}_i - y_i)^2}{n}} \tag{3}$$

$$\text{rRMSE} = \text{RMSE}/\overline{y} \tag{4}$$

where $\hat{y}_i$ was the predicted forest GSV and y_i was the forest GSV observed in the field.

3. Results

3.1. Effects of Moving Window Size on the Precision of GSV

The effect of varying the moving window size on the precision of GSV estimation depended on the algorithm used for variable selection. The R^2 decreased with increasing moving window size when CART, SVM, and RF were used to estimate the GSV. In other words, 3×3 and 5×5 window yielded the highest R^2 and lowest RMSE and rRMSE when using these methods (Table 4). For the ANN method, the R^2 initially decreased with the moving window size and then increased, the highest R^2 values (0.83) was achieved with this model when using a window size of 15×15 in this case (Table 4).

Table 4. The effects of window size on the precision of forest growing stock volume estimation.

Methods	Window Sizes	R^2	RMSE (m³/ha)	rRMSE (%)
	3×3	0.77	36.78	29.43%
	5×5	0.76	36.61	29. 29%
	7×7	0.78	37.71	30.17%
CART	9×9	0.68	42.95	34.36%
	11×11	0.68	42.49	33.99%
	13×13	0.43	60.06	48.03%
	15×15	0.57	50.59	40.47%
	3×3	0.72	40.07	32.05%
	5×5	0.73	39.64	31.71%
	7×7	0.64	45.64	36.51%
SVM	9×9	0.71	41.44	33.16%
	11×11	0.68	43.81	35.05%
	13×13	0.67	44.57	35.65%
	15×15	0.67	44.70	35.76%
	3×3	0.73	39.60	31.45%
	5×5	0.73	39.92	31.94%
	7×7	0.61	49.29	39.43%
ANN	9×9	0.65	46.90	37.54%
	11×11	0.70	42.99	34.39%
	13×13	0.77	38.19	30.55%
	15×15	0.78	35.78	28.63%
	3×3	0.78	36.71	29.37%
	5×5	0.84	35.96	28.77%
	7×7	0.75	42.54	34.03%
RF	9×9	0.72	43.35	34.67%
	11×11	0.66	47.36	37.89%
	13×13	0.61	51.25	41.00%
	15×15	0.74	42.35	33.88%

CART = classification and regression tree, SVM = support vector machine, ANN = artificial neural network (ANN), RF = random forest, R^2 = the coefficient of determination, RSME = root mean square error, and rRMSE = the relative RMSE.

3.2. Performance of CART, SVM, ANN, and RF

To evaluate the performance of CART, SVM, ANN, and RF for predicting the forest GSV when using only textural parameters, four machine learning algorithms were compared in terms of their precision (R^2) and RMSE (Figure 2). When the window size was set as 3 × 3, 5 × 5, and 9 × 9, RF demonstrated the best regression performance in terms of model precision and RMSE. The ANN achieved the highest precision when the window size was 11 × 11, 13 × 13, or 15 × 15 (Figure 2).

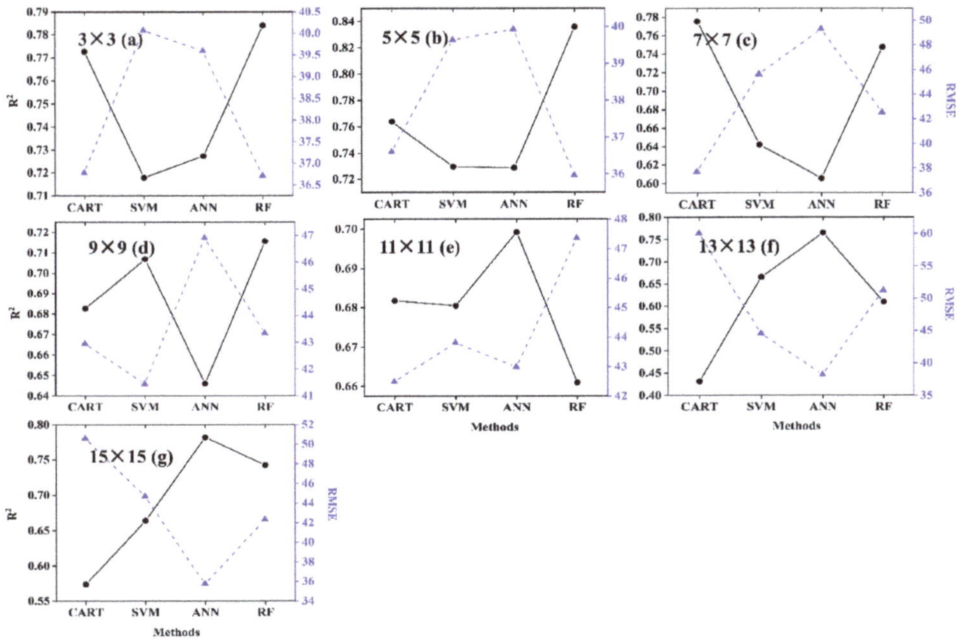

Figure 2. The coefficient of determination (R2) and root mean square error (RMSE) the forest GSV estimated by different regression models (classification and regression tree, CART; support vector machine, SVM; artificial neural network, ANN; random forest, RF) with different textural information calculated with different window sizes; (a) 3 × 3 window size, (b) 5 × 5 window size, (c) 7 × 7 window size, (d) 9 × 9 window size, (e) 11 × 11 window size, (f) 13 × 13 window size, and (g) 15 × 15 window size.

On the other hand, a comparative analysis based on texture parameters and textural parameters together with SVIs demonstrated that RF was the best method for estimating GSV. The R^2 values for models generated using this method ranged from 0.82 to 0.86 (Figure 3). For models using only SVI data as inputs, ANN yielded the highest precision followed by SVM and RF. When using textural parameters together with SVIs as inputs, SVM achieved the second highest precision (Figure 3).

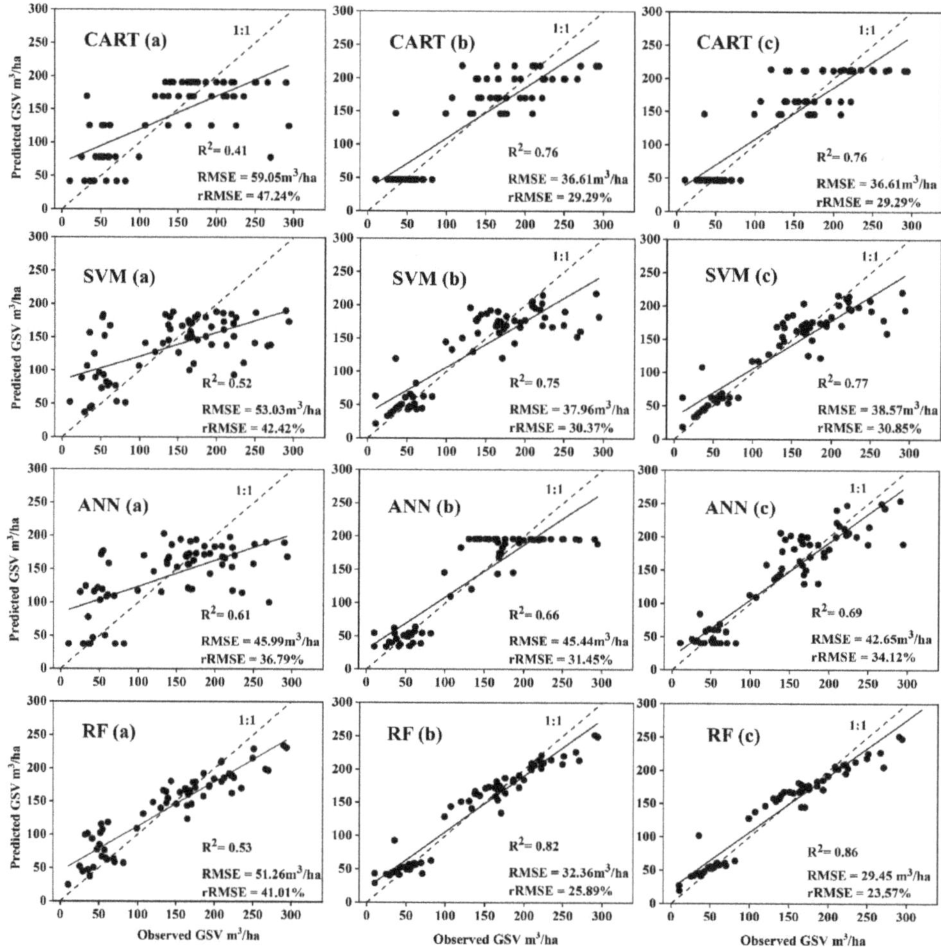

Figure 3. Scatter plots of predicted versus observed growing stock volume (GSV) using classification and regression tree (CART), support vector machine (SVM), artificial neural network (ANN), and random forest (RF) models for (**a**) spectral vegetation indices (SVIs), (**b**) texture features, and (**c**) SVIs plus texture features. R^2 = the coefficient of determination, RSME = root mean square error, and rRMSE = the relative RMSE.

3.3. Performance of Spectral, Texture, and Fusion of Spectral and Texture Information

Regression models using SVIs as inputs achieved worse estimating precision than those using textural parameters independent of the choice of regression method. The R^2 value for the model based on texture data alone (the R^2 values were 0.76, 0.75, 0.66, and 0.82 for CART, SVM, ANN, and RF, respectively) was substantially higher than that for the model based on SVIs alone (Figure 3). Likewise, the RMSE and rRMSE values for the model based on textural parameters alone were considerably lower than those for the model based on SVIs alone. The RMSE values were 36.61 m³/ha, 37.96 m³/ha, 45.44 m³/ha, and 32.36m³/ha, and the rRMSE values was 29.29%, 30.37%, 31.45%, and 25.89% for CART, SVM, ANN, and RF, respectively. Fusing spatial and textural information did not greatly increase the estimating precision relative to that achieved using textural parameters alone, although minor improvements were observed with the SVM and RF methods (Figure 3). The R^2 values achieved with the CART methods when using textural parameters alone as inputs were almost equal to those achieved using textural parameters together with SVIs (Figure 3). Figure 4 indicates the importance of explanatory variables when the RF algorithm is used. The NDVI and MEAN calculated with 3 × 3 and 5 × 5 window sizes, were of higher importance than the other variables.

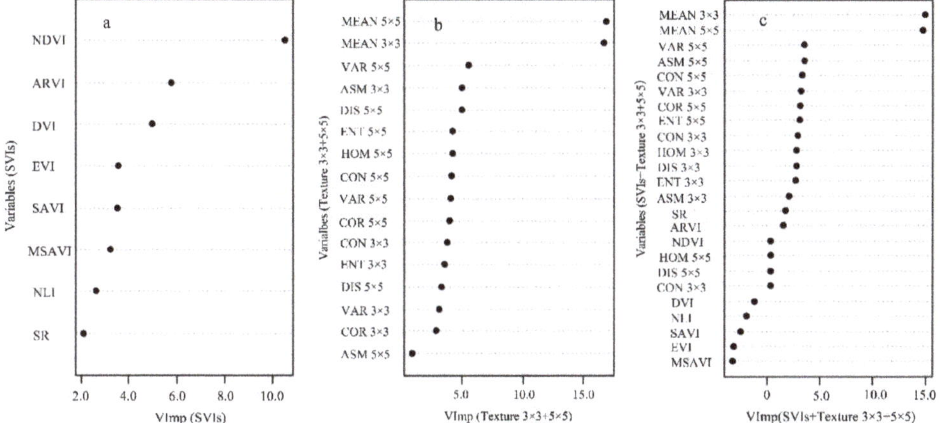

Figure 4. The importance of explanatory variables: (**a**) spectral vegetation indices (SVIs), (**b**) texture features, (**c**) SVIs plus texture features when the Random Forest (RF) algorithm was used. VImp = the variable's importance values, NDVI = Normalized Difference Vegetation Index, ARVI = Atmospherically Resistant Vegetation Index, DVI = Difference Vegetation Index, EVI = Enhanced Vegetation Index, SAVI = Soil Adjusted Vegetation Index, MSAVI = Modified Soil Adjusted Vegetation Index, NLI = Non-linear Vegetation Index, SR = Simple Ratio. Texture (3 × 3 + 5 × 5) was the value of textural parameters when window size was set as 3 × 3 or 5 × 5. MEAN 3 × 3, VAR 3 × 3, DIS 3 × 3, ASM 3 × 3, CON 3 × 3, COR 3 × 3, ENT 3 × 3, and HOM 3 × 3 were Mean, Variance, Dissimilarity, Angular Second Moment, Contrast, Correlation, Entropy, and Homogeneity values when the moving window size was set as 3 × 3. MEAN 5 × 5, VAR 5 × 5, DIS 5 × 5, ASM 5 × 5, CON 5 × 5, COR 5 × 5, ENT 5 × 5, and HOM 5 × 5 were Mean, Variance, Dissimilarity, Angular Second Moment, Contrast, Correlation, Entropy, and Homogeneity values when the moving window size was set as 5 × 5.

4. Discussion

The GSV of *Pinus massoniana* plantations was estimated using four machine learning algorithms (CART, SVM, ANN, and RF) together with spectral and textural information from a SPOT6 image. The effect of varying the window size on the precision of GSV estimation was investigated, and the usefulness of textural parameters from SPOT6 images for forest GSV estimation was assessed.

Image texture is a complex aspect of visual perception [48]. The optimal moving window size for extracting textural parameters from the SPOT6 image generally depends on the choice of regression method. In this study, the CART, RF, and SVM methods performed best with 3 × 3 or 5 × 5 moving windows, while 15 × 15 windows were best for ANN. These results are consistent with those of previous studies [6,49], in which small windows (3 × 3 or 5 × 5) were found to be best when using the multiple linear regression method to estimate forest variables. Small window sizes increase sensitivity to interpixel differences in the proportions of tree crown and shadow and may better detect fine-scale variation in pixel brightness, whereas larger windows may extract texture information inefficiently due to over-smoothing of textural variation [50–52]. Some researchers have suggested that the window size should match that of the sample plots to increase precision [7,53,54]. Moreover, Chryasfis et al. [5] concluded that no single window size could adequately characterize the texture conditions in different phenological scenes when using the bagging least absolute shrinkage and selection operator (LASSO) algorithm. One of the reasons ANN performed well at 15 × 15 windows, while CART, RF, and SVM performed better at 3 × 3 or 5 × 5 windows, may be related to the autocorrelation of texture parameters from different window sizes.

MLAs have fewer assumptions, higher methodological accuracy, and high non-linear adaptation, and can efficiently model the complex non-linear relationships between forest biophysical parameters and RS data [55–59]. Four machine learning regression algorithms, CART, SVM, ANN, and RF were explored in this study. The RF method yielded more precise GSV estimates than the other tested methods (CART, SVM, and ANN) when texture parameters alone, or a fusion of spectral and texture information, were used. This is consistent with the results of Zhao et al. [4], who found that RF achieved the greatest precision when estimating forest variables in black locust plantations on the Loess Plateau. Similarly, Fallah et al. [60] concluded that SVR and RF outperformed k-NN in volume/ha estimation. These results may be due to the fact that RF is more sensitive to overfitting during model training, and can handle high data dimensionality [47,61,62]. The algorithm of the random forest method naturally includes the interactions of variables which are often ignored in other models because of their complexity [63]. However, Zhang et al. [12] found that ANN and SVM were the best methods for estimating sawgrass marsh AGB, with both methods achieving correlation coefficients (r) above 0.9, while ANN was the best method for total biomass estimation (r = 0.94). ANN achieved the best estimating precision independently of the SVIs of input data in this study. This result may be explained by the fact that ANN has advantages in complex pattern learning and generalizing in noisy environments. For CART, it was observed that for larger values of observed GSV, the predicted values of GSV were systematically lower (i.e., always below the 1:1 line), although high R^2 was obtained when texture parameters or a combination of texture parameters and SVIs were used. This may be contributed to by over-fitting. Another reason for underestimation may be due to optical saturation under the conditions of high biomass in dense forests [64]. Overall, CART, SVM, ANN, and RF each have their own advantages. The selection of MLAs may be related to the type of RS information applied. For example, in this study, ANN could be used when applying SVIs alone for GSV estimation, and RF was a good choice when texture parameters or a fusion of textural and spectral information were used. The main drawback of these MLAs was that they did not reveal the functional relationships between the target and predictor variables. MLAs are often referred to as "black box" approaches [7,12,42,65].

Models using textural parameters as inputs achieved greater precision than spectral models, confirming that high-resolution textural information is more useful than spectral data for forest parameter estimation, especially in dense and complex forests characterized by a mosaic of dense and sparse vegetation cells [6,7,66]. Our results indicate that combining spectral and textural parameters yields, at best, marginal improvements in estimating precision compared to using textural parameters alone when estimating forest GSV. This is consistent with the conclusions of Chrysafis et al. [5], based on their work on Mediterranean forests. Textural information was indispensable in the estimation of forest parameters, especially when a high-resolution image was adopted. However, there were many uncertainties that needed to be considered, such as window size. It may be necessary to determine the optimal moving window size for extracting textural parameters.

In this research, forest GSV tended to be overestimated at lower values and underestimated at higher values (Figure 3). This uncertainty may be contributed to the inconsistency between the time of in-site data collection and RS image, although we applied the growth model of *Pinus massoniana* tree for stand volume conversion from 2018 or 2019 to 2015. The growth was calculated from the stand age of forest, which decreased the precision in our study. We suggested that it was better to keep the time of field data surveyed consistent with time of RS acquisition in estimation of forest parameters.

5. Conclusions

Four machine learning regression algorithms including classification and regression tree (CART), support vector machine (SVM), artificial neural network (ANN), and random forest (RF) algorithms were used to estimate the growing stock volume (GSV) of *Pinus massoniana* plantations in the northern subtropical area of China based on spectral vegetation indices (SVIs) and textural parameters extracted from a SPOT6 remote sensing image. The RF method was found to offer the best performance for GSV estimation, although SVM, CART, and ANN also performed reasonably well. Increasing the size of the moving window used when extracting data from the SPOT6 image reduced the predictive performance of the CART, SVM, and RF methods, and the highest coefficient of determination (R^2) values in these cases were achieved with 3×3 or 5×5 windows. Using both spatial and textural parameters as model inputs improved estimating precision relative to using spectral data alone. However, the combined approach did not appreciably outperform models using only textural parameters as inputs.

Author Contributions: Y.D. supervised the work and provided advice for preparing and revising the paper. J.Z. proposed the idea, collected field survey data, organized the writing, and revised the paper. Z.Z. contributed to the programming and experiments, and drafted the manuscript. Q.Z. and Z.H. analyzed the data and prepared figures. J.X. collected field survey data and did image preprocessing. P.W. helped to collect field survey data and revise the manuscript. All authors read and approved the final manuscript.

Funding: The research was funded by the national key research and development plan (grant number, 2017YFC050550404) and project 2662020YLPY020 supported by the fundamental research funds for the central university.

Acknowledgments: We would like to thank Jie Luo from Huazhong Agricultural University for providing constructive comments, which greatly improved the manuscript.

Conflicts of Interest: The authors declare no conflict of interest.

References

1. Xu, Y.; Li, C.; Sun, Z.; Jiang, L.; Fang, J. Tree Height Explains Stand Volume of Closed-Canopy Stands: Evidence from Forest Inventory Data of China. *For. Ecol. Manag.* **2019**, *438*, 51–56. [CrossRef]

2. Ioki, K.; Imanishi, J.; Sasaki, T.; Morimoto, Y.; Katsunori Kitada, K. Estimating Stand Volume in Broad-Leaved Forest Using Discrete-Return Lidar: Plot-Based Approach. *Landsc. Ecol. Eng.* **2010**, *6*, 29–36. [CrossRef]

3. Fang, J.; Chen, A.; Peng, C.; Zhao, S.; Ci, L. Changes in Forest Biomass Carbon Storage in China between 1949 and 1998. *Science* **2001**, *292*, 2320–2322. [CrossRef] [PubMed]

4. Zhao, Q.; Yu, S.; Zhao, F.; Tian, L.; Zhao, Z. Comparison of Machine Learning Algorithms for Forest Parameter Estimations and Application for Forest Quality Assessments. *For. Ecol. Manag.* **2019**, *434*, 224–234. [CrossRef]

5. Chrysafis, I.; Mallinis, G.; Tsakiri, M.; Patias, P. Evaluation of Single-Date and Multi-Seasonal Spatial and Spectral Information of Sentinel-2 Imagery to Assess Growing Stock Volume of a Mediterranean Forest. *Int. J. Appl. Earth Observ. Geoinf.* **2019**, *77*, 1–14. [CrossRef]

6. Zhou, J.J.; Zhao, Z.; Zhao, J.; Zhao, Q.; Wang, F.; Wang, H. A Comparison of Three Methods for Estimating the Lai of Black Locust (*Robinia Pseudoacacia* L.) Plantations on the Loess Plateau, China. *Int. J. Remote Sens.* **2014**, *35*, 171–188. [CrossRef]

7. Zhao, Q.; Wang, F.; Zhao, J.; Zhou, J.; Yu, S.; Zhao, Z. Estimating Forest Canopy Cover in Black Locust (*Robinia Pseudoacacia* L.) Plantations on the Loess Plateau Using Random Forest. *Forests* **2018**, *9*, 623. [CrossRef]

8. Brosofske, K.D.; Froese, R.E.; Falkowski, M.J.; Banskota, A. A Review of Methods for Mapping and Prediction of Inventory Attributes for Operational Forest Management. *For. Sci.* **2014**, *60*, 733–756. [CrossRef]

9. Motlagh, M.G.; Kafaky, S.B.; Mataji, A.; Akhavan, R. Estimating and Mapping Forest Biomass Using Regression Models and Spot-6 Images (Case Study: Hyrcanian Forests of North of Iran). *Environ. Monit. Assess.* **2018**, *190*, 352. [CrossRef]

10. Thamaga, K.H.; Dube, T. Understanding Seasonal Dynamics of Invasive Water Hyacinth (Eichhornia Crassipes) in the Greater Letaba River System Using Sentinel-2 Satellite Data. *Gisci. Remote Sens.* **2019**, *56*, 1355–1377. [CrossRef]

11. Chen, W.; Xie, X.; Wang, J.; Pradhan, B.; Hong, H.; Bui, D.T.; Duan, Z.; Ma, J. A Comparative Study of Logistic Model Tree, Random Forest, and Classification and Regression Tree Models for Spatial Prediction of Landslide Susceptibility. *Catena* **2017**, *151*, 147–160. [CrossRef]

12. Zhang, C.; Denka, S.; Cooper, H.; Mishra, D.R. Quantification of Sawgrass Marsh Aboveground Biomass in the Coastal Everglades Using Object-Based Ensemble Analysis and Landsat Data. *Remote Sens. Environ.* **2018**, *204*, 366–379. [CrossRef]

13. Wang, J.; Xiao, X.; Bajgain, R.; Starks, P.; Steiner, J.; Doughty, R.B.; Chang, Q. Estimating Leaf Area Index and Aboveground Biomass of Grazing Pastures Using Sentinel-1, Sentinel-2 and Landsat Images. *ISPRS J. Photogramm. Remote Sens.* **2019**, *154*, 189–201. [CrossRef]

14. Eckert, S. Improved Forest Biomass and Carbon Estimations Using Texture Measures from Worldview-2 Satellite Data. *Remote Sens.* **2012**, *4*, 810–829. [CrossRef]

15. Pu, R.; Cheng, J. Mapping Forest Leaf Area Index Using Reflectance and Textural Information Derived from Worldview-2 Imagery in a Mixed Natural Forest Area in Florida, Us. *Int. J. Appl. Earth Observ. Geoinf.* **2015**, *42*, 11–23. [CrossRef]

16. Pu, R.; Landry, S. Evaluating Seasonal Effect on Forest Leaf Area Index Mapping Using Multi-Seasonal High Resolution Satellite Pleiades Imagery. *Int. J. Appl. Earth Observ. Geoinf.* **2019**, *80*, 268–279. [CrossRef]

17. Colombo, R.; Bellingeri, D.; Fasolini, D.; Marino, C.M. Retrieval of Leaf Area Index in Different Vegetation Types Using High Resolution Satellite Data. *Remote Sens. Environ.* **2003**, *86*, 120–131. [CrossRef]

18. Frampton, W.J.; Dash, J.; Watmough, G.; Milton, E.J. Evaluating the Capabilities of Sentinel-2 for Quantitative Estimation of Biophysical Variables in Vegetation. *Isprs J. Photogramm. Remote Sens.* **2013**, *82*, 83–92. [CrossRef]

19. Hlatshwayo, S.T.; Mutanga, O.; Lottering, R.T.; Kiala, Z.; Ismail, R. Mapping Forest Aboveground Biomass in the Reforested Buffelsdraai Landfill Site Using Texture Combinations Computed from Spot-6 Pan-Sharpened Imagery. *Int. J. Appl. Earth Observ. Geoinfor.* **2019**, *74*, 65–77. [CrossRef]

20. Zhou, J.; Guo, R.Y.; Sun, M.; Di, T.T.; Wang, S.; Zhai, J.; Zhao, Z. The Effects of Glcm Parameters on Lai Estimation Using Texture Values from Quickbird Satellite Imagery. *Sci. Rep.* **2017**, *7*, 1–2. [CrossRef]

21. Quan, W.; Ding, G. Root Tip Structure and Volatile Organic Compound Responses to Drought Stress in Masson Pine (*Pinus massoniana* Lamb.). *Acta Physiol. Plant.* **2017**, *39*, 258. [CrossRef]

22. Zhang, R.; Ding, G. Seasonal Variation of Soil Carbon and Nitrogen under Five Typical Pinus Massoniana Forests. *Chem. Ecol.* **2017**, *33*, 543–559. [CrossRef]

23. Du, M.; Ding, G.; Cai, Q. The Transcriptomic Responses of *Pinus massoniana* to Drought Stress. *Forests* **2018**, *9*, 326. [CrossRef]

24. Jiang, W. Volume Estimating Model of *Pinus massioniana* Lamb in the Middle of Guizhou Based on 3s Technolog. *For. Inventory Plan.* **2015**, *40*, 13–18. (In Chinese)

25. Jiang, W. The Research of Estimating *Pinus massioniana* Lamb Volume in the Middle of Guizhou Based on Landsat8 Data—Taking Guiyang as an Example. Master's Thesis, Guizhou University, Guizhou, China, 2015. (In Chinese).

26. Jia, X.; Bi, J.; Zhou, Z.; Liu, X.; Gao, D.; Guo, G.; Zhou, H. Water—Holding Capacity of Litter and Soil under Major Pure Plantation in Hilly Region of Central Hubei. *J. Huazhong Agric. Univ.* **2013**, *3*, 39–44. (In Chinese)

27. Ye, Y.; Ling, Y.; Zhuang, E. Compilation of Binary Standing Timber Volume Table of *Pinus massoniana* in Hubei Province. *Cent. South For. Inventory Plan.* **1996**, *3*, 6–8. (In Chinese)

28. Zhang, Z.; He, M. The Growth and Growing Model of *Pinus massoniana* Plantation. *J. Anhui Agric. Coll.* **1992**, *19*, 202–208. (In Chinese)

29. Yuan, J.; Niu, Z.; Wang, X. Atmospheric Correction of Hyperion Hyperspectral Image Based on Flaash. *Spectrosc. Spectr. Anal.* **2009**, *29*, 1181–1185.

30. Haralick, R.M.; Shanmugam, K.; Dinstein, I. Textural Features for Image Classification. *IEEE Trans. Syst. Man Cybern.* **1973**, *SMC3*, 610–621. [CrossRef]

31. Kaufman, Y.J.; Tanre, D. Atmospherically Resistant Vegetation Index (Arvi) for Eos-Modis. *IEEE Trans. Geosci. Remote Sens.* **1992**, *30*, 261–270. [CrossRef]

32. Richardson, A.J.; Wiegand, C.L. Distinguishing Vegetation from Soil Background Information. *Photogramm. Eng. Remote Sens.* **1977**, *43*, 1541–1552.

33. Huete, A.; Justice, C.; Liu, H. Development of Vegetation and Soil Indexes for Modis-Eos. *Remote Sens. Environ.* **1994**, *49*, 224–234. [CrossRef]

34. Qi, J.; Kerr, Y.H.; Moran, M.S.; Weltz, M.; Huete, A.R.; Sorooshian, S.; Bryant, R.B. Leaf Area Index Estimates Using Remotely Sensed Data and Brdf Models in a Semiarid Region. *Remote Sens. Environ.* **2000**, *73*, 18–30. [CrossRef]

35. Rouse, J.W.; Haas, R.H.; Schell, J.A.; Deering, D.W. Monitoring Vegetation Systems in the Great Plains with Erts. *NASA Goddard Space Flight Center Third ERTS-1 Symp.* **1973**, *351*, 309.

36. Gong, P.; Pu, R.; Biging, G.S.; Larrieu, M.R. Estimation of Forest Leaf Area Index Using Vegetation Indices Derived from Hyperion Hyperspectral Data. *IEEE Trans. Geosci. Remote Sens.* **2003**, *41*, 1355–1362. [CrossRef]

37. Huete, A.R. A Soil-Adjusted Vegetation Index (Savi). *Remote Sens. Environ.* **1988**, *25*, 295–309. [CrossRef]

38. Jordan, C.F. Derivation of Leaf-Area Index from Quality of Light on Forest Floor. *Ecology* **1969**, *50*, 663–666. [CrossRef]

39. Sarker, L.R.; Nichol, J.E. Improved Forest Biomass Estimates Using Alos Avnir-2 Texture Indices. *Remote Sens. Environ.* **2011**, *115*, 968–977. [CrossRef]

40. Breiman, L.; Friedman, J.H.; Olshen, R.A.; Stone, C.J. *Classification and Regression Trees*; Wadsworth International Group: Belmont, CA, USA, 1984.

41. Mountrakis, G.; Im, J.; Ogole, C. Support Vector Machines in Remote Sensing: A Review. *Isprs J. Photogramm. Remote Sens.* **2011**, *66*, 247–259. [CrossRef]

42. Were, K.; Bui, D.T.; Dick, Ø.B.; Singh, B.R. A Comparative Assessment of Support Vector Regression, Artificial Neural Networks, and Random Forests for Predicting and Mapping Soil Organic Carbon Stocks across an Afromontane Landscape. *Ecol. Indic.* **2015**, *52*, 394–403. [CrossRef]

43. Rodriguez-Galiano, V.; Sanchez-Castillo, M.; Chica-Olmo, M.; Chica-Rivas, M.J. Machine Learning Predictive Models for Mineral Prospectivity: An Evaluation of Neural Networks, Random Forest, Regression Trees and Support Vector Machines. *Ore Geol. Rev.* **2015**, *71*, 804–818. [CrossRef]

44. Tien Bui, D.; Pradhan, B.; Lofman, O.; Revhaug, I. Landslide Susceptibility Assessment in Vietnam Using Support Vector Machines, Decision Tree, and Naive Bayes Models. *Math. Probl. Eng.* **2012**, *2012*, 974638. [CrossRef]

45. Conforti, M.; Pascale, S.; Robustelli, G.; Sdao, F. Evaluation of Prediction Capability of the Artificial Neural Networks for Mapping Landslide Susceptibility in the Turbolo River Catchment (Northern Calabria, Italy). *Catena* **2014**, *113*, 236–250. [CrossRef]

46. Wang, H.; Lu, K.; Pu, R. Mapping *Robinia pseudoacacia* Forest Health in the Yellow River Delta by Using High-Resolution Ikonos Imagery and Object-Based Image Analysis. *J. Appl. Remote Sens.* **2016**, *10*, 045022. [CrossRef]

47. Breiman, L. Random Forests. *Mach. Learn.* **2001**, *45*, 5–32. [CrossRef]

48. Coburn, C.A.; Roberts, A.C. A Multiscale Texture Analysis Procedure for Improved Forest Stand Classification. *Int. J. Remote Sens.* **2004**, *25*, 4287–4308. [CrossRef]

49. Zhou, J.J.; Zao, Z.; Zhao, Q.; Zhao, J.; Wang, H. Quantification of Aboveground Forest Biomass Using Quickbird Imagery, Topographic Variables, and Field Data. *J. Appl. Remote Sens.* **2013**, *7*, 073484. [CrossRef]

50. Nichol, J.E.; Sarker, M.L. Improved Biomass Estimation Using the Texture Parameters of Two High-Resolution Optical Sensors. *IEEE Trans. Geosci. Remote Sens.* **2011**, *49*, 930–948. [CrossRef]

51. Fuchs, H.; Magdon, P.; Kleinn, C.; Flessa, H. Estimating Aboveground Carbon in a Catchment of the Siberian Forest Tundra: Combining Satellite Imagery and Field Inventory. *Remote Sens. Environ.* **2009**, *113*, 518–531. [CrossRef]

52. Lu, D.; Batistella, M. Exploring Tm Image Texture and Its Relationships with Biomass Estimation in Rondonia, Brazilian Amazon. *Acta Amazonica* **2005**, *35*, 249–257. [CrossRef]

53. Wood, E.M.; Pidgeon, A.M.; Radeloff, V.C.; Keuler, N.S. Image Texture as a Remotely Sensed Measure of Vegetation Structure. *Remote Sens. Environ.* **2012**, *121*, 516–526. [CrossRef]

54. Gómez, C.; Wulder, M.A.; Montes, F.; Delgado, J.A. Delgado. Forest Structural Diversity Characterization in Mediterranean Pines of Central Spain with Quickbird-2 Imagery and Canonical Correlation Analysis. *Can. J. Remote Sens.* **2011**, *37*, 628–642. [CrossRef]

55. Dang, A.T.; Nandy, S.; Srinet, R.; Luong, N.V.; Ghosh, S.; Kumar, A.S. Forest Aboveground Biomass Estimation Using Machine Learning Regression Algorithm in Yok Don National Park, Vietnam. *Ecol. Informa.* **2019**, *50*, 24–32. [CrossRef]

56. Liu, K.; Wang, J.; Zeng, W.; Song, J. Comparison and Evaluation of Three Methods for Estimating Forest above Ground Biomass Using Tm and Glas Data. *Remote Sens.* **2017**, *9*, 341. [CrossRef]

57. Srinet, R.; Nandy, S.; Patel, N.R. Estimating Leaf Area Index and Light Extinction Coefficient Using Random Forest Regression Algorithm in a Tropical Moist Deciduous Forest, India. *Ecol. Informa.* **2019**, *52*, 94–102. [CrossRef]

58. Cooner, A.J.; Shao, Y.; Campbell, J.B. Campbell. Detection of Urban Damage Using Remote Sensing and Machine Learning Algorithms: Revisiting the 2010 Haiti Earthquake. *Remote Sens.* **2016**, *8*, 868. [CrossRef]

59. Pham, L.T.; Brabyn, L. Monitoring Mangrove Biomass Change in Vietnam Using Spot Images and an Object-Based Approach Combined with Machine Learning Algorithms. *ISPRS J. Photogramm. Remote Sens.* **2017**, *128*, 86–97. [CrossRef]

60. Shataee, S.; Kalbi, S.; Fallah, A.; Pelz, D. Forest Attribute Imputation Using Machine-Learning Methods and Aster Data: Comparison of K-Nn, Svr and Random Forest Regression Algorithms. *Int. J. Remote Sens.* **2012**, *33*, 6254–6280. [CrossRef]

61. Shao, Y.; Lunetta, R.S. Lunetta. Comparison of Support Vector Machine, Neural Network, and Cart Algorithms for the Land-Cover Classification Using Limited Training Data Points. *ISPRS J. Photogramm. Remote Sens.* **2012**, *70*, 78–87. [CrossRef]

62. Belgiu, M.; Drăguţ, L. Random Forest in Remote Sensing: A Review of Applications and Future Directions. *ISPRS J. Photogramm. Remote Sens.* **2016**, *114*, 24–31. [CrossRef]

63. Cutler, D.R.; Edwards Jr, T.C.; Beard, K.H.; Cutler, A.; Hess, K.T.; Gibson, J.; Lawler, J.J. Random Forests for Classification in Ecology. *Ecology* **2007**, *88*, 2783–2792. [CrossRef] [PubMed]

64. Avitabile, V.; Camia, A. An Assessment of Forest Biomass Maps in Europe Using Harmonized National Statistics and Inventory Plots. *For. Ecol. Manag.* **2018**, *409*, 489–498. [CrossRef] [PubMed]

65. Mas, J.F.; Flores, J.J. The Application of Artificial Neural Networks to the Analysis of Remotely Sensed Data. *Int. J. Remote Sens.* **2008**, *29*, 617–663. [CrossRef]

66. Rodriguez-Galiano, V.F.; Chica-Olmo, M.; Abarca-Hernandez, F.; Atkinson, P.M.; Jeganathan, C. Random Forest Classification of Mediterranean Land Cover Using Multi-Seasonal Imagery and Multi-Seasonal Texture. *Remote Sens. Environ.* **2012**, *121*, 93–107. [CrossRef]

MDPI
St. Alban-Anlage 66
4052 Basel
Switzerland
Tel. +41 61 683 77 34
Fax +41 61 302 89 18
www.mdpi.com

Forests Editorial Office
E-mail: forests@mdpi.com
www.mdpi.com/journal/forests